Food Safety

This book is dedicated to the memory of two lovely ladies:
My mother
Audrey Shaw
28 November 1928 to 28 May 2009

My mother-in-law
Jeanne Zehms
6 May 1919 to 5 December 2010

They both inspired me in their own special ways.

Food Safety

The Science of Keeping Food Safe

Ian C. Shaw

Director of Biochemistry & Professor of Toxicology
University of Canterbury, Christchurch, New Zealand

WILEY-BLACKWELL

A John Wiley & Sons, Ltd., Publication

Registered Office
John Wiley & Sons, Ltd, The Atrium, Southern Gate, Chichester, West Sussex, PO19 8SQ, UK

Editorial Offices
9600 Garsington Road, Oxford, OX4 2DQ, UK
The Atrium, Southern Gate, Chichester, West Sussex, PO19 8SQ, UK
2121 State Avenue, Ames, Iowa 50014-8300, USA

For details of our global editorial offices, for customer services and for information about how to apply for permission to reuse the copyright material in this book please see our website at www.wiley.com/wiley-blackwell.

Library of Congress Cataloging-in-Publication Data

Shaw, Ian, 1950–
    Food safety : the science of keeping food safe / Ian Shaw.
        p.   cm.
    Includes bibliographical references and index.
    ISBN 978-1-4443-3722-8 (pbk. : alk. paper)   1. Food–Analysis.   2. Food contamination.
3. Food–Safety measures.   I. Title.
    TX531.S5848 2013
    363.19'26–dc23
                                                        2012013710
A catalogue record for this book is available from the British Library.

Wiley also publishes its books in a variety of electronic formats. Some content that appears in print may not be available in electronic books.

Cover image: Bacterial culture – © iStockphoto.com/Linde1; Crop spraying – © iStockphoto.com/fotokostic; Food tasting – © iStockphoto.com/webphotographeer
Cover design by Steve Thompson

Set in 9.5/11.5pt Interstate-Light by SPi Publisher Services, Pondicherry, India
Printed and bound in Malaysia by Vivar Printing Sdn Bhd

1   2013

Contents

A colour plate section falls between pages 52 and 53

Preface

We expect our food to be safe; we certainly don't expect to be ill after eating a meal. However, it is important to remember that this is an affluent Western world expectation and that many undernourished people in the poorer parts of the world simply want to eat – the safety of their food is a secondary, or an even lesser, consideration.

Our desire for safe food, spurred on by food disasters like Mad Cow disease in the UK in the mid 1980s, has led to developed countries introducing legislation to ensure safe food – to make sure that food is fit for purpose.

In order to make food safe, we need to understand what makes it unsafe. Why do some microorganisms (pathogens) in food cause disease in their consumers, while others are harmless – or even beneficial? We need to minimise our exposure to food pathogens in order to minimise consumer risk. We need to understand why chemical food contaminants, like pesticides used in food production, can harm their consumers and we need to know the doses that are harmful so that we can set safe levels for chemical contaminants in food and so further minimise risk.

To store food we often use preservatives, otherwise harmful microorganisms might grow on the stored food; if we use chemical preservatives we must understand their potential toxicity to the consumer and make sure the chemical preservatives don't solve a microbiological problem, but introduce unacceptable chemical toxicity.

As consumers become more pernickety about their food they want it to look and taste exactly right – and by exactly right I mean how *they* think it should look and taste. To achieve this, colours and flavours are added to many pre-prepared foods. But are these additives safe? What are their effects on their consumers? Is using colours and flavours to enhance our food experience an acceptable risk?

Food is inextricably linked to health. If we eat too much fat or sugar we might become obese and our health will be significantly impacted – this might lead to heart disease or diabetes, both serious diseases. Some bacteria (e.g. Listeria) that might contaminate food cause serious diseases, even death. On the other hand, the contaminants and additives present in our food might affect our health in far more esoteric ways following very long-term exposure. For example, some food colours are known to cause cancer in rats at high doses, but what effects might they have on human consumers of infinitesimally tiny doses in food? Are these risks outweighed by the benefits of the chemicals? Is bright red cherryade worth the vanishingly low risk of its consumer contracting thyroid cancer? Do you *need* your cherryade to be bright red? Is any health risk associated with food colour acceptable – however small?

These are all fundamentally important questions – and there are many, many more – to which we should seek answers if we are to make our food safer. We need to understand the science that underpins food safety; we need to tease out the health effects of chemicals in our food and set these risks against their benefits. Is the risk of a bacterium growing in our food greater than the chemical used to kill it? Why is the chemical harmful to its consumer? Could we modify its molecule to make it less toxic, but maintain its bactericidal properties? These are some of the answers we might need to help us to produce and regulate our food and make it as fit for purpose as possible.

Over the last 50 or so years our understanding of food safety has grown to such an extent that we no longer accept food-borne illness as a consequence, albeit rare, of eating. Those responsible for food-borne illness outbreaks can fall foul of strict food legislation and find themselves subjected to heavy fines or, in rare cases, even imprisonment. Just 50 years ago this would not have been thought possible.

My book takes a trip through the world of food safety, from microbiological food pathogens, through chemical contaminants, natural toxins and the chemicals we use to colour, preserve and flavour our food. It grapples with the esoteric prion that causes Mad Cow disease which led to the collapse of the UK beef industry and prevents me as a Brit living in New Zealand from donating blood because of the perceived risk of transferring the prion to my fellow New Zealanders. It uncovers the controversy of 'organic' food and food irradiation. Finally, it looks at the laws that are used to make sure that when we eat our dinner or buy a snack on the street we don't contract a food-borne illness or expose ourselves to chemicals that might compromise our health in the future. This is a long journey flavoured with many examples from around the world; I hope you enjoy it!

Professor Ian C. Shaw PhD, FRSC, FIFST, FRCPath
Christchurch, New Zealand
September 2012

Acknowledgements

I thank Wiley-Blackwell for asking me if I would write a textbook on Food Safety, and, in particular, David McDade for steering the idea to contract and Andrew Harrison for trying to keep me on schedule and for gently prompting me by e-mail when I missed deadlines by a country mile. He never once raised his e-mail voice even though this would have been justified on many occasions. In addition, I am indebted to Alison Nick for her supreme proof reading skills and frightening efficiency.

As with any wide-ranging textbook, there are subjects included that the author is less *au fait* with; in my case food legislation was the subject that I needed expert help with. I thank Keith Zehms and Sharon McIlquham for their advice on US law and John Reeves for his help with the New Zealand food legislation.

When I agreed to write the book and signed the contract on 21 January 2010 I could not, in my worst nightmare, have anticipated the devastation that the 4 September 2010, 22 February and 13 June 2011 Canterbury (New Zealand) earthquakes would bring to my life and environment. Much of this book was written during a period of regular and significant aftershocks, telecommunication failures, lack of internet, uncertainty about the stability of buildings, workmen everywhere, continual battles with the New Zealand Earthquake Commission and our insurers, and deep, deep sadness for the loss of our city and some of its people. Throughout this, my partner, David Zehms, gave me unwavering support and provided some semblance of emotional normality that allowed me to retire to my cracked and crumpled study to write this book – thank you David.

As you read, think of the people of Christchurch and Lyttelton, New Zealand, who have lost so much and have a long, hard road ahead.

I hope you enjoy my book.

Ian C. Shaw
Christchurch, New Zealand

Chapter 1

Introduction

Introduction

Food safety is a relatively recent 'invention'. It was introduced in the developed world to increase confidence in food. In our modern world it simply is not acceptable to have food that might make us ill. Sadly even now a good proportion of the world's people are very much more concerned about getting food and stemming their unrelenting hunger than they are about whether they might get a stomach upset as a result of eating the food. We must always remember these horrifying facts when we study food safety. Food safety and the legislation emanating from it are for the relatively rich countries that have the luxury of having sufficient food to allow them to make rules about what is safe to eat.

A brief history of food safety

Prehistoric times

The risk of eating in prehistoric times was very much more an issue of the dangers of catching the beast to eat than the ill effects suffered after eating it. To survive, cavemen had to eat and their animal instincts dominated their behaviour with respect to food. These instincts, no doubt, made them avoid food they had learned made them sick, but their overriding instinct was 'eat to live'. Some foods, however, might have been so toxic that they threatened the early man's survival. Behaviour that minimised consumption of toxic food would have been selected in because individuals that succumbed to toxins in their food simply did not survive. This is the raw material of Darwinian evolution and could be considered a very early manifestation of food safety issues! Whether this happened or not thousands of years ago is impossible to know, but we do know that modern-day animals avoid toxic plants in their diet. This might be because some of the toxins (e.g. alkaloids) have a bitter taste that warns the would-be consumer of the risk. Prehistoric man probably behaved in exactly this way which is why he was able to survive in such a harsh environment in which every day posed new and unknown food challenges.

Food Safety: The Science of Keeping Food Safe, First Edition. Ian C. Shaw.
© 2013 John Wiley & Sons, Ltd. Published 2013 by John Wiley & Sons, Ltd.

This is hardly prehistoric food safety policy, but it illustrates our inborn survival instinct that extends to the food we eat. We have an innate desire not to eat something that will make us ill. This has not changed over the millennia.

Evolution of cellular protection mechanisms

It is important to remember too that our metabolic systems (and avoidance strategies) evolved during the tens of thousands of years of prehistoric times. Metabolism of toxins from food in order to reduce their toxicity and so make the food 'good' developed over millions of years. There are highly complex metabolic systems 'designed' to detoxify ingested toxins that evolved long before man, but the enzyme systems from the primitive cells

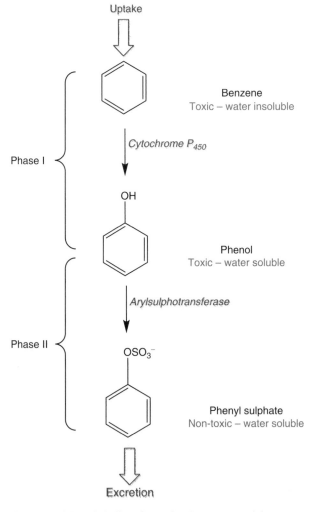

Figure 1.1 Phase I and II metabolism for a simple compound, benzene, showing how the molecule is detoxified, made water soluble and excreted (e.g. in urine).

in which they evolved were selected into the human genome through the evolutionary process and were inevitably expressed by the earliest hominids. These detoxification systems gave man an advantage because he could eat food that contained chemicals which if not detoxified would make the food too toxic to eat. These enzyme systems are now very well understood; they include the cytochromes P_{450} mixed function oxidases (termed Phase I metabolism) and the conjugating enzymes (termed Phase II metabolism) (Figure 1.1).

There are many food toxins that are detoxified by these systems, so making the food safe to eat (this will be discussed further in Chapters 7 and 8); for example, parsnips contain bergapten, a photosensitising toxin that also causes cancer (see Chapter 8, *Furocoumarins in parsnips, parsley and celery*); bergapten is detoxified by Phase I and II metabolism (Figure 1.2)

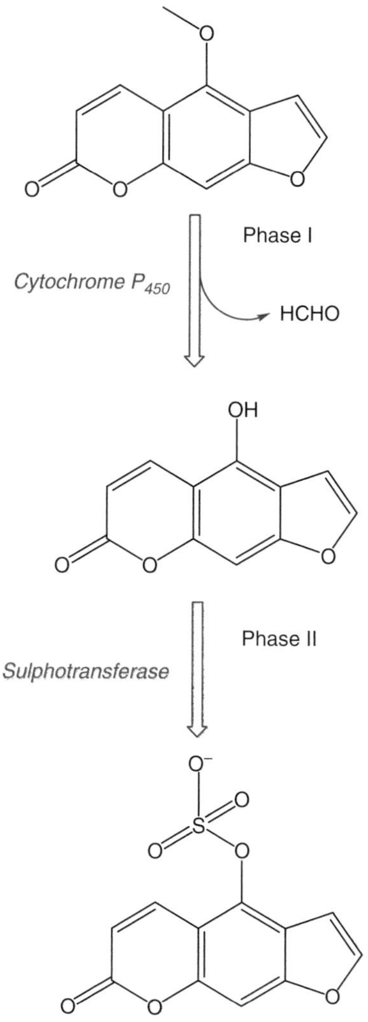

Figure 1.2 A proposed metabolic pathway for bergapten.

thus making parsnips safe to eat. These metabolic processes are the cell's internal food safety mechanisms and broaden the range of foods we can eat without suffering the ill effects that some of their components would cause.

There are significant differences in the susceptibility of different animal species to toxic chemicals; these are due to the evolutionary selective pressures under which the particular species developed. This means that safe foods for some species might be highly toxic to others. For example, the toxin in the swan plant (*Asclepias fruticosa*), labriformidin, is very toxic to birds but harmless to the monarch butterfly (*Danaus plexippus*) (see Chapter 8, *Why produce natural toxins?*).

The monarch butterfly uses this differential toxicity as a means of protection. Its caterpillar eats swan plant leaves and incorporates labriformidin into its body; this makes it toxic and unpalatable to predatory birds. This interesting means of survival is by no means unique amongst animals. Indeed, some plants that are eaten by animals are very toxic to humans. For example, it would only take a few leaves of hemlock (*Conium maculatum*) to kill a person, but the skylark (*Alauda arvensis*) is unaffected by its toxin (Figure 1.3). Indeed, there have been cases of human poisoning in Italy following consumption of skylarks which (strange as it may seem) are a delicacy in that country. The toxin in hemlock is coniine (Figure 1.3) – it is very toxic; about 200 mg would be fatal to a human. Hemlock was the poison used to execute Socrates in 399 BC for speaking his mind in the restrictive environment of ancient Greece.

Tudor England (1485–1603)

In the 1500s I doubt whether many people thought about illness being linked to what they had eaten, but I imagine food-borne illness was prevalent in that rather unhygienic society. In fact spices were introduced into Tudor England to mask the putrid taste of some foods particularly meat – this is a 'head in the sand' approach where masking the bad taste was thought to take away the bad effects. Whether the Tudors thought that masking the taste of putrefying meat stopped them getting ill I cannot know, but they certainly thought that masking the terrible smells of putrid plague-ridden London prevented them catching fatal diseases like the Plague. The gentry used, amongst other things, oranges stuck with cloves, and ornate necklaces with receptacles for sweet-smelling spices and resins (pomanders – derived from the French *pomme d'ambre* meaning apple of amber; ambergris, a sweet-smelling substance produced by sperm whales was often used to scent pomanders) to waft in front of them to take away the evil smells as they walked the streets. This is hardly food safety legislation, but it might just be the beginning of people connecting off-food with illness – a key step in making food safe.

The times of King George III of England (1760–1820)

The Georgian era was a time of great social division. The rich ate well, if not exuberantly, and the poor just about found enough food to keep them alive. The idea that bad smells were associated with disease prevailed as

Figure 1.3 Socrates (469–399 BC), coniine, the poison from hemlock used to execute him, and the skylark (*Alauda arvensis*) which is unaffected by coniine. (Pictures from http://en.wikipedia.org/wiki/File:Socrates_Louvre.jpg, © Sting; http://en.wikipedia.org/wiki/File:Alauda_arvensis_2.jpg, © Daniel Pettersson; photograph of hemlock taken by the author.)

did the naïve thought that if the smell was masked, putrid food was good to eat. Susannah Carter, an American cookery author, described a 'method of destroying the putrid smell which meat acquires during Hot Weather' in her book *The Frugal Housewife, or, Complete Woman Cook*, published in New York in 1803. Some people must have been very ill after eating food prepared under this rather naïve food safety philosophy; i.e. bad smell means high risk and hiding the smell minimises the risk. I wonder if they connected their stomach upset with the food they had eaten? Probably not because such illness would be the norm in the 1700s and people probably simply took it for granted.

Figure 1.4 Louis Pasteur (1822–1895). (Picture from http://en.wikipedia.org/wiki/File:Louis_Pasteur.jpg.)

The 1800s – Pasteur's Germ Theory, Lister's antiseptics and the first refrigerators

In the mid 1800s in Europe there was a significant improvement in the understanding of disease and, in particular, public health. This was the time that the connection between microorganisms and disease was beginning to be understood. Louis Pasteur (1822–1895; Figure 1.4) proposed the Germ Theory of Disease while he was working at the University of Strasbourg in France in the 1860s. He later extended his understanding of 'germs' to propose that heating contaminated broths to a high temperature for a short time would kill the 'germs'. This is the basis of one of today's most important methods of assuring safe food – pasteurisation.

Disinfectants

Joseph Lister (1827–1912) followed Pasteur's work with his discovery of antiseptics. He showed that carbolic acid (phenol; Figure 1.5) killed germs and reduced post-operative infection. This revolutionalised surgery, which was often a sentence of death pre-Lister. The people of Victorian England embraced scientific development – they were fascinated by science and were keen to understand and use it. Lister's antiseptics were modified and developed and became the carbolic and creosote disinfectants that were used to keep Victorian (1837–1901) homes free of germs. There is no doubt that this 'clean' approach to living reduced food-borne illnesses in the kitchens of the Victorian upper classes. The lower classes were still scrambling to get enough food to feed their large families and probably knew nothing of the new-fangled theories of germs and antiseptics. A disinfectant fluid was

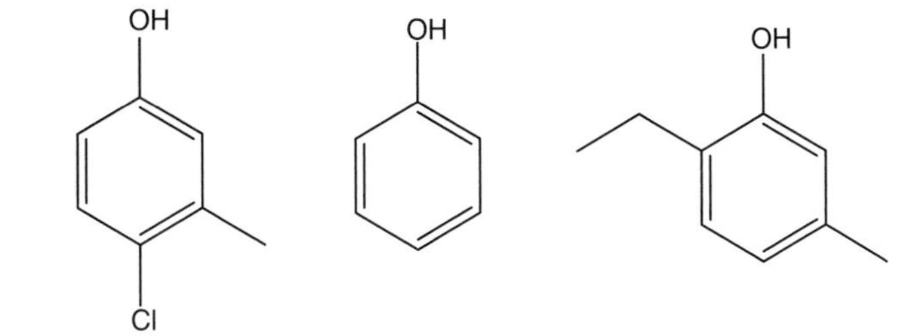

Figure 1.5 Molecular structures of some of the components of Jeyes' Fluid, a very effective disinfectant introduced in Victorian times.

patented by John Jeyes in 1877 in London which was a product of the increased interest in 'germs' and antiseptics and was based on Lister's phenol. Jeyes' Fluid comprises 5% 3-methyl,4-chlorophenol (chloro m-cresol) and 5% alkylphenol fraction of tar acids (these were a by-product of the coal industry; Figure 1.5); it is still used today.

Refrigeration
It has been known for a long time that food keeps better when it is cooled. The Victorians equated this with suppression of the growth of spoilage germs and introduced complicated means of keeping their food cool. Refrigerators, as we know them now, were not introduced until the 1860s, but before then 'iceboxes' were used in which large chunks of ice kept the food cool. The production of ice was not an easy task either – this is a circular problem; without refrigeration it is difficult to produce ice. In the early days, ice was collected during the winter and packed into ice houses, then the ice houses were used for storage of perishable food. With good insulation the ice could be maintained for a good proportion of the year in temperate climates. Later ice was made using cooling chemicals and water. For example, when diethyl-ether evaporates it takes in heat, thus cooling its surroundings; the cooling property of ether was used to freeze water for iceboxes. There is no doubt that the increased availability of iceboxes increased the safety of mid-1800s' food. In the 1860s, the Industrial Revolution was under way; the developed world was enthralled by mechanical devices and commercial, large-scale manufacture. Long-haul transport became important as a means of moving products, including food, around and between nations; this led to a renewed interest in cooling devices both to keep food cold at home, and, perhaps more importantly, to allow food to be transported long distances without spoiling. Since the problem of food spoilage was more acute in hot countries, it is perhaps not surprising that it was a man from Scotland living in Australia who appreciated the need to cool food. This man was James Harrison (1816-1893) and he developed one of the first mechanical cooling devices based on the compression and expansion of a volatile liquid (when liquids evaporate – remember the ether example above – they take up heat). Harrison was granted a patent for the vapour-compression refrigerator in 1855. He used

Figure 1.6 A cow creamer. (Photographed with permission from the collection of Mrs S. Drew, Christchurch, New Zealand.)

this device to manufacture ice for the first attempt to transport meat from Australia to England in 1873. Unfortunately the ice melted before the ship arrived in England and the meat spoiled. It was not until 1882 that the first successful shipment of cooled meat was made from the antipodes to England and it went from New Zealand not Australia.

Refrigeration revolutionised food safety and continues to be used as one of the main ways we keep our food safe in the 21st century.

It is clear that the Victorians were aware of hygiene and its link to health. Mrs Beeton's *Book of Household Management* (published 1861) has many tips on hygiene; she advises suspending chloride of lime (calcium hypochlorite – $Ca(ClO)_2$)-soaked cloths across the room. Chloride of lime slowly liberates chlorine gas which is a powerful antiseptic. Such methods would have killed bacteria and therefore made food preparation more hygienic.

There are some good examples of the Victorians' concern about food hygiene. For example, they loved intricate, delicate china to accompany afternoon tea. Milk was served from creamers (small jugs) sometimes shaped like cows. Cow creamers (Figure 1.6) disappeared in the late 1800s because of concerns about hygiene – it was very difficult to clean them properly because of their intricate design.

Chemical preservatives

Food spoilage and food-borne illness can also be prevented by using naturally produced chemicals to kill bacteria or significantly reduce their growth rate. Some of these methods are very old. For example, fermentation; here 'good' microorganisms are used to produce natural preservatives in the fermented food. Salami manufacture relies upon fermentation. The acid products of the fermentation process (e.g. lactic acid) preserve the meat by inhibiting the growth of pathogens and spoilage bacteria which do not thrive in acid conditions (see Chapter 11, *Antimicrobial food preservatives*). On the other hand, yoghurt is simply milk infected with good bacteria (traditionally *Lactobacillus bulgaricus* and *Streptococcus thermophilus* and more recently *L. acidophilus*); these bacteria colonise the milk so effectively that they prevent harmful bacteria growing. Yogurt production, as a means of preserving milk, has been known for at least 4,500 years and probably began in Bulgaria.

Some chemical preservatives are added to food to prevent food spoilage. Some of these preservatives have been used for thousands of years. Vinegar (acetic acid; ethanoic acid) produced by fermenting ethanol (originally from wine) is a good example; traces have been found in Egyptian urns from 3,000 BC and it is still used today to pickle vegetables (e.g. onions) and make chutneys, etc. The acidity of vinegar inhibits most bacterial and fungal growth, thus preventing food spoilage – the principle is the same as described above for food preserved by fermentation, but, in this case, the acid is added to the food rather than being produced by fermentation of the food (see Chapter 11, *Other organic acids*).

Sugar is also used as a preservative. If the concentration is high enough it too prevents bacterial and fungal growth by scavenging the water that microbes need to survive (sugars form hydrogen bonds with water, thus effectively removing the water from the system). Sugar, either in the form of refined sugar (sucrose) or honey (mainly fructose), has also been used for thousands of years to preserve food. Jam is simply fruit boiled with sugar and bottled aseptically. Sugar can also be used to bottle or can fruit which involves heating the fruit in a strong sugar solution in jars and sealing the jars aseptically. Both bottled fruits and jams will keep for years.

There are also many modern means of preserving food using gases (e.g. nitrogen) and chemicals (e.g. sodium benzoate) to inhibit microorganism growth, or using irradiation (see Chapter 12) to kill them. These techniques are associated with risks to the consumer and therefore are often controversial; we must not forget, however, that the risk of harm following exposure to a food pathogen is likely to be greater than the risk of the method of preserving the food (this will be covered in detail in Chapter 11). However, there is no doubt that pickling with vinegar and preserving in sugar represent a negligible risk to the consumer ... unless, of course, you eat too much of the sugar-preserved food and your teeth decay and you become obese!

Sodium benzoate itself has a very low toxicity – no adverse effects have been seen in humans dosed up to 850 mg/kg body weight/day. However, in the presence of ascorbic acid (vitamin C) sodium benzoate can react to form benzene (Figure 1.7) which is a carcinogen. Since many foods that sodium benzoate might be used to preserve might also contain ascorbic acid, perhaps the risk is not worth the benefit. On the other hand, benzoic acid is present at low concentrations naturally in some fruits (e.g. cranberries) and they contain ascorbic acid too, so you cannot avoid the risk if you choose to eat these foods. Sometimes 'natural' is not good (See Chapter 8 for many more examples), but whichever way you look at it the risk is very low indeed (see Chapter 2).

For cats, the risk of cancer following benzene exposure via foods preserved with benzoate is significant because cats have very different routes of metabolism to humans and are unable to detoxify benzoate efficiently and so benzoate itself is toxic to cats. For this reason, the allowable level of sodium

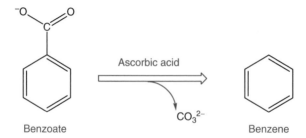

Figure 1.7 The oxidation of benzoate by ascorbic acid to form highly toxic benzene.

benzoate in proprietary cat foods is significantly lower than the corresponding level for foods intended for human consumption.

The influence of religion on food safety

Many religions are strict about what foods can be eaten and how they should be prepared. There is often little rationale for this except that it was decreed thousands, or more, years ago by the prophets or gods of the religion concerned. It is tempting to speculate that the reason that the food rules were originally introduced was because they constituted a simple means by which food was made safer to eat. There are good examples that illustrate this from Judaism and Islam.

The *Old Testament* prohibits the Jews from eating pork:

And the swine, because it divideth the hoof, yet cheweth not the cud, it is unclean unto you: ye shall not eat of their flesh, nor touch their dead carcase. (*Deuteronomy* 14:8)

Similarly the Koran forbids pork consumption:

He has only forbidden you dead meat, and blood, and the flesh of swine ....'

Banning pork was a very sensible food safety rule for a warm climate thousands of years ago. Pigs can be infected by the parasite *Trichinella* (see Chapter 5, *Trichinella sp.*) and it is likely that many more pigs were infected then than are infected now.

Trichinella is a roundworm (nematode) that infects pigs and spreads quickly via its eggs in infected animals' faeces. Consumption of undercooked *Trichinella*-infected pork can lead to human infection which leads to severe fever, myalgia, malaise and oedema as the *Trichinella* larvae infest the host's muscles. Modern meat production hygiene operated in most developed countries has reduced the incidence of human trichinellosis to very low levels – in the USA there were only 25 cases between 1991 and 1996, whereas in Asia and parts of eastern Europe there are still thousands of cases annually. Since the animal husbandry and meat production hygiene were primitive in the times of Christ and Allah it is very likely that most pigs were *Trichinella*-infected and therefore the risk of disease from eating pork was great. So what better food safety legislation than to ban pork consumption through the religious statutes?

The impact of space travel on food safety

The biggest impetus to make absolutely certain that food is safe was the introduction of space travel in 1960s USA. Astronauts must eat, but they simply cannot become ill while floating around in space, primarily because they usually do not have a doctor on board to treat them, and if they did the 'hospital' facilities would be rudimentary at best. There is a rather more pressing and pragmatic reason for not getting food-borne illness in the confines of a space craft orbiting the earth – most food-borne illnesses are associated with diarrhoea and vomiting and this is out of the question in a spaceship at zero gravity for obvious reasons. The developers of the US space programme realised the potential problems associated with unsafe food in space and therefore they formulated a series of extremely strict rules to ensure that the food consumed by astronauts would not make them ill. Producers of food for space travel had to ensure that it was sourced from reliable producers, that it was prepared under ultra-hygienic conditions, that it was cooked properly (to kill any pathogenic organisms that might be present) and packaged in a way that prevented later contamination (Figure 1.8). In addition, they developed a testing regime to check that astronauts' food was not contaminated with potential human pathogens. The system worked – as far as I am aware there has not been a serious incident of food-borne illness on any space mission so far.

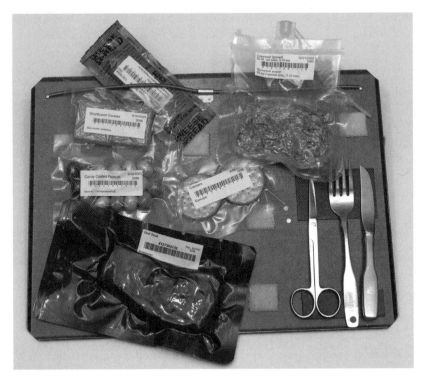

Figure 1.8 Space food used by US astronauts. It is sterilised and vacuum packed to prevent food-borne illness in space. (Picture from http://en.wikipedia.org/wiki/File:ISSSpaceFoodOnATray.jpg.)

The system that the US Space Agency formulated is the basis of modern food safety principles and has been adopted as the Hazard Analysis and Critical Control Point (HACCP) approach to minimising food-associated risk.

It is clear that making food safe by preventing the growth of spoilage and pathogenic organisms has been practised for a very long time. This is important because it allows food to be stored for times when it is less plentiful. We still use ancient food preservation techniques today to make some of our finest delicacies, including salami, yoghurt and cheeses. The idea that 'germs' in food might make the consumer ill is a much more recent (within the last 150 years) leap in understanding and the concept of chemical contamination causing illness is even more recent; these two facets of food safety form the basis of food legislation (see Chapter 16) in most countries.

In the following chapters we will explore what makes food unsafe, the processes that are used to make food safe and the laws that are in place to make it an offence to sell unsafe food. Food is safer now than it has ever been. Read on and you'll find out why.

Chapter 2
Food Risk

Introduction

Everything we do is associated with risk – nothing is risk free. There is a risk you will be killed crossing the road – in fact this is high relative to many of life's other risks, like dying from the ill effects of food. It is important that any risk is kept in perspective and compared to the general risks of everyday life if we are to assess it appropriately and determine how much money and time we spend minimising it. Governments through their regulatory authorities protect their countries' populations by introducing legislation to minimise risks. For example, most countries have laws that make the wearing of seat belts compulsory when travelling in a car; this significantly reduces the risk of dying following a car accident. In addition, most countries have laws that set rules for driving safely (e.g. stopping at a red light); these rules significantly reduce the number of accidents. So combining the road rules, which reduce accidents, and the seat belt laws, which reduce deaths following car accidents, results in a significant reduction in fatalities as a result of travelling in cars. Fines and even prison sentences are applied to 'persuade' people to follow the rules. This is an excellent example of successful risk management – identify the risk and reduce it.

Food risk has to be managed too because it is simply not acceptable for people to die as a result of eating. Acceptability (i.e. what level of risk we will accept) of becoming ill, or even dying, as a result of eating depends on an individual's perspective. A starving person would accept a far greater risk than someone who had too much food to eat; the benefit of the food to the former is survival and to the latter is pure pleasure. Clearly, benefit is an important consideration when determining what level of risk is acceptable.

This chapter will explore the following:

- What is risk?
- How is risk determined?
- Can risk be acceptable?

Food Safety: The Science of Keeping Food Safe, First Edition. Ian C. Shaw.
© 2013 John Wiley & Sons, Ltd. Published 2013 by John Wiley & Sons, Ltd.

- Managing risk
- Food-related risk
- Food risk assessment
- Risk versus benefit
- How regulators minimise food risks

What is risk?

Risk is the probability of something going wrong. It is a word used in everyday conversation; a businessman might find a particular investment too risky (i.e. he fears losing his money), you might exclaim to a friend who asks you to dive off the highest board in the swimming pool, 'Wow, that's too risky for me!' In the latter example it is possible that if your diving skills are not very good you will injure yourself or even die if you dive off the high board. On the other hand, if you are an excellent swimmer who has dived off high boards many times before, your dive is likely to be an enjoyable experience. When you are standing on the high board (or perhaps before you climb the ladder), you make a risk assessment; 'Do I know what I am doing?' 'Will I hurt myself if I do this?' 'Do I want to do this?' The answers to these questions will determine whether you dive or not.

Similarly when you cross the road you assess the risk. You stand at the curb and check whether a vehicle is approaching; if it is, you decide whether you can get across the road without being hit. Most of us include a significant safety margin in this risk assessment to make absolutely certain we don't become one of our country's road traffic accident statistics.

This logical approach to assessing risk might be acceptable to a layman, but if we are to assess risk properly we need some numerical measurements of risk to base our decision on. This is termed quantitative risk. For example, 87% of lung cancer deaths in the UK (2002) were smoking related. This means that you are very much more likely to die prematurely if you smoke. This means that any sensible person deciding whether to smoke or not would regard the high risk of death from smoking as unacceptable and therefore would decide not to take that risk and would not smoke. This is a very simplistic approach to risk assessment; however, not only is there a need to express the risk in numerical terms to enable us to consider it properly, but also it is necessary to consider what factors determine risk and to measure them too.

The factors that contribute to risk

Humans have taken account of risk in their day-to-day decisions since they first walked the plains of the Serengeti about 2.5 million years ago. Arguably all animals are risk averse. My dog will not jump out of the back of my car, presumably because she thinks she might hurt herself if she does – she waits for me to help her down. Even though 'instinctive risk assessment' had been a part of human life since man evolved, it was not until the early 1500s that the Swiss philosopher and scientist Phillippus Aureolus Theophrastus Bombastus von Hohenheim (known as Paracelsus; 1493–1541) (Figure 2.1) defined risk in terms that we still use today.

Figure 2.1 Paracelsus (1493–1541) – the scientist who first defined risk. (From http://commons.wikimedia.org/wiki/File:Wenceslas_Hollar_-_Paracelsus_%28State_2%29.jpg.)

In his tome, *The Four Treatises of Theophrastus von Hohenheim, called Paracelsus*, he wrote:

Alle Dinge sind Gift	All things are poisons
und nichts ohn Gift	and nothing is not a poison
alien die Dosis macht,	it is the dose that makes
daβ ein Ding kein Gift ist	a thing safe

This was inspired thinking for its time because it recognised that the amount of poison ingested determines the effect – the dose makes the poison. Therefore even the most poisonous chemical will not cause harm if ingested at a low enough dose. For example, potassium cyanide (KCN) is very toxic – it would take only 100 mg to kill a person. Nevertheless drinking a glass of 0.0000001 M KCN (aq.) would cause no harm at all because the KCN dose is very, very low (i.e. in a 300 mL glass of 0.0000001 M KCN (aq.) there would be only 0.0065 mg of KCN). This principle applies to any risk situation. Therefore, returning to the example of crossing the road, the lower your 'dose' of car, the safer the crossing! In this situation, we, of course, aim for a dose of zero. The 'thing' we are exposed to is termed the hazard – the car and KCN are hazards. Risk is defined as follows:

$$RISK = HAZARD \times EXPOSURE$$

Exposure is used in place of *dose* because it applies to everything, whereas *dose* applies only to chemicals.

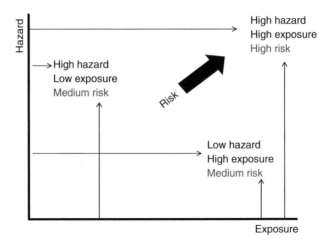

Figure 2.2 The relationship between hazard, exposure and risk – a low level of exposure to a high hazard or a high level of exposure to a low hazard both result in medium risk. The highest risk can only result from high-level exposure to a high hazard.

Therefore to determine the risk of a particular situation we need to know the hazard and measure the exposure to it; the risk associated with a particular hazard goes up with the exposure (Figure 2.2).

Measuring hazard

Hazard is an intrinsic property of something. If the 'something' is a chemical, hazard is a measure of its toxicity; if it is a pathogenic microorganism (e.g. a virus) the hazard would be less well defined, but would be a measure of how harmful the virus could be. For example, Ebola virus results in death and therefore has a very high hazard, whereas Norovirus (see Chapter 4) causes an unpleasant bout of gastroenteritis which rarely causes death and therefore it is a low–medium hazard. The risk associated with both chemical and microbiological hazards is determined by the exposure level. For example, if exposure to Norovirus is high (i.e. millions of viral particles) it is very likely that severe but short-duration gastroenteritis will result, but if exposure is very low (i.e. a few tens of viral particles) the body's immune system is likely to prevent infection and therefore gastroenteritis will not develop.

To measure chemical hazard, groups of animals are exposed to the chemical at different doses and the dose at which a toxic effect occurs is noted. If the toxic effect (end point) measured is death, the dose that kills 50% of one of the groups is the LD_{50} (lethal dose for 50% of a population). LD_{50} tests are rarely carried out now because they are considered inhumane and therefore LD_{50} has been replaced by the No Observable Adverse Effect Level (NOAEL). The NOAEL is determined using a non-lethal end point; for example, the effect of a test chemical on the liver which can be measured by a change in the serum activity of the liver enzyme glutamate pyruvate aminotransferase (SGPT). The NOAEL is the dose given to the group of animals immediately below the group showing the effect (e.g. raised SGPT) (Figure 2.3).

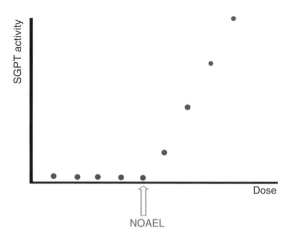

Figure 2.3 Determination of NOAEL using serum glutamate pyruvate aminotransferase (SGPT) activity as a marker of liver damage in an animal experiment. The NOAEL is the dose given to the dose group immediately below the group that shows elevated SGPT (i.e. effect).

Acceptable Daily Intake (ADI)

LD_{50} and NOAEL tests are carried out in animals (often rats) and there is significant debate about whether these hazard measurements can be extrapolated to humans. Because of the uncertainty of this, the Acceptable Daily Intake (ADI) is often calculated for chemicals that might be ingested by humans (e.g. chemical contaminants in food). The ADI is based on the NOAEL following long-term dosing.

$$ADI = \frac{NOAEL}{SF}$$

where SF = Safety Factor, usually 1,000.

The ADI includes a large safety margin which takes account of the variability of toxic effects between different animal species. The ADI is the dose at which the risk is theoretically zero for a lifetime exposure.

Maximum Residue Level (MRL) and Maximum Limit (ML)

The MRL is the maximum residue of a chemical (e.g. a pesticide) that will occur in a crop following use of a particular agrochemical (e.g. a pesticide) in accordance with Good Agricultural Practice (GAP). GAP comprises a set of rules covering the use of agrochemicals on farms; it covers issues such as when it is appropriate to spray a crop (e.g. to prevent spray drift crops should not be sprayed on windy days) and when it is appropriate to harvest the crop for human or animal consumption (e.g. withdrawal times are used to ensure that the crop is not harvested when residues of the agrochemical will be at their highest). The MRL is therefore not toxicologically based but is simply the maximum concentration of the agrochemical that will be present if the chemical is used properly. It is a trading standard that ensures that exports of agricultural products (e.g. wheat) have been grown in accordance with internationally agreed standards. Countries monitor imports and if MRLs are exceeded they will often refuse to allow the product to enter their country. It would be very serious in a political sense if a country exported a product that exceeded the MRL for a particular residue.

Maximum limits (MLs) are similar to MRLs but relate to environmental chemical contaminants (e.g. lead) rather than chemicals that have been used intentionally as part of normal farming practice (e.g. pesticides). For example, lead levels in prunes (dried plums) are sometimes rather high – MLs are used to ensure that prunes are not sold with unacceptable lead levels (see *Food risk assessment – case examples; Lead in prunes*).

MRLs and MLs are set by international committees (e.g. Codex Alimentarius, a committee of the FAO and WHO). FAO is the Food and Agriculture Organisation of the United Nations (UN). Its major role is to defeat hunger by ensuring food supply, particularly to developing countries. The World Health Organisation (WHO) is also an agency of the UN and is the coordinating authority on international public health. In 1963 the FAO and WHO joined forces and created the Codex Alimentarius Commission (usually referred to as 'Codex'). Codex develops and polices food standards (e.g. MRLs) worldwide as a means of ensuring food is safe and that internationally agreed standards are met when food is traded internationally.

It is important not to regard MRLs and MLs as measures of hazard. The popular press often does because it does not understand how these values were determined and assumes they have a toxicological basis – they don't.

Tolerable Daily Intake (TDI) and Tolerable Weekly Intake (TWI)

The TDI and TWI are the amounts of a substance in food that if consumed every day or every week for an entire lifetime would not be expected to cause harm. They are calculated in the same way as ADIs – i.e. from toxicity data with applied safety factors (see *Acceptable Daily Intake (ADI)*). TDIs and TWIs are used for chemical food contaminants that are not incurred as a normal part of food production, processing or preserving (e.g. heavy metals).

Determining risk

Armed with a measure of hazard (e.g. NOAEL) and an estimate of exposure level (i.e. dose), risk can be determined. Aspirin (acetylsalicylic acid) is a good example to illustrate risk. The oral LD_{50} for aspirin in the rat is 200 mg/kg body weight. Assuming this hazard level can be extrapolated to humans, it would require a dose of about 12 g of aspirin to kill a human (average human weight = 60 kg). The recommended dose of aspirin is two tablets; each tablet contains 300 mg aspirin, therefore the dose is 600 mg, which is only 5% of the lethal dose – clearly this represents a very low risk of death to the consumer. However, if 30 aspirin tablets were taken at once (i.e. 9 g) this would result in a dose that is 75% of the lethal dose and therefore presents a very high risk of death. The hazard remains the same (i.e. the LD_{50} of aspirin), but the risk changes with dose.

Zero risk

No activity has zero risk; the risk might be vanishingly small, but it cannot be zero. At the other end of the spectrum, even high hazards can result in near zero risk. Take the example of a ferocious animal that has not been fed for several weeks; it is likely to attack anyone who approaches it – it is a high hazard. However, if that ferocious animal is locked in a strong metal cage, it would be safe to approach the animal because it could not bite you. In this case the exposure is zero and the hazard is very high, which makes the risk apparently zero.

$$HAZARD \times EXPOSURE = RISK$$
$$High \, HAZARD \times zero \, EXPOSURE = zero \, RISK$$

However, I said that zero risk was not possible – even in this scenario it is possible (although very unlikely) that the vicious creature could force the lock on its cage open, then the risk would be far from zero! (Figure 2.4).

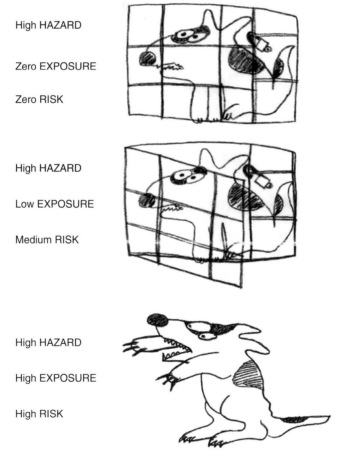

High HAZARD

Zero EXPOSURE

Zero RISK

High HAZARD

Low EXPOSURE

Medium RISK

High HAZARD

High EXPOSURE

High RISK

Figure 2.4 A high hazard ferocious animal apparently has zero risk when it is locked in a strong cage ... but it is just possible that it will break out and then the risk will be far from zero! (Illustration from Shaw (2005), *Is it Safe to Eat?* Springer, Berlin.)

Consider the aspirin example; the therapeutic dose is a very small proportion of the LD_{50}, nevertheless some people are allergic to aspirin and a dose in the microgram (μg; 10^{-6} g) range could lead to a fatal anaphylactic allergic reaction.

The food situation is often difficult for consumers to accept (I will discuss this later in this chapter) because the concentrations of chemical contaminants in food are usually very low (i.e. low exposure), but often they are high hazard chemicals (e.g. pesticides; see Chapter 7), but, of course, this results in low risk. It is anathema for people who do not fully understand the concept of risk to accept that a high hazard chemical contaminant can result in a low risk. For example, if I gave you a cake and told you that there was cyanide (high hazard) in it, you might not want to eat it. Now that you know more about risk assessment you might be happier if I told you that the contamination level was extremely low and that I only have one small cake for you to eat (extremely low exposure), which means that the risk is extremely low. Armed with this information you should be happy to eat the cake, but you might still be a little concerned when you took the first bite. In fact, almonds contain cyanide and therefore cakes containing almonds also contain cyanide – I bet you have eaten such foods in the past and not thought anything of it!

Almonds are the kernels from the fruit of the tree *Prunus dulcis*. There are different genotypes of the *P. dulcis* that produce kernels with different degrees of bitterness and sweetness. The bitter taste is caused by the cyanogenic glycoside amygdalin, which is broken down by enzymes in the kernels (emulsin) or the mammalian intestine (β-glucosidase) to generate cyanide which has a bitter taste (Figure 2.5). The more cyanide, the more bitter the flavour. In Italy, bitter almonds are used to make the liqueur Amaretto which has a strong almond flavour. Just a few very bitter almonds could be fatal to a human because of their high cyanide content – there is about 5 mg of hydrogen cyanide (HCN) per bitter almond and the lethal oral dose of HCN in humans can be as low as 1.5 mg/kg (for a 70 kg human lethal dose=105 mg); therefore just 20 bitter almonds could be fatal.

So are we dicing with death eating a traditional Dundee cake? Providing sweet almonds were used to make it, there is no toxicological issue whatsoever, even if the almonds were bitter; since HCN boils at 25.6°C it is very likely that most of it will boil off during baking (a Dundee cake is baked for more than an hour at 180°C).

Risk assessment for a Dundee cake

A typical recipe for a Dundee cake is as follows:

Butter	150 g
Sugar	150 g
Eggs	3 (50 g each)
Flour	225 g
Baking powder	4.5 g
Currants	175 g
Sultanas	175 g
Glace cherries	50 g
Candied peel	50 g

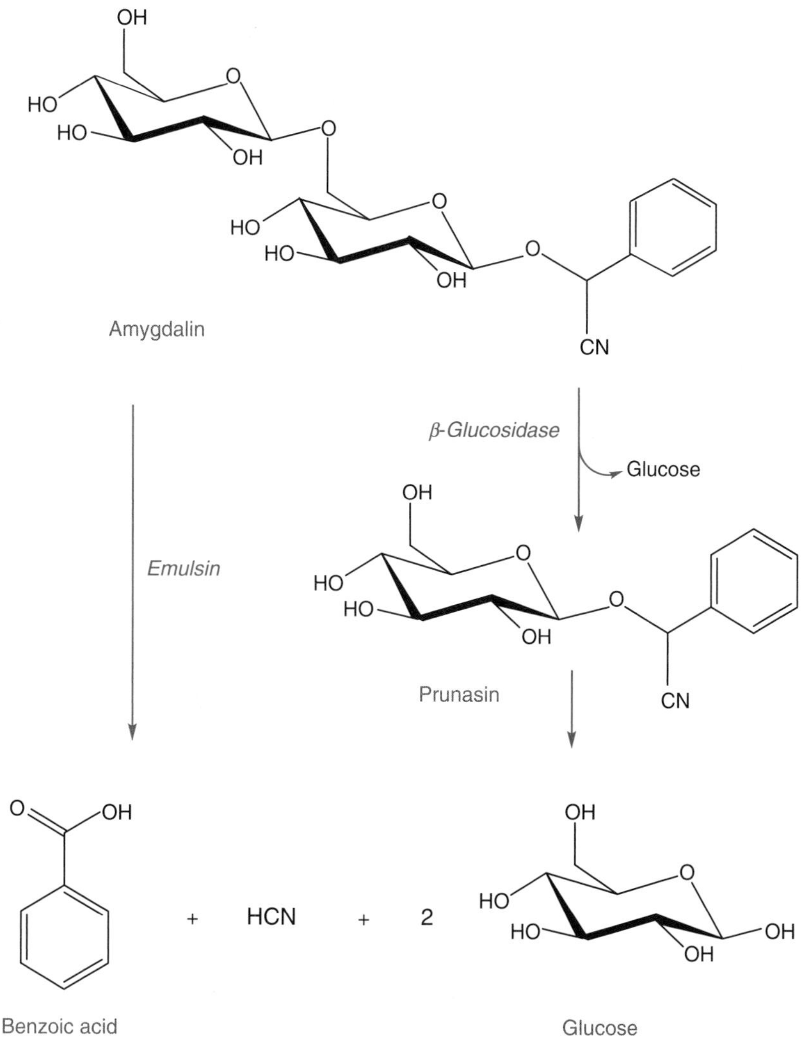

Figure 2.5 A mixture of enzyme-catalysed and chemical breakdown of amygdalin (a cyanogenic glycoside) releases cyanide and benzoic acid in almonds.

Ground almonds   15 g
Orange zest
Lemon zest
Blanched almonds   50 g

The concentration of HCN in sweet almonds is in the range 0–10.5 mg/100 g, so taking the worst case example (i.e. 10.5 mg/100 g) the cyanide level in the raw Dundee cake is calculated as follows:

[HCN] in raw almonds   = 10.5 mg/100 g
Mass of almonds in Dundee cake   = 65 g

Mass of HCN in these almonds	= 6.8 mg
Approximate total mass of cake	= 1.2 kg
[HCN] in raw cake	= 0.06 mg/100 g
Approximate mass of slice of cake	= 120 g
Absolute dose of HCN	= 0.072 mg
Dose of HCN (70 kg human)	= 0.001 mg/kg body weight

The lethal oral dose of HCN in humans is approximately 1.5 mg/kg body weight; therefore the dose of HCN in a slice of Dundee cake is only 0.07% of the lethal dose.

The FAO and WHO have both set the ADI for cyanide at 0.05 mg/day. This means that a daily dose of 0.05 mg cyanide every day for a lifetime would have no adverse effect on the consumer. Our slice of Dundee cake results in a dose of HCN below the ADI (2% of ADI). In fact, it is very likely that the HCN levels in the baked cake would be significantly lower than those we have calculated for the raw product because of the low boiling point of HCN. Therefore on all counts it is safe to eat Dundee cake – now that is a relief!

Other kernels from fruits related to almonds have cyanogenic glycosides; some have very much higher concentrations, indeed concentrations that make the risk of foods containing them greater than most people would be prepared to take.

The French bake a tart containing the kernels of apricots. The apricot tree (*Prunus armeniaca*) is very closely related to the almond. We eat only the kernel of the almond and usually eat only the fruit (pericarp) of the apricot. However, the apricot stone is very like the almond and has a similar kernel, but the apricot kernel has much more amygdalin than sweet almonds and therefore is able to generate more HCN (Table 2.1).

The high concentration of HCN in the apricot kernel gives it its very characteristic, pungent, almond-like flavour which is why it is used to make the delicious French apricot tart. A recipe for *tarte aux abricots* follows:

Shortcrust pastry	250 g
Apricots	12
Ground almonds	100 g
Crème fraîche	100 mL
Egg yolks	3
Sugar	4 tablespoons

Table 2.1 Concentration of HCN in sweet almond and apricot kernels. (Data from Dicenta *et al.* (2002) *Agricultural and Food Biochemistry,* **50**, 2150; New Zealand Food Safety Authority, Cyanogenic Glycosides Information Sheet, available at http://www.nzfsa.govt.nz/science/chemical-information-sheets/fact-sheet-cyanogenic-glycosides.pdf.)

Kernel	mg HCN/100 g
Sweet almond	0–10.5
Apricot	8.9–217

Apricot kernels contain 90–2,170 mg HCN/kg and a kernel weighs approximately 0.7 g. This means that 12 apricots would have approximately 8.5 g of kernels; therefore the *tarte aux abricots* would contain 0.77–18.4 mg HCN. Taking the worst case (i.e. 18.4 mg HCN/tart) and assuming that a tart serves six people, the approximate dose per helping would be 3.1 mg HCN. Assuming an average person weighs 70 kg this means that the dose = 0.04 mg/kg body weight. Remember, the human lethal dose for HCN is approximately 1.5 mg/kg body weight and the FAO ADI is 0.05 mg/day; therefore the worst case HCN dose is worryingly close to the ADI, but only 2.7% of the human lethal dose. This means that consumption of the tart every day for an entire lifetime could have an effect on the consumer, but that it is very unlikely to cause death. You would have to eat six worst case tarts to receive a fatal HCN dose. This, of course, is highly unlikely which means that the risk of harm from an occasional slice of *tarte aux abricots* is low. Remember also that some of the HCN would be lost during baking, but the cooking period is very short and therefore much of the HCN is likely to remain in the tart.

A report in the *Annals of Emergency Medicine* in 1998 records a case of a 41-year-old woman in the USA who consumed 30 apricot kernels in one sitting and suffered the characteristic acute toxic effects of HCN within 30 minutes. She collapsed on her bathroom floor, but fortunately was able to telephone her friend before this and so the emergency services were able to administer a cyanide antidote (amyl nitrate and sodium thiosulphate) and she survived. Thirty (approximately 21 g) worst case apricot kernels could contain 45 mg HCN; this is a dose of approximately 0.65 mg/kg body weight which is 43% of the human HCN lethal dose, which explains why the woman was so ill.

Acceptable risk

Whether a particular risk is acceptable or not depends upon the associated benefit and depends on one's personal perspective. The French apricot tart discussed above is a good example. The flavour of the tart is wonderful and therefore many people would accept the risk of harm due to its cyanide content simply because it tastes so good and gives them great pleasure. This is an acceptable risk.

Smoking – an acceptable risk?

Many people choose to smoke cigarettes even though it is well known that the activity is associated with lung diseases including bronchitis, emphysema and cancer. The WHO has estimated that smoking causes 10 million deaths worldwide per year and that the average smoker's life expectancy is reduced by 12 years. A 50-year-old man who has smoked 20 cigarettes a day since he was 18 years old has a 1% risk of developing lung cancer during the next 10 years of his life (you can calculate your own risk using the Sloan-Kettering Lung Cancer Risk Assessment tool at http://www.mskcc.org/mskcc/shared/forms/Nomograms/flash/load.cfm?type=lung&width=585&height=445&title

=LungCancerRiskAssessment). Not all cases of lung cancer are smoking-related. Indeed approximately 10% of lung cancer sufferers are non-smokers; the risk of a 50-year-old non-smoking man developing lung cancer during the next 10 years of his life is about 0.1%.

Despite the horrifying health statistics, 1.3×10^9 people worldwide smoke – clearly they either don't understand or they accept the risks. Many smokers are from developing countries where the persuasive powers of advertising still exist; however, 22% of US (i.e. well-informed people in a developed country) citizens still smoke (WHO data for 2003). The most likely reasons for such a large number of people accepting the risks of smoking is because they are addicted to the nicotine content of tobacco and therefore find it difficult (and an unpleasant experience) to give up, or they enjoy smoking and don't want to give up; i.e. they accept the risk.

Toxic fugu sashimi – tasty, but potentially lethal

Fugu (Figure 2.6), a puffer fish used in the preparation of sashimi (Figure 2.7) in Japan, contains highly toxic tetrodotoxin (Figure 2.8):

LD_{50} [mouse] $= 10\,\mu g/kg$
Lethal dose for a human based on mouse toxicity $= 0.07\,mg$

Fugu sashimi is an expensive delicacy in Japan and is served in specialist restaurants (Figure 2.9) where the chefs are specially trained in its preparation. Tetrodotoxin is present in the liver and bile of the fugu and the chef cuts the fish in such a way as to contaminate the flesh with the toxin. He does this because when consumed tetrodotoxin causes a tingling sensation on the lips

Figure 2.6 Fugu on a Japanese fishmonger's slab. (From http://en.wikipedia.org/wiki/File:Fugu.Tsukiji.CR.jpg. This Wikipedia and Wikimedia Commons image is from the user Chris 73 and is freely available at //commons.wikimedia.org/wiki/File:Fugu.Tsukiji.CR.jpg under the creative commons cc-by-sa 3.0 license.)

which the sashimi consumer enjoys. In fact to some people the benefit of the experience of eating fugu sashimi outweighs the very real risk of death from overconsumption of the toxin. I must admit that when I visited Japan recently and was offered fugu sashimi my very rapid risk assessment resulted in my saying 'No thank you', much to the surprise of my Japanese friends who relished the experience. Clearly the risk to us all was the same, but our assessment of the benefit was different, which determined whether we ate or refused the sashimi.

Every year in Japan four or five people die of tetrodotoxin poisoning; indeed the Japanese government has recently introduced legislation that requires fugu sashimi chefs to be specially trained so that they do not overcontaminate their sashimi with lethal tetrodotoxin. This is an attempt at risk minimisation.

If this were a drug for the treatment of a life-threatening cancer the considerable risk associated with the drug would be offset by the 50% cure rate. The risk of death after taking the drug is less than the risk of dying from the disease that the drug is being used to treat. On the other hand, if the drug had

Figure 2.7   Fugu sashimi. (From http://en.wikipedia.org/wiki/File:Fugu_ sashimi.jpg.)

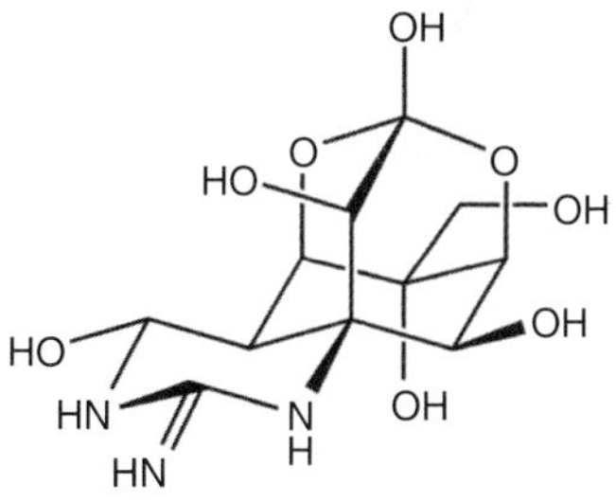

Figure 2.8   The molecular structure of tetrodotoxin – the lethal toxin from fugu.

Figure 2.9 A fugu sashimi restaurant in Kyoto, Japan – the fish hanging above the entrance is a dried fugu. (Photograph by Shaun Hendy.)

been developed for relieving the symptoms of the common cold, the risk would not be acceptable because the risk of harm following medication is greater than the lasting harm caused by the disease the drug is being used to treat. This example illustrates that the risk associated with exposure to a particular hazard can be acceptable in some circumstances, but not in others.

Risk versus benefit

Determining what is an acceptable risk depends on the benefit. For example, consider a new medicine that is being assessed by a regulatory authority as part of the process of deciding whether or not it should be granted marketing approval. The hypothetical drug has the following toxicity profile, therapeutic dose and efficacy:

LD_{50} [rat]=25 mg/kg body weight
Fatal dose to human based on rat toxicity=1.5 g
NOAEL [rat]=0.4 mg/kg body weight
Salmonella typhimurium His⁻ test positive (i.e. the drug is mutagenic)
Therapeutic dose=5 mg/day
Efficacy: 50% cure rate

If the drug was proposed for the treatment of the common cold – which carries a very low risk of death and a near 100% recovery rate – it would not be acceptable because its toxicity profile is not outweighed by the benefit of the treatment. On the other hand, if the drug was proposed for the treatment of a particular

Acceptable risk

Unacceptable risk

Figure 2.10 If the benefit outweighs the risk the risk is acceptable (top); if the benefit does not outweigh the risk the risk is not acceptable (bottom).

cancer with a 95% mortality rate, the risk is acceptable because despite its harsh toxicity profile there is a 50% chance of cure, which is better than not being treated at all. So the benefit determines whether a risk is acceptable or not. We are prepared to take greater risks for larger benefits (Figure 2.10).

In the context of food it is more difficult to assess the risk benefit balance. A Japanese person is much more likely to take the risk of eating fugu sashimi because in their culture it is considered a great delicacy, whereas a New Zealander (like me) is much less likely to take the risk. Therefore risk benefit assessments are not absolute, they depend on the circumstances (e.g. country in which the assessment is being made). Risk assessment is not always a precise science.

Risk perception

Perception plays a key role in an individual's risk-related decisions. One person's acceptable risk is another's unacceptable risk; the fugu sashimi and cigarette smoking examples illustrate this well. In addition, individuals will rank risks differently and will change their views on how they rank risks with time. Vaccination is an excellent example of this. The measles, mumps and rubella (MMR) vaccine when first introduced (late 1960s) was received by mothers with great enthusiasm because they knew very well from first-hand experience how serious measles, mumps and rubella could be to their child (in the case of rubella, to their unborn child). The development of a vaccine significantly reduced the risks associated with these diseases. Most mothers probably did not consider the risk associated with the vaccine because this was much lower than the risks of the diseases themselves. In time (about 20 years) mothers forgot about the severity of the diseases because the success of vaccination programmes meant that the diseases became very rare, so most mothers had not experienced mumps, measles or rubella first hand. At this time, one or two cases of adverse reaction (autism) to the vaccine were reported. This became the key risk from the mother's perspective and outweighed the benefits of vaccination. This is because the mother's perspective had changed; she ranked the risk of her child developing autism as a result of vaccination higher than the risk of him or her contracting one of the high-risk diseases that the vaccine had so successfully almost eradicated. The outcome of this changed perception is that some mothers decided that the risk of vaccination was too great and so they elected not to vaccinate their child. For a vaccination programme to be successful high vaccination compliance is essential, otherwise reservoirs of disease develop in unvaccinated communities. That is exactly what is happening now; measles in particular is reappearing. It won't be long before mothers who personally experience measles will be able to assess the risk of vaccination in the context of the seriousness of the disease and decide to vaccinate their child.

In the UK in 1996 92% of children were vaccinated against measles (using MMR vaccine). Shortly after this the controversy over MMR being linked to autism began and by 2002 vaccination coverage had reduced to 84%. During the period of high vaccination coverage (1998) there were 56 cases of measles in the UK; by 2006 when vaccination had declined significantly there were 450 cases of measles (including one death; this was the first measles death in the UK since 1992) in the first 5 months of the year – so, in just 5 months of a low vaccination year there had already been eight times the number of measles cases that had occurred in a complete high vaccination year. Clearly the controversy over perceived vaccine risk resulted in a very much greater risk from the disease itself.

Risk perception can also be influenced by the media. This happened during the UK's bovine spongiform encephalopathy (BSE) crisis (see Chapter 6). The risk of contracting variant Creutzfeldt-Jakob disease (vCJD) from BSE-contaminated beef in the UK in the 1990s was very low – there were 172 cases of vCJD in the UK between 1995 and 2010 which means that the risk of contracting vCJD was very approximately 1 in 5.5 million. This was estimated by normalising the vCJD rate to the mean of 11 cases/year over the 15 years since its discovery, and using the 2008 UK population of 61 million. One in 5.5 million is a low risk when compared to the acceptable risk of dying in a road traffic accident. In the UK during the same period the risk of dying in a car accident was 1 in 35,714 (equivalent to 154 in 5.5 million), but still a large proportion of the UK population rated the vCJD risk greater than travelling in a car. Their perception of the risk had been distorted by the huge furore in the news media which resulted in their wrongly assessing the risk (Figure 2.11). The media's misrepresentation of the risk and the consumers' assimilation of this led to a decline in beef consumption which eventually led to the collapse of the UK beef industry.

As the media reported the major issues after the discovery of BSE in 1986, consumption of beef declined as the consumer's perception of the risk became more and more inflated. Eventually (1997) the low demand for beef resulted in a significant reduction in its price which in turn appears to have led to increased consumer interest. At this point the benefit of economy out-weighed the risk associated with BSE for some consumers.

The BSE/ CJD saga has led to many erroneous risk-based decisions being made around the world. For example, there is a shortage of donated blood for transfusion in New Zealand, but the New Zealand authorities will not allow anyone who travelled to or lived in the UK during the height of the

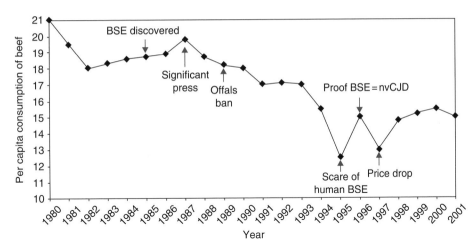

Figure 2.11 The consumption of beef in the UK before and during the BSE saga. (Data from UK Agriculture Committee (2002). From Shaw (2005), *Is it Safe to Eat?* Springer, Berlin.)

BSE crisis to give their blood. The risk of these people transferring the BSE prion (the causative agent of BSE and vCJD) via donated blood is miniscule, but still the authorities persist with the rule, probably because the public perceive the risk as being greater than it really is. Yet arguably the authority's role is to protect their population from real risk, not perceived risk!

If you were in a situation in which you were bleeding to death in New Zealand and the only blood available for transfusion was from someone who had lived in the UK for 6 months in 1991, would you say no? Of course you wouldn't – your risk assessment would rank the risk of your situation (i.e. bleeding to death) as high and the risk of the solution (i.e. transfusion) as low.

Risk perception in relation to food is a key factor in the choices consumers make. Often, novel foods are regarded with greater scepticism (i.e. high risk) because the consumer is not provided with all of the information necessary to make a risk assessment. On the other hand, familiar foods are likely to be regarded as low risk even if they are not. A good example of this thinking is the consumer's response to genetically modified food (GMF) (see Chapter 10). There are significant concerns about the environmental impact of GMF production, but these are unarguable and in my opinion must be dealt with if we are to accept genetically modified (GM) crops. However, the health implications of GMF are far less certain and most scientists, while accepting that GM crops are not biochemically identical to their conventional counterparts, do not expect them to pose a significant risk to the consumer from a food safety perspective. On the other hand, it must be noted that there are a small number of scientists who would vehemently disagree with this statement. GMF is new and therefore consumers have little information upon which to make a sensible risk assessment, and so they become risk averse and assume that unknown risk is high risk. This has led to some countries (e.g. New Zealand) not permitting GMF to be included in their country's food chain. Again this decision has been driven by an uninformed public rather than a proper risk assessment. It is important that consumers are given as much information as possible to help them to make their risk assessment before an attempt is made to introduce new and novel foods. In this case the consumer is applying the precautionary principle (see below) which is a valid way of dealing with uncertain risk even if the uncertainty is due to misperception of risk.

The precautionary principle

If you are uncertain about anything you will proceed with caution; you don't know what might be around the corner. If you attempt to cross the road on a blind bend you will be far more cautious than if you can clearly see approaching traffic. The same applies to any uncertain risk situation; the GMF example illustrates this very well.

Some toxicologists and regulators frown upon the precautionary approach to risk assessment because they say it impedes development of new products. When toxicological risk assessment was first seriously and methodically applied to medicines licensing in the UK in the 1980s (under the Medicines Act 1981) it became almost impossible for pharmaceutical companies developing anticancer drugs to get them approved for use in patients. This was because almost all anticancer drugs have horrific side effects, since they are designed to be toxic – this is how they kill cancer cells. Clearly the precautionary approach to medicines licensing did not work for these compounds. A major debate ensued in the early 1980s amongst toxicologists and legislators about how best to deal with this. This is when the precautionary principle began to be regarded as inhibitory to progress. In order to facilitate anticancer drug development, the toxicity testing requirements for such drugs were changed in the UK and the interpretation of toxicology findings modified to allow the toxicity associated with their mode of action.

Nowadays the precautionary principle is rarely applied by regulatory authorities, but is often argued by protest groups. For example, protesters against GMF want such foods to be banned on the grounds that they might be harmful to consumers even though a considerable body of research suggests that they are safe.

A good example of the importance of the precautionary principle in food risk decision making relates to pesticide residues in food. Pesticides are used worldwide in most modern agricultural systems; as a result their residues remain in the crops that have been treated with pesticides during their production. It is inevitable that the consumer of food made from these crops (e.g. bread made from pesticide-treated wheat) will ingest very small quantities of pesticides. The question is, will such low doses of pesticide cause harm to the consumer? Toxicity studies suggest the risk is so low that it is negligible. However, the risk calculations relate to individual pesticides and most consumers ingest complex cocktails of pesticide residues, albeit at infinitesimally low doses. Those who protest against the use of pesticides use the precautionary principle in their argument, saying, 'but we don't know what effect cocktails of pesticides will have on the consumer'. They are right, but is this reason enough to ban pesticide use when without them farmers would be unable to produce food efficiently and economically? If you have ever tried to grow cabbages or Brussels sprouts without pesticides you will

Figure 2.12 A Brussels sprout plant grown without the use of pesticides in the author's garden; it is clear that insects have severely damaged the crop. (Photo by the author.)

understand the problem – they become infested with cabbage white butterfly (*Pieris rapae*) caterpillars which makes them inedible (Figure 2.12). Spraying with a pyrethroid insecticide solves the problem and means that cabbage and Brussels sprouts can be on the menu. If we apply the precautionary principle there would be no pyrethroid insecticides and perhaps no commercially available cabbage or Brussels sprouts, or at least not at a price most of us would be prepared to pay.

Clearly, if we were to apply the precautionary principle as a matter of course to all risk assessment situations, very little progress would be made. However, it has an important part to play in the risk assessment of novel products or processes, because it ensures that regulators don't approve them too quickly before their risk to the consumer is well understood and accepted (by regulators).

Food risk assessment

There are many hazards associated with food and eating – we have considered several above (e.g. tetrodotoxin in fugu sashimi). There are physical hazards, like bones in fish, that present a risk of choking which might lead to death if first aid is not at hand, and there are chemical hazards relating either to natural chemical components of food (e.g. HCN in almonds) or man-made chemical contaminants of food. The latter fall into two broad categories: those we intentionally add to food (e.g. chemical preservatives) and those we did not intend to contaminate food (e.g. residues of pesticides used in food production). Assessing the acceptability of the risks of each of these contaminants of food chemical hazards depends upon their benefits, but determining the benefit to the consumer of a particular risk is often not easy.

Consider a food preservative hazard; the preservative is added to the food to prevent food spoilage and the possibility of pathogenic microorganisms infecting the food which in turn would have health implications for the consumer. There is a clear benefit to the consumer associated with the risk of the preservative. On the other hand, the risk to the consumer of pesticide residues in food is very much more difficult to weigh up with respect to benefit because on this occasion the risk is to the consumer and the benefit is to the farmer – the use of pesticides helps the farmer to grow his crops and so maximise his income. However, the consumer also gets the benefit of being able to eat the crop (or food made from it) and buy it at a reasonable price (it might be more expensive if the farmer's pesticide-aided productivity was lower). There is significant debate about the risk of pesticide residues to the consumer simply because the consumer is not the recipient of all of the benefits and therefore is not prepared to accept the risk.

It is difficult to determine whether a particular risk is acceptable. If I determined the risk of a particular food contaminant as low (e.g. a chemical preservative) and the benefit as high (i.e. reducing food-borne illness), it is still difficult for someone about to eat the preserved food to decide whether or not they want to eat it. For this reason, we usually have regulatory committees composed of experts who make risk assessments on the consumer's behalf. In order for this to be successful we must trust the committees' judgements. As an aid to communicating risks to consumers, the risk is often set in the context of an everyday risk like crossing the road or flying in an aeroplane (relative risk) – this puts the food risk in perspective.

This all went wrong during the BSE (Mad Cow disease) crisis in the UK in the late 1980s and early 1990s. The public were justifiably concerned about the risk of eating beef that might have originated from BSE cattle for fear of contracting the human form of BSE, CJD (see Chapter 6). Initially they accepted government and experts' assessment that the risk was low, but as it became clear that consumption of BSE beef could cause CJD, they lost faith in government and expert advice – this is illustrated well by the UK beef consumption statistics (see *Risk perception* and Figure 2.11).

The public's lack of faith in government food advice and risk decisions led to the UK reorganising its food safety system by introducing a Food Standards Agency (FSA) led by an acknowledged independent expert, Sir John Krebs, with minimal if any conflicts of interest (e.g. the government could not modify what he had to say because he did not report to a government department). This was an attempt to restore the consumer's faith in food risk assessment, and it worked. Acceptance that a country has robust and reliable food risk assessment systems is very important in relation to the export of food too. Importing countries rely upon the integrity of food risk assessments of exporters as a means of ensuring that their populations are not exposed to unacceptable risks. To verify the validity of exporters' risk assessment processes, importing countries often police imports by carrying out random analysis of product samples to check that the exporter's claims are valid. This adds to their country's consumers' faith in the safety assurance processes that are intended to protect them.

The UK abolished the FSA in 2011; presumably the need for public assurance of the safety of their food had diminished as the BSE saga faded into history. Similarly, the New Zealand Food Safety Authority (NZFSA) was closed in 2011 and its function moved into the Ministry of Agriculture and Forestry (later re-named the Ministry for Primary Industries).

Relative risk and risk ranking

In order to assess a particular risk it often helps to compare it to another risk. If the comparator risk is accepted and the two risks are similar this is good reason to accept the new risk under consideration. We discussed BSE and vCJD above and compared the risk of contracting vCJD in the UK to the risk of dying in a car accident. The latter is very much greater than the former and therefore it is arguable that we should accept the former risk. Indeed banning cars in the UK would have a greater impact on health than not eating beef – this, of course, does not take account of the benefits of having a car. It is often not appreciated how small food risks are when compared to other risks we take for granted (Figure 2.13) – relative risk can be used to illustrate this.

Crossing the road and driving a car are good examples of risks we accept and take for granted. If we compare food risks to these acceptable risks in a developed country, the food risks pale in comparison. This is interesting and perhaps a useful way of making the point that in the context of life's other risks, food is associated with relatively low risk while providing a huge benefit. However, just because a risk (e.g. of food) is low relative to the risk of an everyday activity (e.g. crossing the road), it does not mean that we should automatically accept it. Every risk, however small, adds to life's overall risk. It is our duty to minimise all risks rather than accepting them, even if they are small. So even though the risk of dying from campylobacteriosis (caused by the food-borne pathogenic bacterium *Campylobacter jejuni*; see Chapter 3) is very low (one person a year dies from campylobacteriosis in New Zealand, which has a population of four million) in comparison to dying in a car accident (331 people died in car accidents in New Zealand in 2008), arguably it is not acceptable that anyone dies from a food-related illness. There is, of course, great debate around such issues, and the overriding philosophy is that we must strive to reduce risk as much as possible.

Since campylobacteriosis is a particular problem in New Zealand (Figure 2.14), risk management strategies are an important means of reducing the risk of contracting campylobacteriosis and so reducing the risk of the disease to New Zealand consumers.

Risk management

Remember, RISK=HAZARD×EXPOSURE: therefore, if we reduce exposure to a particular hazard we will reduce the risk. If we minimise exposure, we will minimise the risk. The latter situation relies upon accepting a particular level of risk because it is not possible to achieve zero risk.

In the campylobacteriosis example discussed above, the hazard is the pathogenic bacterium *C. jejuni* which contaminates food (usually meat, often chicken) and thus the risk can be reduced by reducing exposure, or minimised by minimising exposure. Therefore to manage the risk of campylobacteriosis we must reduce exposure to *C. jejuni*. Risk management requires a detailed understanding of exposure routes; New Zealanders enjoy barbequing in their gardens during their long warm summers and chicken is a very popular meat to cook in this way. *C. jejuni* is a gut bacterium in chickens that during the

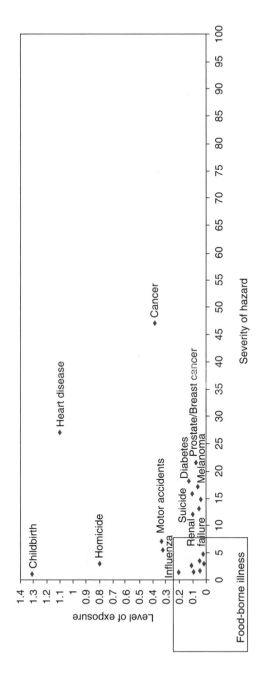

Figure 2.13 Some common food risks (e.g. food-borne illness due to viral and bacterial pathogens) in the context of other risks of living (e.g. cancer). The food risks are very low in comparison to some risks we accept. (From an unpublished study by I.C. Shaw and E.A. Harris (2002).)

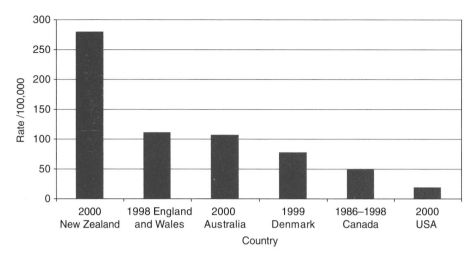

Figure 2.14 The incidence of campylobacteriosis compared to other countries. New Zealand's incidence of the disease is higher than other countries, which is why, even though the risk is low relative to accepted everyday risks, New Zealand strives to reduce the number of cases. (Data provided by Dr Andrew Hudson, ESR Ltd, New Zealand.)

slaughter process contaminates the chicken meat – it is very difficult to avoid this contamination. If the chicken is cooked properly (i.e. the cooking temperature is high enough) the *C. jejuni* will be killed and therefore the risk of contracting campylobacteriosis from a properly cooked piece of chicken is virtually zero (i.e. the ideal risk situation). It follows that, since the campylobacteriosis rate in New Zealand is high, New Zealanders are consuming *C. jejuni*-infected chicken. This suggests that they do not cook their chicken properly and so allow some of the pathogen to survive and infect its consumer. This hypothesis, linked to New Zealanders' penchant for barbequing, suggests that barbequing might be responsible for undercooking. If this were the case, the risk management protocol would be to teach New Zealanders to barbeque their chicken properly.

Studies (see *Further reading*) showed that barbequing chicken is an effective way of killing *C. jejuni* because the bacterium is very temperature sensitive and it is not difficult to attain a sufficiently high temperature to kill the bacteria on a piece of chicken being barbequed. However, careful observation of people barbequing showed that they use the same tongs to put the raw chicken on to the barbeque as they did to serve the cooked chicken. This is where the problem lies; the *C. jejuni* infects the outside of the chicken and is transferred to the tongs when the raw chicken is put on to the barbeque. It remains alive on the tongs because they do not attain a high enough temperature to kill the bacteria during the cooking process. When the infected tongs are used to serve the barbecued chicken the chicken is contaminated with live *C. jejuni* which are able to infect the consumer and cause campylobacteriosis a few days later. For this reason, an appropriate risk management approach would be to educate consumers how to barbecue their chicken properly, making sure they understand the cross contamination possibilities of the barbeque tongs.

The NZFSA is responsible for managing food risk in New Zealand. It has produced a series of television infomercials that explain how and why chicken should be properly cooked. This is likely to be an effective risk management strategy because a large proportion of the population watch television and they will learn how they can reduce the risks associated with barbequing chicken. It will be interesting to see if this education strategy has an effect on New Zealand's campylobacteriosis incidence.

Risk management strategies are a very important means of reducing risk. Risk communication is a key component in the toolkit of risk mangers because if consumers are not informed of the risks they cannot modify their behaviours to reduce them.

Risk communication

Risk communication is simply telling people about risks and discussing their consequences. Most consumers understand the everyday use of the word 'risk', but don't know what risk means in a scientific sense. The everyday use of the word does not allow an understanding of how risk might be reduced except by not partaking in the risky activity. Indeed the dictionary defines 'risk' as 'hazard' which is why the two words are often used interchangeably in common usage.

> *Shorter Oxford English Dictionary* definition of risk:
> Hazard, danger, exposure to mischance or peril

Explaining that 'hazard' and 'risk' have different meanings, that RISK= HAZARD×EXPOSURE and that reducing one's exposure to the hazard will reduce the risk makes sense to most people. Illustrating this with an example makes it even clearer – if the hazard is a cigarette and smoking is the exposure route, it is clear that smoking less reduces the risk. But when you explore this further the hazards are actually the carcinogens (i.e. chemicals that cause cancer, e.g. benzo[a]pyrene; Figure 2.15) in cigarette smoke, so it would, in theory, be possible to reduce the exposure to the hazard by reducing the concentration of carcinogens in cigarette smoke instead of reducing exposure by smoking less; this is the idea behind low-tar cigarettes because benzo[a] pyrene is a component of the tar mixture.

Food risk communication is an important part of risk management. Campylobacteriosis associated with chicken consumption is a good example (see *Risk management*) – here the NZFSA is using risk communication via a series of television infomercials to increase public understanding in order to reduce exposure to *C. jejuni* and in turn reduce the risk of contracting campylobacteriosis with its consequent risk to life (i.e. four deaths/year in New Zealand).

Quantitative risk assessment

Quantitative risk assessment is putting a numerical value on risk, for example:

- The chance of dying from campylobacteriosis in New Zealand is 4 in 4,000,000.
- 1% of 50-year-old smokers who have smoked 20 cigarettes/day for 20 years die of lung cancer.

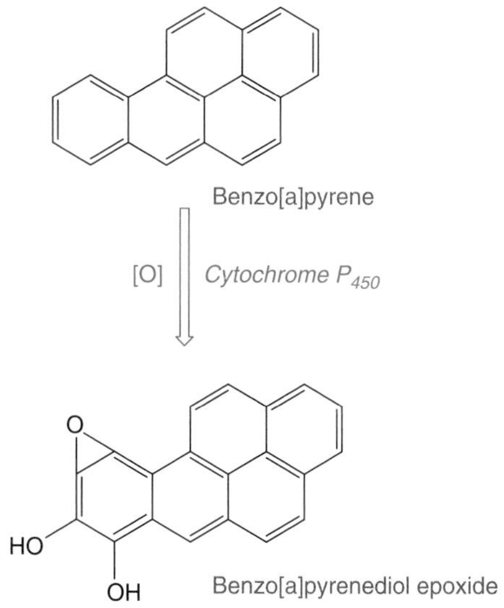

Figure 2.15 Benzo[a]pyrene, a component of the tars present in tobacco, is absorbed via the lungs from cigarette smoke and metabolised by cytochrome P_{450} in lung cells to highly carcinogenic benzo[a]pyrenediol epoxide. This is one of the causes of lung cancer in smokers.

The above are quantitative risk statements. They allow us to understand what risk actually means; what are the chances of succumbing? A quantitative risk statement is more meaningful if you know what the chances of succumbing are if you are not exposed to the hazard in question. In the smoking example, knowing the quantitative risk of dying of lung cancer if you have never smoked will add meaning to the smoker statistics; how much greater is the risk of lung cancer if I choose to smoke?

Of the approximately 220,000 people who died of lung cancer in the USA in 2009, 198,000 (90%) were cigarette smokers and 22,000 (10%) were not smokers. From these statistics it is clear that the risk of dying from lung cancer is very much greater if you smoke (data from National Cancer Institute, USA; www.cancer.gov).

Quantitative risk data are important in assessing risk versus benefit. For example, a 50-year-old man who has smoked since he was a young man can decide whether the benefit (i.e. the pleasure) of smoking outweighs the risk. Is it worth a 1% risk of dying to enjoy a cigarette? Exposure to the hazard in this example is within the control of the person taking the risk, but many risks are outside the consumer's control. For example, the risk of pesticide residues in food; the only way to reduce the theoretical risk is to eat only organic produce which has been grown without the use of pesticides and therefore theoretically does not contain pesticide residues – in reality this is not the case because a great deal of organic produce does contain pesticide residues, albeit at very low concentrations (see Chapter 7). Risks that are outside the control of the consumer are determined and assessed by regulatory committees and the experts on these committees advise government ministers whether it is acceptable to subject the population to such risks.

Regulatory committees

In most of the developed world regulatory committees charged with assessing risk and advising ministers whether or not a particular risk is acceptable are composed of independent experts and usually have an independent chairman. It is important that a large proportion of the committee's membership is independent and that the chair is independent so that conflicts of interest do not influence the decision-making and advisory process. Consumers are less likely to accept decisions from a regulatory committee composed of ministry employees because their opinions might be influenced by their government's (and employer's) desires. For example, until the UK's BSE saga the food regulatory committees were within the Ministry of Agriculture, Fisheries and Food (MAFF; now replaced by the Department of Environment, Food and Rural Affairs – DEFRA) and were composed mainly of MAFF employees. The public seriously questioned the integrity of their advice because they thought that the committee would make decisions that its ministry wanted to hear. For example, decisions relating to the use of pesticides in food production and the risks associated with eating food containing pesticide residues made by a committee of the ministry (i.e. MAFF) that represented farmers might be biased. Whether biased decisions were made is not known, but the UK government changed its committee structure by removing its advisory committees from the control of the ministries and populating them with independent experts and an independent chair. In addition, they created an independent body (the FSA) to oversee food safety issues. This gave advice and decisions great credibility and gave the consumer greater confidence in the safety of UK food.

Some countries in the developed world still run ostensibly government food risk decision-making processes in committees composed mainly of government employees housed in the ministry that is also responsible for farming – this is a significant conflict of interest. Surprisingly the NZFSA (formerly an independent agency) has recently been moved into the Ministry of Agriculture and Forestry (MAF), the ministry responsible for farmers; will this reduce consumers' and importers' acceptance of NZFSA food risk advice?

Food risk assessment – case examples

Lead in prunes

In May 2005 routine monitoring of food imported into New Zealand revealed what appeared to be high concentrations of lead (Pb) in prunes (Table 2.2). The question was asked, is the risk of the lead intake as a result of eating prunes acceptable to the consumer?

The ML for Pb in fruit is 0.1mg/kg, but prunes are dried plums and therefore a correction, or processing, factor ($\times 3.5$) is applied to take account of this – the concentration in dried fruit will be greater because the water has been removed, but the Pb has stayed behind in the dried fruit. This means that the ML for prunes is $0.1 \times 3.5 = 0.35$ mg/kg. Samples 5, 7, 9, 11 and 13 (Table 2.2) therefore exceed the ML. From a trading perspective it was 'illegal' to export those prunes that exceeded the 'corrected ML' and the New Zealand authorities had a serious conversation with the regulatory authority of the exporting country.

This is all well and good, but would a consumer of the high Pb concentration plums suffer any harm following their consumption? In order to determine this we must estimate how many prunes an average consumer eats. Such data

Table 2.2 Concentrations of lead in 13 samples of prunes (dried plums) measured by the New Zealand Food Safety Authority. (From http://www.nzfsa.govt.nz/importing/monitoring-and-review/summary-lead-in-plums-2009.htm. Survey of Lead in Plums. With permission from MAF.)

Sample number	Lead concentration (mg/kg)
1	0.25
2	0.045
3	0.023
4	0.32
5	1.2
6	0.089
7	0.54
8	0.073
9	1.3
10	0.069
11	0.56
12	0.33
13	0.53

are usually obtained from dietary intake statistics. Since prunes are not a dietary staple it is less likely that specific data for a particular country will be available. Indeed New Zealand has no data on prune consumption and therefore I will use data from the USA for this risk assessment. In 2001 in the USA prune consumption was approximately 550 g/person/year (data from http://aic.ucdavis.edu/profiles/prunes.pdf); this is approximately 1.5 g/person/day. Using the worst case residue example (i.e. sample 9 in Table 2.2) this means that the average consumer of prunes with this concentration of Pb would take in $1.3 \times 1.5 \times 10^{-3} = 1.95$ mg Pb/day.

The ADI for Pb is 0.0035 mg/kg body weight/day; therefore our average consumer (assuming they were also average weight, i.e. 60 kg) would have a Pb dose of 0.0325 mg/kg body weight/day, which is ten-fold greater than the ADI and therefore if consumed at this level for a lifetime would be expected to cause harm. Of course it is very unlikely that anyone would be unfortunate enough to eat a plum with a high Pb concentration every day for their entire life. Despite this, exceeding the ADI is taken very seriously by food toxicologists and would trigger further action; for example, a detailed dietary intake study based on consumption of, in this case, prunes, as part of a dietary survey (this is what New Zealand did in the case example given).

Lindane in milk
If, on the other hand, the lead contamination had been in a dietary staple (e.g. bread or rice) that would be expected to be consumed by most people every day, the situation would be much more serious and would elicit a major response from the regulatory authorities. A situation like this occurred in the UK in 1995 (Figure 2.16) when γ-hexachlorocyclohexane (γ-HCH, Lindane;

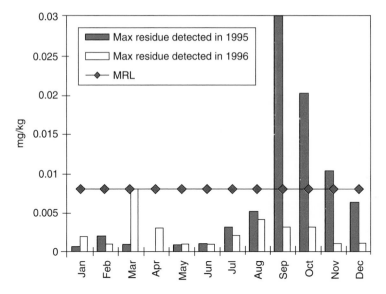

Figure 2.16 Lindane levels in UK milk in 1995 and 1996 showing that the MRL was exceeded in September, October and November of 1995. To exceed the ADI (8 μg/kg/ day) it would have been necessary to consume about 10 L milk/day. (Reproduced from the Annual Report of the Working Party on Pesticide Residues: 1996, with permission of MAFF, London. Crown Copyright. Reproduced from http://www.nationalarchives. gov.uk/doc/open-government-licence/open-government-licence.htm.)

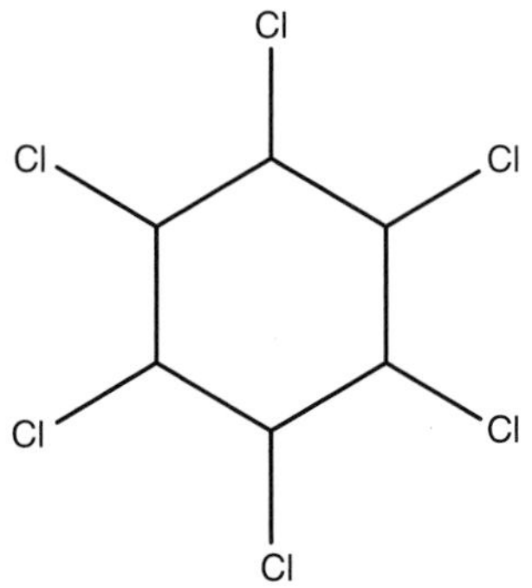

Figure 2.17 Hexachlorocyclohexane; the γ-isomer is Lindane, an insecticide banned in most parts of the world because of concerns about its persistence in biological systems and its disputed link to breast cancer.

Figure 2.17) was found in cow's milk (see Chapter 7 for a full discussion of this example) at concentrations very close to Lindane's ADI. Indeed in September 1995 the concentration of Lindane in nine milk samples analysed as part of a routine monitoring scheme would have exceeded the ADI (FAO/WHO ADI [Lindane]=0.008 mg/kg body weight) based on national milk consumption data. Samples analysed in the months before September had very much lower Lindane concentrations and would not have resulted in ADI exceedances, and therefore the regulatory risk assessors decided that it was unlikely that anyone would consume milk resulting in ADI exceedance for long enough to cause them harm (remember ADI is the acceptable daily intake for a life-

Table 2.3 Some foods consumed (mean g/week) by men by age group from the UK National Diet and Nutrition Survey 2002 (www.statistics.gov.uk). It is clear that some foods, e.g. pizza, burgers and kebabs, and pasta, are favoured by younger men whereas the consumption of some dietary staples, e.g. white bread, is favoured by older men. These data are very important to allow food residue intakes to be calculated.

Food	Age group/years			
	19-24	25-34	35-49	50-64
White bread	600	610	639	629
Wholemeal bread	*	450	376	363
Whole milk	*	997	1,190	996
Semi-skimmed milk	983	1,163	1,435	1,502
Pasta	425	463	382	356
Pizza	479	422	322	233
Burgers and kebabs	295	292	232	*

*Number of consumers <30 and too small to calculate a reliable mean.

time exposure). They decided to step up their monitoring and revisit the risk assessment in the following months. In October [Lindane]=0.02 mg/kg and in November=0.01 mg/kg and therefore it was concluded that the risk was declining and that no further action was necessary except to continue monitoring to make sure that Lindane concentrations did not rise again.

Determining exposure

Exposure to a food-borne contaminant depends on how much of a particular food is consumed. This varies greatly from country to country. In China, rice is a staple part of the diet, whereas in New Zealand or the UK it is not; in these countries bread and potatoes are the staple carbohydrate sources. Clearly, to assess the risk of a particular food residue we must use consumption data from the country in which the exposure occurs, or if data are not available from a country with dietary habits similar to the country in which the exposure occurred. For example, New Zealand could use UK dietary data since both countries have similar dietary habits.

Which food people eat is determined by dietary surveys (Table 2.3), and which hazards they are exposed to (e.g. pesticide residues) is determined by either total diet surveys or surveillance schemes.

Dietary surveys

Dietary surveys are used to find out what people eat. They involve sampling a population and asking individuals to note exactly what and how much of each component they eat over a prescribed time (e.g. a week). The people surveyed are asked to note the contents of each meal and to weigh the individual components. This gives very useful information on the diet of a nation and is also important in determining dietary trends.

For example, the UK's National Diet and Nutrition Survey (NDNS) published in 2002 surveyed approximately 2,500 randomly generated residential

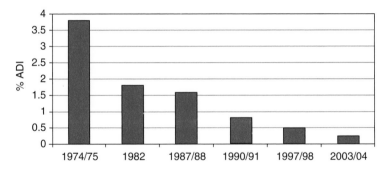

Figure 2.18 Total DDT residues intake trend in 19- to 24-year-old men from the New Zealand Total Diet Survey 2005 – the decline is due to government advice to farmers not to graze dairy cattle on land contaminated with high levels of DDT. This shows how TDSs can be very effectively used to police government attempts to minimise food residues intake. (Data from the 2003/04 New Zealand Total Diet Survey, NZFSA, Wellington, 2005. With permission from MAF.)

addresses including only people between the ages of 19-64 years (i.e. the adult population) in the survey. The survey was complex including questions on dietary supplement consumption, use of artificial sweeteners and dieting, but also included basic data on amounts of different foods consumed. Such surveys are expensive and therefore as much information as possible is gathered as part of the process. From the food safety perspective it is the consumption of different foods that is important. The UK 2002 NDNS gives a unique insight into that country's diet and the different foods consumed by different age groups. This is crucial for food risk assessment.

Total diet surveys

Total diet surveys (TDS) are used to determine the intake of contaminants in food (Figure 2.18) – they provide a snapshot view of the residues a population is consuming with its food. They involve the same sort of statistical randomisation of participants used for dietary surveys, but instead of determining the amounts of different foods consumed, a TDS involves asking the participants to prepare duplicate meals: one to eat and one for the survey. The survey meal is homogenised and analysed for a range of pre-specified contaminants (e.g. pesticides) which gives an idea of how much of a particular contaminant an average consumer takes in each day. New Zealand carries out a TDS every 4 years and therefore is able to track changes in intake of particular residues. The TDS also enables a country to determine whether intake of a particular residue is acceptable or not, e.g. does it exceed the ADI? Of course the data are retrospective and so nothing can be done to change the exposure that has already occurred, but changes can be recommended to prevent the trend continuing.

In 2003/04 the New Zealand TDS showed that intake of cadmium (Cd) was increasing and that this was most likely due to consumption of shellfish (e.g. oysters) (Figure 2.19 and Plate 2.1, Figures 2.20 and 2.21) grown in environments where Cd concentration is naturally high – NZ is a volcanic country and Cd is a component of volcanic emissions and is leached into the silts of estuaries and the sea where filter feeding mussels and oysters are grown commercially. As they filter their food the shellfish take in and retain Cd-contaminated silt.

Cd in food in New Zealand

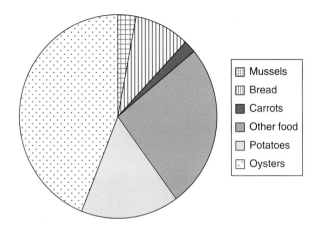

Figure 2.19 The contribution of Cd from different foods to the diet of a 25+ year-old New Zealand male, showing that oysters contribute more Cd (44% of dietary exposure) than any other food. (Data from the 2003/04 New Zealand Total Diet Survey (2005), New Zealand Food Safety Authority, Wellington; see www.nzfsa.govt. nz. With permission from MAF.) (To see a colour version of this figure, see Plate 2.1.)

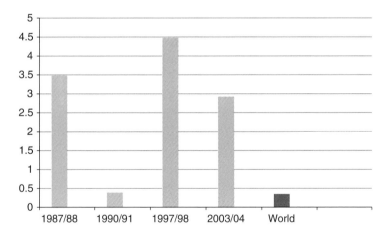

Figure 2.20 Cd (mg/kg) concentrations in New Zealand oysters harvested at different times showing high Cd levels (except in 1990/91 which is difficult to explain) when compared to the WHO average world oyster Cd concentration. (Data from the 2003/04 New Zealand Total Diet Survey (2005), New Zealand Food Safety Authority, Wellington; see www.nzfsa.govt.nz.)

When they are consumed the consumer gets a dose of Cd too. Cd is toxic and is one of the few non-organic carcinogens; therefore its intake must be controlled.

Food surveillance

Food surveillance is a more precise (and continuous) way of determining food residues intake. It involves a long-term rolling programme of food analysis for particular residues (e.g. Lindane). There are many different designs for food

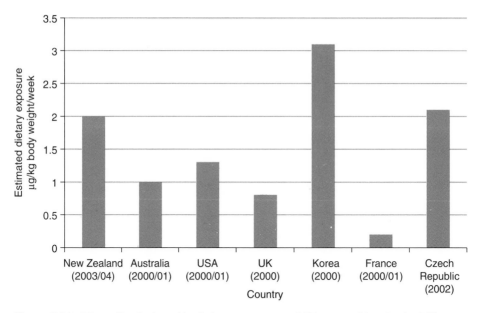

Figure 2.21 The estimated weekly dietary exposures of 25+ year-old males to Cd in various countries. New Zealand's relatively high intake is likely to be explained by Cd in soils and silts that results in contamination of foods grown in these environments. (From the 2003/04 New Zealand Total Diet Survey (2005), New Zealand Food Safety Authority, Wellington; see www.nzfsa.govt.nz. With permission from MAF.)

surveillance programmes, but generally the food staples (e.g. bread) are ana-lysed for a suite of residues every year. The analytes that comprise the resi-dues suite change so that over a 5- or 6-year period a broad range of residues are looked for. Less commonly consumed foods (e.g. Brussels sprouts) are analysed less frequently (e.g. on a 5-year cycle) and rarely consumed foods (e.g. prunes) are analysed very much less frequently (e.g. on a 10-year cycle). Special surveys are carried out if there is concern about a particular residue or food. Food surveillance programmes are expensive.

The UK conducts a very broad ranging food surveillance programme (Table 2.4) including residues of pesticides, veterinary medicines, radio-activity and toxic anions (e.g. nitrate – NO_3^-) and cations (e.g. Cd^{2+}), which gives a full picture of UK residents' exposure to these residues in food. Trends can be seen from food surveillance programmes because they are conducted continuously – this is an important facet of their design.

Decision-making/advisory process

As discussed in *Regulatory committees*, the membership of such committees is very important in determining the acceptability of advice to consumers. There is a significant food lobby in many countries that constantly questions government decisions on food safety issues because they understand the conflicts of interest that government ministers often face when making their decisions. For example, the NZFSA is part of MAF, the ministry responsible for farming. Since agricul-tural products (e.g. milk powder – New Zealand has the largest dairy company in the world; Fonterra) are an important export for the country and a key facet of NZ's Gross National Product (GNP), it is not surprising that lobby groups question

Table 2.4 Bread surveillance data from the UK Food Surveillance Programme. It shows some surprising results – white bread is less likely to have residues than brown bread, and, even though organic bread has the lowest percentage residues contamination, it still has residues. These results are explained by the use of post-harvest pesticides in grain storage to prevent insect damage. They are absorbed onto the testa (bran) of the wheat grain. White bread is made from flour devoid of bran, whereas brown, wholemeal and multi-grain breads include the bran. Organic wheat should be grown and stored and the flour made without the use of pesticides and therefore if all parties are being honest there should be no residues! (Data from the Annual Report of the Working Party on Pesticide Residues, MAFF, London 1996.)

Type of bread	Percentage of samples analysed containing residues
White	12
Brown	23
Wholemeal	27
Multi-grain	24
Organic	11

the ministers' decisions when the alternative decision might affect dairy exports. Such lobby groups, while tiresome to ministers and government officials, are important to maintain their honesty.

The advisory process is simple: the regulatory committees advise government ministers and the minister makes the decision – even if he has little or no knowledge of the science involved. Good ministers, of course, listen to and heed the advice of their experts when making their decisions, but they don't have to!

Take home messages

- Risk = hazard × exposure.
- Everything we do is associated with risk (e.g. crossing the road).
- Food risks are lower than many other of life's risks (e.g. being killed in a car accident).
- Risk can be minimised by reducing exposure to hazards. In a food context this means reducing exposure to chemical and microbiological contaminants.
- Safety parameters (e.g. ADIs) for food contaminants are set by governments and international bodies (e.g. Codex) to minimise exposure to food contaminant hazards.

Further reading

Gilbert S, Lake R, Hudson A & Cressey P (2007) *Risk profile: shiga-toxin producing* Escherichia coli *in uncooked comminuted fermented meat products*. ESR, Wellington, New Zealand. http://www.foodsafety.govt.nz/elibrary/industry/Risk_Profile_Shiga_Toxin_Producing_Escherichia-Science_Research.pdf.

Hudson JA, Whyte R, Armstrong B & Magee L (2003) Cross contamination by *Campylobacter* and *Salmonella* via tongs during the cooking of chicken. *New Zealand Journal of Environmental Health*, March: 13–14.

Slovik P (2001) *The Perception of Risk*. Earthscan, London.

Chapter 3
Bacteria

Introduction

Bacteria are everywhere. If we were to remove everything on the earth except bacteria we would still see an eerie outline of the world as we know it because everything harbours bacteria. They cover our skin, they live in the soil, on rocks, at the top of the highest mountain and at the depths of the deepest ocean; they cover plants, our pets and wild animals, and even the furniture and ornaments in our homes, and they live in and on our food. The vast majority of bacteria are beneficial; they break down waste products and recycle their component chemicals in natural systems - without bacteria we would quickly disappear under a huge pile of waste. They fix nitrogen from the atmosphere, so making our soils nutritious and able to support crops to feed us. We utilise bacteria to our significant benefit, particularly in food storage and manufacture; when milk is infected with particular bacteria it clots - this is the first stage of cheese making. Meat can be preserved by bacteria-generated acids (e.g. salami), and yoghurt is simply milk infected with billions of harmless bacteria.

A very small proportion of bacterial species or strains are harmful. They cause disease - some terrible diseases like dysentery that claims millions of lives worldwide every year and others far less threatening like strep throat. The bacteria that cause disease are termed pathogenic (i.e. generators of disease) bacteria and there are some pathogenic bacteria that infect food and cause illness in millions of consumers each year.

There is a microbiological battle underway on food as it sits in our refrigerators or in our lunch boxes in our work bags. Food has a natural bacterial ecology (natural flora - bacteria are classed as plants, hence 'flora') which covers the surface of food or permeated liquids (e.g. milk). When a pathogenic bacterium lands on a piece of food the rapid and successful growth of the natural flora prevents the pathogen gaining a stronghold - the natural flora simply outgrow the pathogenic interloper. This is an important function of the natural flora and explains why food-borne bacterial illness is not a great deal more common.

Food Safety: The Science of Keeping Food Safe, First Edition. Ian C. Shaw.
© 2013 John Wiley & Sons, Ltd. Published 2013 by John Wiley & Sons, Ltd.

Despite the protection that the natural flora provide sometimes pathogens do take hold and grow on our food and, as a result, consumers become sick, sometimes very sick, and occasionally die.

This chapter is about the relatively small number of food bacterial pathogens and the diseases they cause.

The discovery of bacteria

The word 'bacteria' was first used in 1838 by the German naturalist Christian Gottfried Ehrenberg (1794–1876). It comes from the Greek word *bakterion* meaning small staff or rod because the first bacteria to be seen through the early microscopes were rod-shaped. We now know that there are other forms of bacteria (Figure 3.1).

The concept of minute infectious particles has been with us since Giralamo Fracastoro (1478–1553) (Figure 3.2), a physician of Verona in Italy in the 1540s, put forward his theory of *contagium vivum* (i.e. contagious life – the idea of disease-causing organisms). In 1671 Athanasius Kircher (1601/02–1680) (Figure 3.3), an Italian monk from Rome, wrote that using a simple, low magnification lens set-up he had observed peculiar 'worms' in the blood of people suffering from plague – this must have been a product of a rather fertile imagination because he could not have seen bacteria with the lens system he had used. Nevertheless, Kircher had the right idea that the infectious agent of bubonic plague (the bacterium *Yersinia pestis*; Figure 3.4) is very small, indeed far too small to see without a microscope. Shortly after this a Dutch draper, Antonie Leeuwenhoek (1632–1723) (Figure 3.5), whose hobby was making lenses, constructed the first microscope at his home in Delft, and in 1674 he described 'animacules' that were present in the scrapings from between his teeth and in rain water from his roof. This was the first sighting of bacteria and was made possible by Leeuwenhoek's microscope

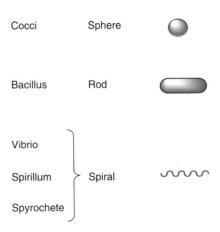

Figure 3.1 Bacteria are classified according to their shape.

Figure 3.2 Giralamo Fracastoro (1478–1553) introduced the idea of disease-causing organisms. (From http://en.wikipedia.org/wiki/File:Fracastoro.jpg.)

Figure 3.3 Athanasius Kircher (1601/02–1680) reported that he had seen the causative agents of bubonic plague in human blood. (From http://en.wikipedia.org/wiki/File:Athanasius_Kircher.jpg.)

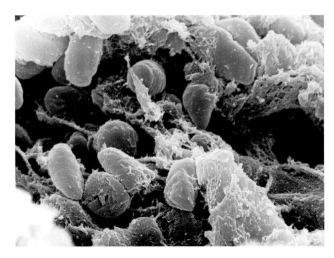

Figure 3.4 A scanning elect.on micrograph of *Yersinai pestis*, the bacterium that causes bubonic plague. (From http://en.wikipedia.org/wiki/File:Yersinia_pestis.jpg.)

Figure 3.5 Antonie Leeuwenhoek (1632–1723), the inventor of the microscope. (From http://en.wikipedia.org/wiki/File:Antoni_van_Leeuwenhoek.png.)

which had the capacity to magnify 300 times – at such magnifications bacteria are visible as tiny dots and dashes.

There was very great interest in Leeuwenhoek's discovery; indeed its importance was recognised by his election to the fellowship of the Royal Society (London) in 1680. There was a proliferation of experiments on Leeuwenhoek's animalcules in Europe over the next 70 or more years. The concept of

Figure 3.6 John Needham (1713–1781), the British Roman Catholic priest who proposed the theory of spontaneous generation. (From http://en.wikipedia.org/wiki/File:John_Turberville_Needham.jpg.)

'spontaneous generation' was thought to explain the origins of these strange little creatures – they simply had to come from nowhere!

In 1750 John Needham (1713–1781) (Figure 3.6) gave a lecture at the Royal Society in London. He described an experiment in which he had heated meat broth (a nutrient-rich liquid) to a temperature he thought would kill all living things. He left the broth at room temperature in a sealed vessel. When he looked at the broth several days later under a simple microscope he found the liquid was swarming with animalcules. This, Needham thought, proved spontaneous generation. In reality it simply proved that he had not heated the broth to a high enough temperature to sterilise it.

An Italian priest, Lazzaro Spallanzani (1729–1799) (Figure 3.7) disproved the theory of spontaneous generation in 1768 by showing that if the meat broth was boiled for several minutes and the vessel in which it was boiled was sealed tightly nothing grew, but if the vessel was either not heated to boiling point or was left unsealed the animalcules appeared as Needham had observed. Spallanzani concluded that the animalcules developed from small numbers of creatures already in the broth, and that boiling killed them. This was a leap in understanding and paved the way for the discovery of bacteria.

In 1856 the French chemist Louis Pasteur (1822–1895) (Figure 3.8) began to get interested in microbiology. His interest was sparked by the French wine industry in and around the city of Lille in northern France which was having problems with their wine production; for reasons they did not understand fermentation had stopped. In the process of solving their problem Pasteur showed that yeast (microscopic single-celled fungi similar to bacteria) were the agents of fermentation; it was not a purely chemical process as had been proposed by the German chemist Justus von Liebig (1803–1873) several years before. This led to Pasteur's Germ Theory of Disease which revolutionised thinking about

Figure 3.7 Lazzaro Spallanzani (1729–1799), the Italian priest who disproved the theory of spontaneous generation. (From http://en.wikipedia.org/wiki/File:Spallanzani.jpg.)

Figure 3.8 Louis Pasteur (1822–1895), the French chemist turned microbiologist who revolutionised our thinking when he proposed his Germ Theory for Disease. (From http://en.wikipedia.org/wiki/File:Louis_Pasteur.jpg.)

the cause and treatment of disease. The concept of bacteria was introduced and the science of bacteriology born.

The biology of bacteria

Bacteria are prokaryotes (without nuclei) and are the only members of the domain bacteria (domain is one of the terms used by biologists as part of the hierarchy of classification of life; see Table 3.1). They have a cell wall that surrounds a cell membrane. The cell wall restricts the cell's size and so prevents water uptake beyond the point that the cytoplasm fills the space enclosed by the rigid cell wall. The cell wall is permeable to most molecules. The cell membrane controls influx and efflux of molecules from the cell by diffusion or facilitated diffusion linked to carrier proteins and membrane channels often associated with molecular pumps. For example, adenosine triphosphate (ATP) is used as an energy source (it is broken down to adenosine diphosphate (ADP) plus phosphate with the release of energy from the phosphate bond) to drive the uptake of Na^+. Similarly membrane channels are formed by trans-membrane proteins to provide a conduit for absorption of useful molecules (e.g. glucose) and ions (e.g. Ca^{2+}) and the efflux of waste products and toxins. In this respect bacteria are like any other cell (Figures 3.9 and 3.10). Similarly, they have a vast array of enzymes, either in their cytoplasm or associated with membrane structures in the cytoplasm, that catalyse thousands of metabolic reactions which in turn drive cellular functions. Again, bacteria are like any other cell in this respect; indeed a great deal of our understanding of cell biochemistry (metabolism) has been derived from experiments on bacteria

Table 3.1 The classification system for life used by biologists.

Classification hierarchy	Example 1 *Escherichia coli*, an important bacterium found in the intestine of mammals	Example 2 Human beings
LIFE ↓		
DOMAIN ↓	Bacteria	Animalia
PHYLUM ↓	Proteobacteria	Chordata
CLASS ↓	Gamma Proteobacteria	Mammalia
ORDER ↓	Enterobacteriales	Primates
FAMILY ↓	Enterobacteriacea	Hominidae
GENUS ↓	*Escherichia*	*Hominini*
SPECIES	*Escherichia coli*	*Homo sapiens*

Cd in food in New Zealand

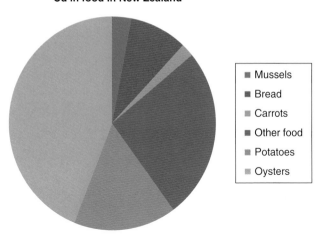

- Mussels
- Bread
- Carrots
- Other food
- Potatoes
- Oysters

Plate 2.1 The contribution of Cd from different foods to the diet of a 25+ year-old New Zealand male, showing that oysters contribute more Cd (44% of dietary exposure) than any other food. (Data from the 2003/04 New Zealand Total Diet Survey (2005), New Zealand Food Safety Authority, Wellington; see www.nzfsa.govt.nz. With permission from MAF.)

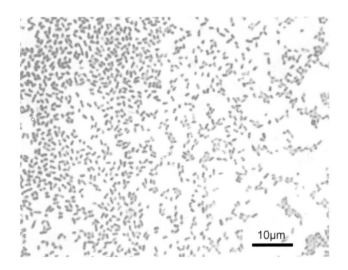

10µm

Plate 3.1 Gram negative *Pseudomonas aeruginosa* will stain pink because fuchsin, a component of Gram's stain, is soluble in the outer bacterial membrane of Gram negative bacteria. (From http://en.wikipedia.org/wiki/File:Pseudomonas_aeruginosa_Gram.jpg.)

Food Safety: The Science of Keeping Food Safe, First Edition. Ian C. Shaw.
© 2013 John Wiley & Sons, Ltd. Published 2013 by John Wiley & Sons, Ltd.

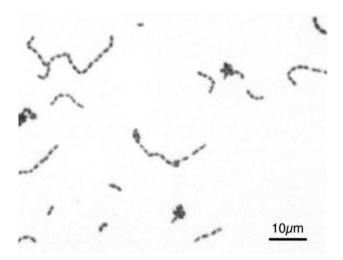

Plate 3.2 Gram positive *Streptococcus mutans* will stain purple because the crystal violet in the Gram's stain binds to the outer polysaccharide cell wall of Gram positive bacteria. (From http://en.wikipedia.org/wiki/File:Streptococcus_mutans_Gram.jpg.)

Plate 3.3 Bread left in a warm place soon becomes a substrate for bacterial and fungal colonies. The colonies on this crumpet (a bread-like teatime treat) are *Penicillium* sp. (Photograph taken by the author.)

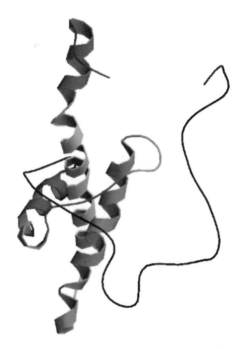

Plate 6.1 The molecular structure of human PrP^C showing its predominant α-helix protein conformation. (From Ilc *et al.* (2010) *Plos One*, **5**, e11715–e11715, downloaded from the Protein Data Bank at http://pdbbeta.rcsb.org/pdb/explore/ explore.do?structureId=2KUN.)

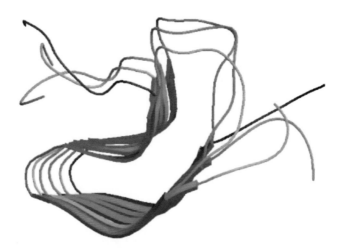

Plate 6.2   The molecular structure of a PrP^{SC} conglomerate forming a fibril, showing its predominant β-pleated sheet conformation – each of the different coloured structures is a PrP^{SC} molecule; they have aligned to form a water-insoluble fibril. (From Van Melckebeke *et al.* (2010) *Journal of the American Chemistry Society*, **132**, 13765–13775, downloaded from the Protein Data Bank at http://pdbbeta.rcsb.org/pdb/explore/explore.do?structureId=2KJ3.)

Plate 8.1 A Monarch (*Danaus plexippus*) caterpillar (above) and adult butterfly (below) resting on a swan plant (*Asclepias fruticosa*) in my garden in New Zealand; its bright colours warn of the toxin (labriformidin) within. (Photograph by the author.)

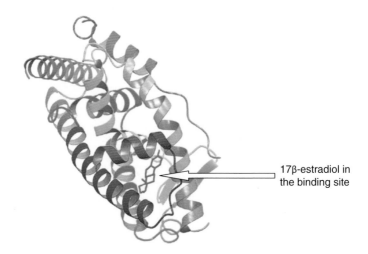

Plate 9.1 The human ER estrogen binding domain with 17β-estradiol bound. (Created by Lisa Graham in Shroedinger 2008 using published X-ray crystallographic data from Brzozowski *et al.* (1997) *Nature*, **389**, 753–758.)

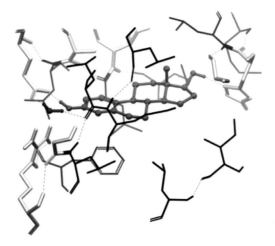

Plate 9.2 17β-estradiol in the binding domain of the human ER. The ER has been simplified so that only the amino acid residues important in the binding of estradiol are shown. The dotted lines are hydrogen bonds and the blue molecule is water which has a key role in the binding of estradiol. You can see that estradiol is bound by hydrogen bonds between its hydroxyl groups and specific amino acid residues in the binding domain. (From Graham and Shaw (2011), SAR QSAR. *Environmental Research*, **22**, 329–350. Reprinted with permission.)

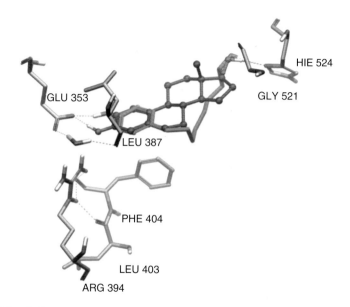

Plate 9.3 4-(9-Hydroxynonyl)phenol in the ER binding domain (purple) with 17β-estradiol (green) superimposed – this shows that 4NP is estrogenic via its refolded hydroxyl-metabolite. (From Graham and Shaw (2011), SAR QSAR. *Environmental Research*, **22**, 329–350. Reprinted with permission.)

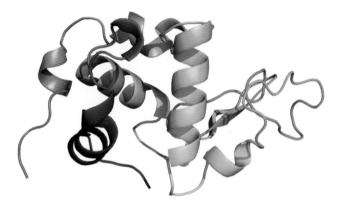

Plate 15.1 The tertiary structure of α-lactalbumin showing the importance of folding in determining its shape: molecular weight $=14$ kDa; concentration in cow's milk ≈ 1 g/L. (Molecular structure from http://upload.wikimedia.org/wikipedia/commons/6/69/Protein_LALBA_PDB_1a4v.png.)

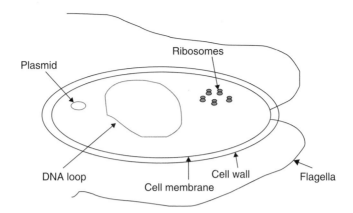

Figure 3.9 Structure of a typical bacterium.

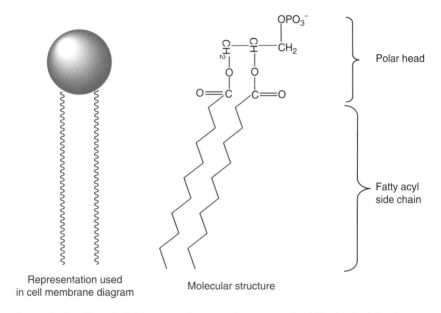

Figure 3.10 Phospholipids are an important component of the bacterial cell membrane.

(especially *Escherichia coli*). Bacteria do not have the complex structures (e.g. mitochondria, nuclei, endoplasmic reticulum) associated with higher life forms, but they do carry out many of the metabolic processes that go on in these organelles (Figure 3.11).

The bacterial cell wall and Gram's stain

Bacteria have a cell wall with a molecular structure unique to bacteria. It is composed of a polymer of two sugar derivatives – N-acetylmuramic (NAM) acid and N-acetylglucosamine (NAG) (Figure 3.12). The cell wall polymer has alternating NAM-NAG units and the polymer chains are held together by

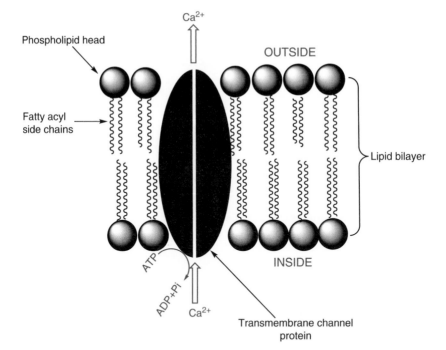

Figure 3.11   A representation of a section of cell membrane (lipid bilayer) with the transmembrane protein Ca^{2+} ATPase used to pump Ca^{2+} across cell membranes showing how the protein forms a channel through which the Ca^{2+} is pumped. Energy to pump the Ca^{2+} out of the cell is derived from the ATP γ-phosphate bond. The bacterial cell membrane has a complex array of carrier proteins and transmembrane proteins that facilitate influx and efflux of myriad important molecules.

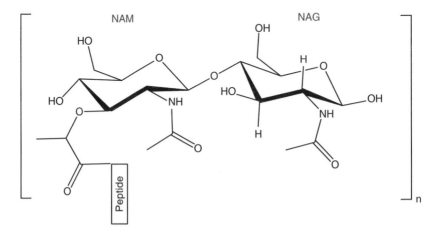

Figure 3.12   The repeating NAG-NAM dimer which forms the backbone of the bacterial cell wall showing the position of attachment to the peptide cross-linking unit.

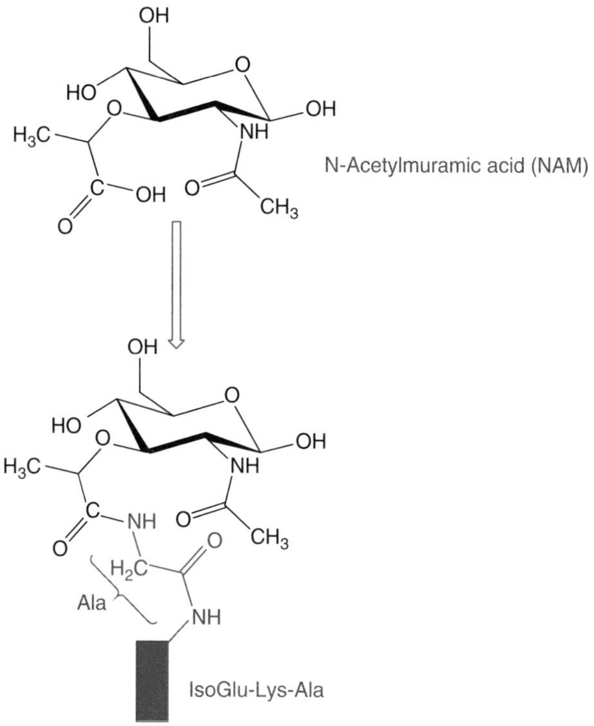

Figure 3.13 NAM is linked to specific short-chain peptides in the bacterial cell wall – this example is for Gram positive bacteria.

short peptides (e.g. Ala-Lys-Glu-Ala) (Figure 3.13). The exact form of the cell wall structure determines whether the bacterium is Gram positive or Gram negative because particular dyes (e.g. crystal violet) are bound to the specific forms of the polymer and therefore stain the bacterium (Gram positive). The Gram staining method (see *Gram's stain*) was invented by the Danish scientist Hans Christian Gram (1853–1938) in 1882 and is still used today as part of the identification and classification of bacteria.

Gram positive bacterial cell walls have a pentaglycyl bridge between the Ala-IsoGlu-Lys-Ala tetrapeptide cross link, whereas Gram negative bacteria have a direct link between the tetrapeptide cross links. The arrangement of the cell wall and cell membrane in Gram positive and negative bacteria differs too. In Gram negative bacteria the cell wall is sandwiched between two lipid bilayers whereas in Gram positive bacteria the cell wall is on the outside of a single lipid bilayer membrane (Figure 3.14). It is the polysaccharide of the cell wall that is stained by crystal violet (a component of Gram's stain) and therefore because Gram positive bacteria have their cell wall on the outside of their membrane it stains. Gram negative bacteria have a membrane on the outside of the polysaccharide cell wall, thus preventing the crystal violet gaining access to stain the cell wall. Gram negative bacteria are stained pink by the Fuschin present in Gram's stain because it is lipid soluble and they have a lipid bilayer membrane on the outside of their polysaccharide cell wall (Figures 3.15 and 3.16; Plates 3.1 and 3.2).

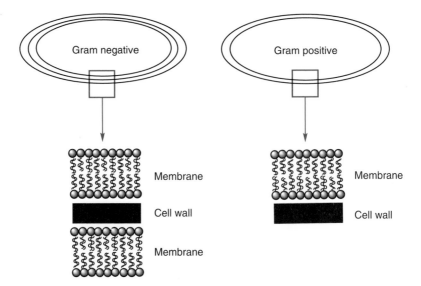

Figure 3.14 The differences in cell membrane/cell wall organisation of Gram positive and Gram negative bacteria.

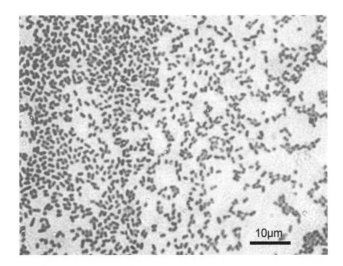

Figure 3.15 Gram negative *Pseudomonas aeruginosa* will stain pink because fuchsin, a component of Gram's stain, is soluble in the outer bacterial membrane of Gram negative bacteria. (From http://en.wikipedia.org/wiki/File:Pseudomonas_aeruginosa_Gram.jpg.) (To see a colour version of this figure, see Plate 3.1.)

Gram's stain

Gram's stain is carried out as follows:

- Heat fix bacteria onto microscope slide.
- Apply crystal violet solution [primary stain] – 2% in a mixture of ethanol/1% ammonium oxalate (aq) (20:80 v/v).

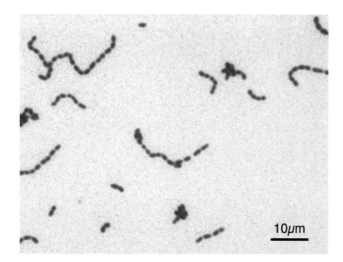

Figure 3.16 Gram positive *Streptococcus mutans* will stain purple because the crystal violet in the Gram's stain binds to the outer polysaccharide cell wall of Gram positive bacteria. (From http://en.wikipedia.org/wiki/File:Streptococcus_mutans_Gram.jpg.) (To see a colour version of this figure, see Plate 3.2.)

- Wash with water.
- Apply Gram's iodine – 1% in 2% potassium iodide (aq).
- Decolourise with acetone.
- Apply safranin (or fuchsin) [counter stain] – 0.5% (aq).
- Wash with water.
- Dry.
- Observe under microscope:

   Gram positive – PURPLE.
   Gram negative – PINK.

Bacterial nucleic acids, transcription and translation

Bacteria are prokaryotes and therefore do not have nuclei (i.e. nucleic acids encapsulated by a nuclear membrane). Instead they have DNA loops that float freely in the cytoplasm. The primary loop is large and includes most of the genes that code for the proteins and enzymes that the bacterial cell needs to live. In addition, there are smaller loops (plasmids) (Figure 3.9) that have the genes for specific properties (e.g. antibiotic resistance). The plasmids are transferred during bacterial conjugation (see *Conjugation*) which is the mechanism of transfer of bacterial resistance and pathogenicity from one strain (or sometimes species) to another.

Transcription (i.e. synthesis of messenger RNA (mRNA) from the DNA template) is carried out on ribosomes in the cytoplasm. Ribosomes are huge enzyme complexes that possess the entire apparatus for attaching, reading and translating (to protein sequence) the mRNA template. The plasmid genes are translated in the same way; in the case of antibiotic resistance the proteins encoded are often enzymes that degrade antibiotics.

Conjugation

Bacteria reproduce by simple cell division (mitosis) generating daughter cells that are identical to the parent cell. This is their 'normal' means of reproduction; however, they are also able to conjugate and exchange genetic material (i.e. plasmids) through a conjugation tube that links the two bacterial cells. This is the means by which plasmids encoding for pathogenicity or antibiotic resistance are transferred from one bacterium to another. Once transferred, mitosis results in multiplication of the plasmid-containing strain. Conjugation usually occurs between cells of the same species, but can occur between different species. The latter is important in transferring pathogenicity and was responsible for the formation of a strain of *E. coli* that is able to synthesise a highly potent toxin, shiga toxin, normally found in members of the bacterial genus *Shigella*, a highly pathogenic genus responsible for some serious and life-threatening diseases, e.g. *S. dysenteriae* causes dysentery. Shiga toxin is named after Professor Shiga Kiyoshi, the Japanese scientist who discovered *Shigella* in 1897. When the shiga toxin plasmid was transferred to *E. coli* a very important food-borne pathogen was born, namely *E. coli* 0157 (the number refers to the strain or serotype) or STEC (shiga toxin-producing *E. coli*) which has resulted in food poisoning epidemics throughout the world (see *E. coli* 0157/STEC later) (Figure 3.17).

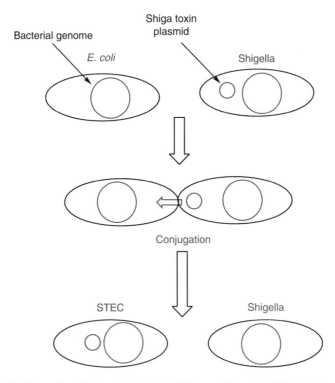

Figure 3.17 Transfer of shiga toxin plasmid from pathogenic *Shigella* to non-pathogenic *E. coli* forming the highly pathogenic shiga toxin-producing STEC/ *E. coli* 0157, an important food-borne pathogen.

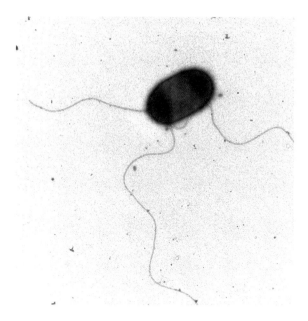

Figure 3.18 *Listeria monocytogenes* (×35,000). (Electron micrograph kindly provided by Phillipa Rhodes.)

Bacterial motility

Some bacteria have cell-surface devices to facilitate their movement. This is very important in the context of food contamination because it allows bacteria to move over the surface of the food, thus infecting a larger area. There are two types of motility devices, flagellae and cilia. Flagellae (Figure 3.18) are long thrashing outgrowths of the bacterial cell – there are usually only one or two per cell. Cillia are small hair-like outgrowths that usually cover the surface of the bacterial cell and beat in waves which results in movement of the cell through its aqueous environment.

The important food-borne pathogen *Lysteria monocytogenes* (see *Listeria*) which causes potentially fatal listeriosis has flagellae (Figure 3.18), and water-borne *Vibrio cholerae* (see *Vibrio*) which causes cholera has thousands of cilia.

Bacterial spores

Some bacteria are able to undergo a significant biochemical, physiological and structural change in response to adverse conditions, to form a resilient, encapsulated form that can withstand the most extreme conditions (e.g. drying and high temperatures) – this highly protected, in limbo, bag of bio-chemicals is termed a spore, or more correctly an endospore (*endo* from the Latin for inside) to distinguish it from the reproductive spores (exospores) produced by other life forms (e.g. fungi).

When a spore-forming bacterium (e.g. members of the genera *Bacillus* or *Clostridium*) encounters adverse environmental conditions (e.g. high

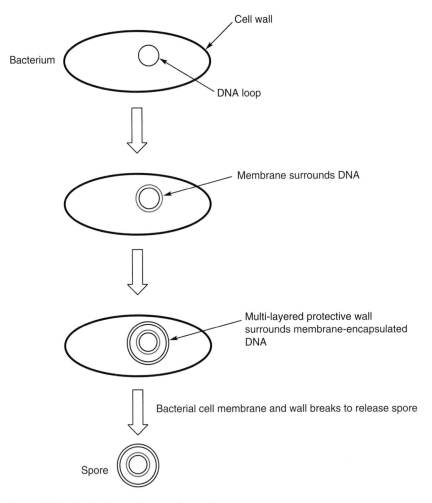

Figure 3.19 Bacterial endospore formation.

temperature) it replicates its DNA and forms a membrane around it (the septum); the membrane is then surrounded by a series of strong walls. All of this happens inside the bacterial cell, hence the term *endo*spore. The bacterial cell breaks down and releases the multi-walled spore, which can lie dormant for a very long time (Figure 3.19). When conditions are favourable the spore germinates; this involves hydration of the spore, activation of the nucleic acid replication apparatus and switching on of the spore's biochemistry. A new bacterial cell emerges from the spore which takes advantage of the favourable environment and replicates.

In the context of food, spores are very important because they are able to withstand the adverse conditions associated with some cooking and food preservation processes. When the bacterial pathogen spores find themselves in more harmonious conditions (e.g. in a warmed-up sauce) they germinate and reproduce just in time to be consumed by an unsuspecting human.

The bacterial ecology of food

Food, like most other things on earth, is covered with bacteria - this is the natural bacterial ecology of food. If food is cooked before being eaten the bacteria will be killed, but if the food is eaten raw, live bacteria will be consumed. Most of the bacteria that are part of food's natural bacterial ecology cause no ill effects when eaten; they are not pathogenic. Indeed they are beneficial and play an important role in preventing pathogens growing on food; the natural food bacteria colonise food and occupy the ecological niche provided by a particular food and prevent pathogenic contaminants taking hold and growing because the natural food bacteria outgrow the pathogen imposters.

As discussed at the beginning of this chapter, we live in a microbial world - bacteria are everywhere and will take advantage of any environmental niche they might find themselves in. Therefore if a bacterium falls onto an apple growing in an orchard, or a piece of meat hanging in the slaughter house, providing there are the right nutrients present the bacterium will grow and colonise the food's surface. Since there are millions of airborne bacteria per litre of air, it is very likely that a complex array of microbes will fall onto and form an equally complex mix of bacteria growing on the food's surface. We can't be precise about the species of bacteria that naturally grow on food because this depends on what is present in the environment in which the food was grown or processed (e.g. slaughter house or packing station). This complex array of bacterial species is the bacterial ecology of food. Not only do bacteria grow on food, but also other microorganisms such as yeast and other fungi, viruses and protozoa take their place in food's microbial ecology. Some of these organisms will be discussed further in Chapters 4 (viruses), 5 (parasites) and 6 (natural toxins - some natural toxins are produced by fungi living on food).

As mentioned above, bacteria in the natural microbial ecology of food might inhibit the growth of human food pathogens that might also find an ecological niche on the surface of food by simply outnumbering and outgrowing the pathogen; however, some bacteria and fungi might produce inhibitory substances (antibiotics) that prevent the growth of other organisms that might be competing for their environment. Indeed this is how antibiotics were discovered by Alexander Fleming (1881-1955) at St Mary's Hospital in London in 1928. He was culturing bacteria on agar plates and some spores of the fungus *Penicillium notatum* settled on the agar and grew. Fleming noticed that the bacterial colonies were killed by adjacent *P. notatum* colonies; this was the discovery of the first antibiotic, penicillin (Figure 3.20), which revolutionised medicine in the 1930s. Other fungi (e.g. *Streptomyces griseus* produces streptomycins) and many bacteria also produce inhibitory substances which are likely to be important in reducing the growth of pathogens on food. Indeed *Staphylococcus xylosus* growing on cheese inhibits the growth of the cheese pathogen *Listeria* monocytogenes (see later in this chapter). A piece of bread left in a warm place quickly shows signs of bacterial and fungal colonisation (Figure 3.21; Plate 3.3). Members of the genus *Penicillium* often grow on bread; they produce penicillins which inhibit the growth of would-be bacterial colonisers.

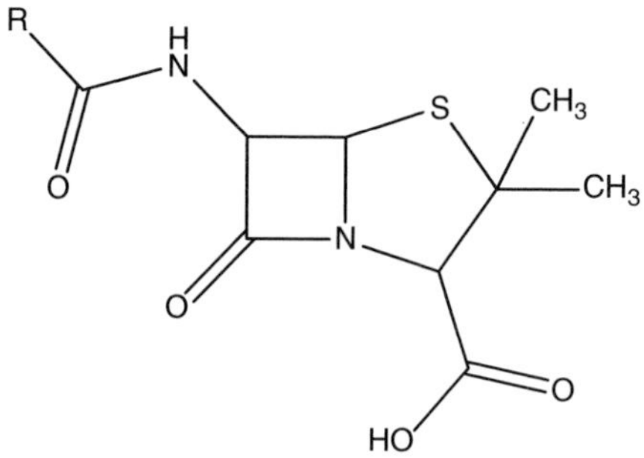

Figure 3.20 β-lactam antibiotics are produced by members of the fungal genus *Penicillium* (R varies according to the specific penicillin; e.g. R=⬡— in benzylpenicillin).

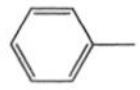

Figure 3.21 Bread left in a warm place soon becomes a substrate for bacterial and fungal colonies. The colonies on this crumpet (a bread-like teatime treat) are *Penicillium* sp. (Photograph taken by the author.) (To see a colour version of this figure, see Plate 3.3.)

Human bacterial pathogens on food

With the exception of a single human viral pathogen (see Chapter 4, *Norovirus*) found on food, bacteria are without doubt the major cause of food-borne illness worldwide. There are many species of pathogenic bacteria associated with food, and each has its own preferences with respect to the food it infects, the conditions it prefers to grow in, its geographical distribution and the severity of human disease it causes. Some food-borne pathogenic bacteria can be fatal (e.g. botulism caused by *Clostridium botulinum*), others can cause severe diseases (e.g. campylobacteriosis caused by *Campylobacter jejuni*) and others simply result in a few days of severe discomfort (e.g. *Staphylococcus aureus*).

Food-borne pathogenic bacteria often produce toxins and it is the toxin that is responsible for the diseases they cause (e.g. botulinum toxin from *Clostridium botulinum*). Sometimes the toxins are produced by the food-borne bacteria growing in the human gut (e.g. shiga toxin–producing *Escherichia coli* (STEC)), whereas other bacteria secrete toxins into the food they infect and the consumer of the food succumbs to the toxin (e.g. staph toxins produced by *Staphylococcus aureus*). Bacterial pathogens are ingenious synthesisers and deliverers of some of the nastiest toxins imaginable – they are the bacterial terrorists of the food world!

I will deal with the most important food-borne bacterial pathogens below, but it is important to know that there are many more bacterial pathogens than those discussed in this chapter, and, more importantly there are new pathogenic strains being created as you read this. Bacteria are an evolutionary success, they are constantly evolving into strains and species that are better suited to their diverse environments and so are more successful in an ecological and evolutionary sense (viz Charles Darwin's concept of 'survival of the fittest'). A very good ecological niche for bacteria to occupy is the nice warm, nutrient-rich human body and what better way to get there than in food? For these reasons alone we are likely to encounter many new food-borne pathogens over the coming years (e.g. the new *E. coli* strain that resulted in serious illness and deaths in Germany in 2011).

Gastroenteritis

Gastroenteritis is also known as gastric or stomach flu – although it is not caused by the influenza virus (i.e. the virus that causes influenza or 'flu'). It is inflammation of the gastrointestinal tract, particularly the stomach and small intestine, and is usually caused by bacterial or viral infection, parasites, toxins or an adverse reaction to other chemicals (e.g. medicines). It is almost always associated with acute diarrhoea.

The term gastroenteritis is very commonly used to describe the clinical condition caused by food-borne pathogens.

Food-borne pathogenic bacteria

Aeromonas

Aeromonads are flagellate and therefore motile, rod-shaped, Gram negative bacteria that have been associated especially with travellers' gastroenteritis – approximately 1.8% of gastroenteritis in travellers to India is associated with *Aeromonas* infection. Aeromonads cause mild diarrhoea to life-threatening cholera-like diseases.

The most commonly encountered food-borne bacterium in this class is *Aeromonas hydrophila*, but *A. caviae* and *A. sobria* are also quite common. Members of the genus *Aeromonas* produce a protein toxin which is thought to be responsible for the symptoms they cause.

The presence of aeromonads in food (especially seafood) is quite common. In a survey of New Zealand seafood (2003), aeromonads were found in 66%

of shellfish and 34% of finfish. Despite this, human aeromonad gastroenteritis is uncommon; in New Zealand there were no recorded cases of *Aeromonas* gastroenteritis between 1997 and 2003 (data from Bremmer *et al.* (2003), New Zealand Institute of Crop and Food Research). This might be because aeromonads are not particularly pathogenic or because the symptoms they cause are so mild that sufferers do not go to their doctors and therefore cases are not recorded.

Bacillus

The most important food-borne member of this genus is the Gram positive, spore-forming flagellate (i.e. motile), rod-shaped *Bacillus cereus*. It causes two distinct types of gastroenteritis: one associated with vomiting and the other with diarrhoea. The two diseases are caused by different toxins produced by distinct strains of *B. cereus*, namely emetic (i.e. vomiting) toxin and diarrhoeagenic toxin. Both emetic and diarrhoeal diseases are not very severe and usually clear up within 24 hours.

B. *cereus* lives naturally in soil and it is likely that food contamination originates from soil contact either directly or indirectly (e.g. via hands). It is present naturally on many foods.

Diarrhoeagenic toxin and diarrhoeal enteritis

Diarrhoeagenic toxin comprises three separate protein toxins (haemolysin BL, non-haemolytic enterotoxin and cytotoxin K) of molecular weight 38,000–46,000 Da; they are destroyed by heat (56°C for 30 minutes) and digested by proteases (e.g. trypsin found in the stomach). Since the toxins are destroyed by stomach enzymes, ingestion of food contaminated with them does not cause disease. To be pathogenic, the bacterium itself has to be ingested. It then grows in the intestine where the growth conditions are just right and there are no proteases to break down the toxins; the toxins produced cause diarrhoea. The incubation period for *B. cereus* diarrhoeal enteritis is 6–15 hours because it takes the bacteria time to get to the intestine, grow and produce toxins.

Diarrhoeagenic toxin is coded for by a gene sequence on the bacterial genome. It is a β-barrel protein which means that it has a structure with a hole through its centre which inserts into cell membranes (e.g. of intestinal cells) and disrupts the potential difference across the membrane, disrupting cell activity and eventually resulting in cell death.

Emetic toxin and vomiting

Emetic toxin (also called cereulide) is a cyclic dodecapeptide (12 amino acid residues) of molecular weight 1,152 Da. It is resistant to heat (126°C for 90 minutes), proteases and extremes of pH. Its physical and chemical stability means that toxin present in food contaminated with *B. cereus* is able to survive the extreme conditions of the stomach (pH 1-3 plus proteases) and results in vomiting. The mechanism by which it induces vomiting is thought to be via cereulide's interaction with gastric serotonin (5-hydroxytryptamine) receptors which leads to afferent vagus nerve stimulation which, in turn, causes vomiting.

Emetic toxin is coded for by a plasmid and therefore, in theory, could be transferred to other bacteria with which *B. cereus* might conjugate.

Foods and conditions associated with *B. cereus* food-borne illness

The diarrhoeal disease is associated with a vast array of foods, but the emetic disease is almost always linked to alkaline carbohydrate foods (e.g. rice) that have been cooked, stored and reheated. Such foods provide the ideal growing conditions for the bacterium and the storage time allows the bacteria to grow and synthesise emetic toxin. Toxin-contaminated food will not be made safe by cooking because of the toxin's heat stability.

B. cereus food-borne illness case example

On 21 July 1993 a chicken fried rice lunch was served to 82 children (age <6 years) and nine staff in a Virginia (USA) day care centre. Of the 67 individuals who ate the lunch, 14 (21%) became ill, with nausea and/or vomiting (71%), abdominal cramps or pain (36%) and diarrhoea (14%) within approximately 2 hours of eating the food. None of the 13 children or staff who did not eat the lunch became ill. The symptoms were gone within 4 hours of onset.

The rice used in the chicken fried rice lunch had been cooked the night of 20 July and cooled to room temperature before being stored in a refrigerator overnight at a central facility. On the day of the lunch the rice was pan fried in oil with pieces of pre-cooked chicken. It was delivered to the day care centre at 10.30 am on 21 July, stored at room temperature and served without reheating at 12.00 noon.

Analysis of food samples showed >10^5 *B. cereus*/g which is consistent with *B. cereus* food-borne illness. The rapid onset, short duration of illness and predominance of nausea/vomiting is consistent with emetic toxin disease.

Rice is an alkaline carbohydrate food and is a common source of *B. cereus* (emetic) infection. The rice was probably contaminated with *B. cereus* spores which survived the first cooking (boiling); they germinated, grew and synthesised emetic toxin while the rice was cooled and stored overnight. Because the emetic toxin is heat stable it was not denatured by the frying process on the day the food was served, but the bacteria would have been killed by the high frying temperatures. The toxin remained in the food during the storage time at the day care centre immediately before the food was eaten for lunch. It is possible that bacterial spores survived frying and germinated and produced more toxin during the storage time at the day care centre. This is supported by the high *B. cereus* culture count found in samples of the food eaten.

Reference to case report: http://www.textbookofbacteriology.net/B.cereus_2.html.

Brucella

Brucellae are non-motile, non-spore forming, rod-shaped, Gram negative bacteria. They cause diseases in animals and people. The most important species are *Brucella abortus* which infects cattle and *B. melitensis* which infects goats; both species can be transferred to humans via raw (unpasteurised) milk, cream and milk products (e.g. cheese). Other species infect pigs (*B. suis*) and sheep (*B. ovis*) and can also infect humans, but are much less common. *Brucella* spp. are killed by pasteurisation.

When humans are infected with *Brucella* sp. they develop septicaemia (i.e. the bacterium enters the blood stream); this is different to most other food-borne bacterial pathogens which infect only the alimentary tract (e.g. intestine) and do not get into the blood stream.

Brucellosis is a serious disease and for this reason most countries in the developed world have programmes to control *Brucella* infection in farm animals, especially cows in order to minimise the risk to the consumer of dairy products. Such programmes involve regular testing of milking herds, isolating farms with infection and culling infected cattle. Control of *B. abortus* in cattle has been very successful, so successful in fact that the most common cause of brucellosis in humans is now *B. melitensis* via goat's milk products and not *B. abortus* from cow's milk as was the case at the beginning of the 20th century.

Brucellosis

The disease was first described by military doctors in the Crimean War in the 1850s, but it was not until 1887 that the causative bacterium was isolated by the Scottish microbiologist Sir David Bruce (1855–1931) – the bacterium was named after him.

Brucellosis is characterised by fever associated with muscular pain and sweating (the sweat often smells like wet hay). The disease can be successfully treated with antibiotics (e.g. doxycycline/rifampicin), but if not treated can become chronic with localisation of *Brucella* in bones resulting in severe, chronic complaints like spondylodiscitis (infection of the vertebral discs).

Brucellosis is uncommon in countries with good farm control measures in place and more common where there is likely to be less control. For example, in the UK (population 61 million) where there are stringent controls there were 33 cases of human brucellosis between 1999 and 2003 (i.e. mean=7 cases/year), whereas in Iran (population 66 million) there were 469 cases (i.e. mean=78 cases/year) between 1997 and 2002 (data from Purcell *et al.* (2007) *Brucellosis in Medical Aspects of Biological Warfare*, Office of the Surgeon General, Department of the Army, USA, and Borden Institute, Washington, USA).

Brucellosis case example

A 16-year-old boy from a village in Turkey who regularly consumed unpasteurised dairy products was admitted to hospital with fever (temperature=39.9°C), abdominal pain, vomiting, diarrhoea and a skin rash. A full investigation including blood biochemistry and faecal microbiology ruled out many possible causes of the symptoms (e.g. typhoid), but it was not until blood immunology revealed an increased *Brucella* titre (i.e. the boy was making antibodies against the bacterium present in his blood) and blood culture grew *B. melitensis* that a diagnosis of brucellosis was made. The boy was treated with rifampicin and doxycycline for 6 weeks and made a full recovery.

Reference to case report: Erbay *et al.* (2009) *Journal of Infection in Developing Countries*, **3**, 239–240.

Campylobacter

Campylobacters are spiral (from Greek *kampylos*=bent), flagellate (therefore motile), Gram negative bacteria (Figure 3.22). They are natural inhabitants of the intestinal flora of many farm animals (e.g. poultry) – where they cause no

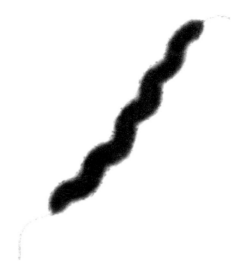

Figure 3.22 *Campylobacter jejuni* (electron micrograph, magnification×25,000). (Provided by Manfred Ingerfeld, University of Canterbury, New Zealand.)

ill-effects to their hosts – and are commonly found in rivers and streams. There are two species important in food-borne illness, namely *C. jejuni* and *C. coli*. When animals are slaughtered, contamination of their flesh with intestinal contents can lead to food-borne *Campylobacter*, particularly in poultry (see *Foods associated with* Campylobacter *contamination*).

Campylobacters* are difficult to culture in the laboratory because they require very specific conditions and for this reason they are not thought to multiply on food. However, when they infect the human intestine the conditions are perfect for their growth and a small number of bacteria ingested soon multiply to numbers capable of causing harm.

Infection with *C. jejuni* or *C. coli* causes campylobacteriosis. Following ingestion of bacteria with food (or drink, e.g. contaminated water) it takes 2–5 days for the symptoms of campylobacteriosis (see below) to appear; during this time the small number (as low as 500–800 cells) of ingested bacteria multiply in the intestine to numbers sufficient to damage the intestine lining. Their mechanism of pathogenesis involves adhesion to the mucosa of the intestine and production of an enterotoxin (cytolethal distending toxin) which prevents cells dividing and stops them eliciting an immune response against the invading bacteria, so allowing the *Campylobacter* to survive for longer and cause more damage to surrounding cells.

Campylobacter serotypes

There are many different strains of pathogenic *C. jejuni* and *C. coli*. These strains elicit an immune response in their mammalian hosts that results in the production of antibodies specific to the different strains – these are termed serotypes (sometimes called serovars). The antibody response is directed against specific cell surface antigens present on the bacteria which are indicators of the different *Campylobacter* strains.

Serotyping can be carried out to determine the strain of *Campylobacter* (and other bacteria) and is very useful when investigating a food-borne infection because it enables the microbiologist to link the *Campylobacter* isolated from food with the isolate (e.g. from faeces) from the infected person. Serotyping can be carried out in the laboratory by looking for specific genetic sequences on DNA from the bacterium – the sequences are associated with the genes that code for the bacterial cell surface markers that result in host immune responses.

Campylobacteriosis

The symptoms of campylobacteriosis include severe diarrhoea (often bloody), abdominal pain, cramps and fever (temperature can reach 40°C); the symptoms last from 2–10 days. Most people recover fully, but in rare cases, in people with impaired immune systems, septicaemia can develop which can be fatal.

C. jejuni can also cause a rare latent autoimmune (i.e. causes antibody production against the infected person's own cells) disease of the nerves of the legs termed Guillain-Barré syndrome; the symptoms are ascending paralysis and dysaesthesias (loss of feeling) below the waist. It can progress to respiratory failure and death if not treated.

Campylobacteriosis is a notifiable disease in many countries which means that cases must be reported to the public health authorities. Campylobacteriosis incidence information from national records is very useful in investigating the epidemiology of the disease and its incidence variability around the world. There is significant variability in incidence of campylobacteriosis from country to country.

Foods associated with *Campylobacter* contamination

Campylobacters can contaminate meat from different animal species. This is not surprising because *Campylobacters* are present in the intestines of animals, so, during the slaughter, boning and cutting process it is inevitable that intestine contents will contaminate the meat. Fortunately *Campylobacters* are fragile bacteria and are easily killed; moderate heat (55°C), drying and freezing are often enough to kill them. It only takes 300–800 live *Campylobacter* cells to result in human campylobacteriosis; a smaller number of cells – even just one – might cause infection, but the risk is correspondingly lower.

Chicken is by far the most important source of *Campylobacter* – 89% of chicken meat tested in New Zealand was contaminated with *Campylobacter* in a 2003/04 survey (Table 3.2). The reason why chicken is so contaminated is the subject of much research and there is no answer to the question as yet, but it might be because poultry have a higher intestinal load of *Campylobacter* and the slaughter process leads to greater contamination.

International incidences of campylobacteriosis

The incidence of campylobacteriosis varies around the world (see Figure 2.14 in Chapter 2). This is likely to reflect differences in the contamination of meat with *Campylobacter* during processing and differences in cooking and preparation procedures – remember *Campylobacter* is easily killed, but not many live cells are needed to cause campylobacteriosis.

New Zealand has a high incidence of campylobacteriosis compared to the rest of the world and therefore a great deal of research is under way there to

Table 3.2 The incidence of *Campylobacter* (*C. jejuni* +
C. coli) contamination in meat in New Zealand from a
national retail survey carried out in 2003 and 2004. (Data
from Lake *et al.* (2007) *Risk Profile: Campylobacter jejuni/
coli in Red Meat*. Institute of Environmental Science &
Research, Christchurch, New Zealand, www.nzfsa.govt.nz.)

Meat	Percentage positive for *Campylobacter* (number tested)
Beef	3.5 (230)
Veal	10 (90)
Lamb/mutton	6.9 (231)
Pork	9.1 (230)
Chicken	89.1 (230)

Table 3.3 *Campylobacter* contamination in chicken
samples from different countries analysed as part of
national surveillance schemes. (Data from Lake *et al.*
(2007) *Risk Profile: Campylobacter jejuni/coli in Poultry
(Whole & Pieces)*. Institute of Environmental Science &
Research, Christchurch, New Zealand, www.nzfsa.govt.nz.)

Country	Percentage of chicken samples positive for *Campylobacter*
New Zealand	89
United Kingdom	59–75*
Taiwan	68
Germany	50
Japan	46
Mexico	36
Denmark	25–40*

*Range of values from several studies.

try to understand why. One possibility is that New Zealand chicken is more
highly contaminated with *Campylobacter* than chicken from other countries
(Table 3.3). Indeed New Zealand chicken does appear to have higher levels of
Campylobacter contamination than chicken meat from other countries.

For a consumer to contract campylobacteriosis she/he must be exposed to
live *Campylobacter* sp. Which means that even if a piece of meat is highly contaminated, providing it is cooked well the bacteria will be killed and therefore
the risk will be near zero. For this reason, when considering the high incidence of the disease in a particular country (e.g. New Zealand) it is important
to consider food preparation (e.g. cooking) methods with a view to determining whether there might be a preparation method peculiar to that country
that allows live bacteria to contaminate the final food product. As discussed
in Chapter 2, this might be the case in New Zealand where barbequed food is

very popular. Tongs are used to put the raw meat (e.g. *Campylobacter*-contaminated chicken) onto the barbeque grill, the meat is then cooked well enough to kill the surface bacteria. However, the tongs that were used to pick up the raw meat might be contaminated with *Campylobacter* from the raw chicken and if they are used to transfer the cooked meat to the consumer's plate they might infect the cooked meat and, in turn, infect the consumer. This possible route of food contamination is not proven, but it illustrates well how a particular food preparation technique might result in food-borne illness (see Hudson *et al.* (2003) in *Further reading*).

Campylobacteriosis case example

Fresh pre-cooked cocktail sausages were purchased from a butcher's shop by a family in New Zealand. The sausages were frozen and the next day some of the sausages were defrosted in a refrigerator and eaten without reheating or cooking by three members of the family. Two days after eating the sausages one member of the family began to feel ill; 5 days after eating the sausages another member of the family (a child) became ill, and 9 days after the sausages had been defrosted another member of the family (another child) became ill, but he had not eaten any sausages. Interestingly the child who became ill but had not eaten any sausages had been bathed with the child who had consumed sausages and became ill. The symptoms were diarrhoea, stomach cramps and fever. The family went to their doctor who took faecal samples and the case was referred to the regional Public Health Laboratory.

Campylobacter jejuni was cultured from the faecal and the sausage samples. In addition, the *C. jejuni* isolates were serotyped and all samples were found to be Penner Type 4. Further studies showed that *C. jejuni* was only present on the outside of the sausages (i.e. *C. jejuni* could only be cultured from saline rinses of the sausages and not from the sausage meat itself). This suggested external contamination of the sausages with *C. jejuni*.

The butcher's shop from which the sausages were purchased was investigated and it was found that the retailer purchased bulk packs of cocktail sausages, opened them and repackaged them into smaller packs for sale. The repackaging was carried out on the same bench used to cut and pack chicken on.

This was a clear case of food-borne campylobacteriosis that was very likely caused by the butcher contaminating the outside of the cocktail sausages because he used the same bench to handle raw chicken (remember chicken is often contaminated with *Campylobacter*) and repack the sausages. Interestingly, one of the family who contracted campylobacteriosis had not eaten a cocktail sausage, but had been bathed with his infected brother; this suggests that he picked up the bacterium from the bath water which was likely to be contaminated with his brother's infected faecal material.

The infected family members all recovered within a week.

Reference to case report: Graham *et al.* (2005) *Australian and New Zealand Journal of Public Health*, **29**, 507–510.

Clostridium

Clostridia are Gram-positive, rod-shaped bacteria capable of producing endospores. They are natural inhabitants of soil. There are several highly pathogenic *Clostridia* associated with life-threatening diseases, including *C. botulinum* which causes botulism, a severe, often fatal disease; *C. difficile*, a natural human intestine inhabitant that can grow uncontrollably during antibiotic therapy (the antibiotic kills many of the other gut microflora which allows *C. difficile* to grow) causing a severe colitis; *C. perfringens* which causes food-borne illness; *C. tetani* which is responsible for tetanus; and *C. sordelli* which can cause severe complications and death following childbirth. Other members of the genus cause far less severe diseases in humans or are apparently harmless natural inhabitants of the environment, including the human intestine.

The most important *Clostridia* associated with food-borne illness are *C. botulinum* and *C. perfringens*. Both bacteria cause disease by producing toxins coded for by plasmids which can be exchanged during conjugation.

The potential for the lethal botulinum toxin to be transferred to other *Clostridia* is very worrying indeed, especially as some food preservation procedures are designed specifically to stop *C. botulinum* growing, but might allow other, harmless, *Clostridia* to survive. If a 'harmless' *Clostridium* sp. acquired the ability to make botulinum toxin this would have the potential to lead to a food safety disaster of unprecedented scale.

Clostridium botulinum

Clostridium botulinum is the most serious of all food-borne bacteria. The disease it causes – botulism – is often fatal. The bacterium prefers basic conditions (i.e. >pH7), but will grow in low acid (pH >4.5) conditions and tolerates high sodium chloride (NaCl) concentrations. Indeed *C. botulinum* was first isolated in 1896 by the Belgian bacteriologist Émile van Ermengem (1851-1932) from a piece of cured ham that had poisoned three people. Cured ham is both low acid and salty (i.e. high NaCl concentration) and therefore provided the *C. botulinum* with the growth conditions it likes.

C. botulinum produces a group of incredibly potent neurotoxins – botulinum toxins – with estimated LD_{50}s in humans in the low ng/kg body weight range (i.e. >100 ng would be fatal to a human). This is incredibly toxic and is responsible for the very high risk associated with *C. botulinum*-contaminated food. It is the toxins not the bacterium *per se* that result in pathogenicity, and usually the toxin contaminates the food due to bacterial growth in the food, but the bacteria can also infect the consumer of contaminated food and produce the toxins *in situ* (e.g. in the consumer's gut).

Botulinum toxins

As discussed above, this group of toxins is incredibly potent; they are fatal to humans in the nanogram exposure range. Indeed they are amongst the most toxic chemicals known.

Botulinum toxins (sometimes referred to as BT or BTX) are proteins (molecular weight in the region of 150,000 Da). They are classified as BTA, B, C, etc., and some are further divided into subgroups (e.g. BTC1). The magnitude of toxicity of the specific toxins varies (Table 3.4), but all are neurotoxins.

The botulinum toxins are made up of two protein subunits: a higher molecular weight subunit (100,000 Da; heavy chain) which chaperones the lower molecular weight subunit (50,000 Da; light chain) which is responsible

Table 3.4 Toxicity of some botulinum toxins.

Botulinum toxin type	LD$_{50}$
A	40–50 ng/kg body weight
C1	Approx. 32 ng/kg body weight
E	3,200 ng/kg body weight

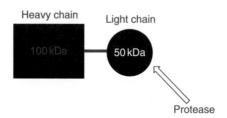

Figure 3.23 The molecular arrangement of the botulinum toxins.

for the neurotoxicity (Figure 3.23). The toxins work by interacting with the presynaptic membrane preventing release of neurotransmitter. This results in inhibition of neurotransmission causing paralysis.

How botulinum toxins inhibit neurotransmission

Before we can understand how botulinum toxins work, we must first have an appreciation of how nerves transfer information to and from the brain.

Nerves are composed of myriad neurones (nerve cells) which stretch from the brain to muscles (motor neurones), or from a sensory site (e.g. skin) to the brain (sensory neurones). Motor neurones conduct electrical messages from the brain to the muscles to initiate contraction (resulting in movement), whereas the sensory neurones take messages (e.g. pain) to the brain. So a sensory message in response to touching a hot object would result in a motor message to move away from the hot object, so preventing one being burned.

A single neurone does not stretch from the brain, via the spinal cord, to either the motor or sensory site, but instead a series of end-to-end-connected neurones make the 'wire' that connects the brain to the sensory or motor site. The individual neurones interact with each other via gaps across which the electrical nerve impulse must jump in order for it to take the message to or from the brain. These gaps are called synapses.

The nerve impulse (action potential) comprises a flow of changed membrane potential caused by Na$^+$ being pumped in to the neurone across the cell membrane followed very shortly afterwards by K$^+$ being pumped out to return the potential across the membrane to zero (Figure 3.24). This flow of changed potential along the neurone membrane is the nerve impulse.

The nerve impulse cannot jump the synapse and so a chemical (called a neurotransmitter) is used to transfer the charge from one neurone to the next. There are a large number of neurotransmitters, e.g. acetylcholine which is used in many motor and sensory systems (Figure 3.25).

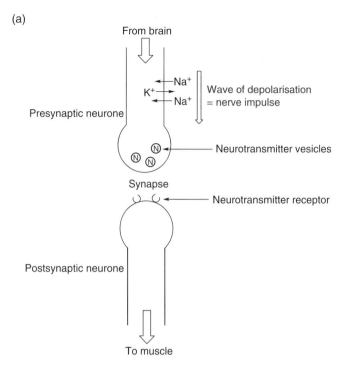

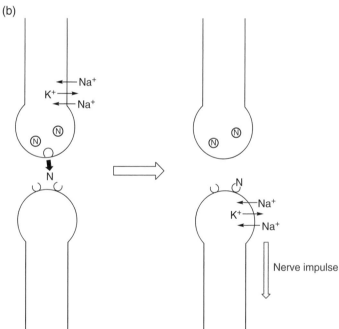

Figure 3.24 (a) Neurones on either side of the synapse showing the wave of membrane depolarisation that causes the nerve impulse. The neurotransmitter vesicles are ready to release their neurotransmitter (N) into the synapse. (b) When the action potential reaches the presynaptic nerve terminus the neurotransmitter vesicles fuse with the presynaptic membrane and release the neurotransmitter into the synapse (left). The neurotransmitter then binds to the postsynaptic receptor and initiates a new action potential (right). Botulinum toxins prevent the neurotransmitter vesicles fusing with the presynaptic membrane.

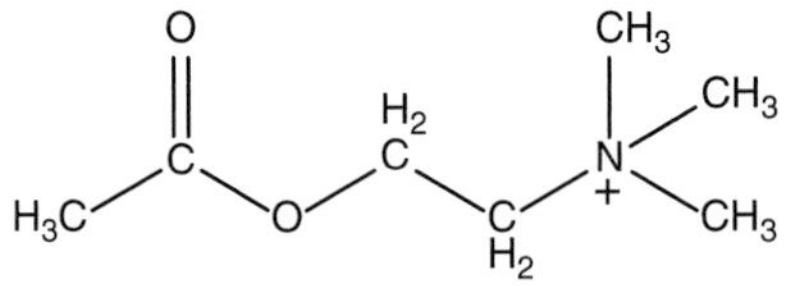

Figure 3.25 The molecular structure of acetylcholine – the positive charge on the nitrogen carries the action potential charge across the synapse.

The neurotransmitter is made in the presynaptic neurone and is stored in membrane-bound vesicles in the cytoplasm. When the action potential reaches the presynaptic terminus (i.e. the part of the neurone next to the synapse) it causes the neurotransmitter to be released into the synapse by fusing the cytoplasmic vesicle containing the neurotransmitter with the presynaptic membrane. The neurotransmitter then crosses the synapse and binds to a postsynaptic receptor located on the postsynaptic membrane. The receptor is a large protein molecule that recognises the shape of a specific neurotransmitter. When the neurotransmitter molecule is docked into the binding site of the postsynaptic receptor the sodium pump in the postsynaptic receptor is activated, resulting in Na^+ efflux which initiates another nerve impulse (Figure 3.24). In order to prevent a neurotransmitter molecule initiating more than one new postsynaptic action potential, the neurotransmitter is either removed (by specific carrier proteins) or destroyed by an enzyme, e.g. acetyl-cholinesterase.

Botulinum toxins prevent the presynaptic neurotransmitter vesicles being released into the synapse and so stop the nerve impulse crossing the synapse. They do this because the light chain of the botulinum toxin molecule is a protease which destroys the membrane proteins to which the neurotransmitter vesicle binds before it fuses with the presynaptic membrane to release its neurotransmitter into the synapse.

Botox
Botulinum toxin A (BTX-A) is used under the trade name Botox for cosmetic purposes. It is injected at very low concentrations usually into the muscles of the face. The BTX-A inhibits neurotransmission to the muscles and causes relaxation which removes lines and wrinkles and apparently makes people look younger ... an interesting application of a very toxic molecule.

Botulism
Botulism is the disease caused by botulinum toxin. The toxin can be present in food contaminated with *C. botulinum* or be produced by *C. botulinum* growing in the consumer's gut. The symptoms include muscle weakness, including the muscles associated with chewing, facial expression (e.g. drooping eyelids), eye movement (often results in double vision) and swallowing. As the disease progresses the muscles of respiration become involved resulting in difficulty in breathing and poor oxygenation of the blood, and eventually respiratory failure leads to coma and death. All of these symptoms relate to botulinum toxins' inhibition of neurotransmission at the synapse.

Botulism is rare, but when it occurs it is very serious. In 11 incidents of botulism between 1922 (the first recorded case) and 1998 in the UK there were

58 people affected of whom 19 (33%) died – this is an extremely high death rate for a food-borne illness (data from Brett (1999) *Eurosurveillance*, **4** (1), pii=45, http://www.eurosurveillance.org/ViewArticle.aspx?ArticleId=45).

Botulism case example

On Thursday 8 June 1989 a 47-year-old woman was admitted to Blackpool Hospital in the UK with suspected botulism. Her son was in intensive care at the nearby Royal Preston Hospital (RPH) with a diagnosis of Guillain-Barré syndrome (also known as Landry's paralysis – there's a significant clue in this name!). There was a second patient admitted to the RPH with the same symptoms and the next day two further patients were admitted to Blackpool Hospital, and two more were in Manchester Hospital 40 miles (75 km) south of Blackpool. Within a day or two there were 27 affected people in hospitals in the north-west of England. After considerable discussion, a diagnosis of botulism was agreed upon for all of the patients – remember an important symptom of botulism is muscle weakness/paralysis, hence the misdiagnosis of Guillain-Barré syndrome.

It was found that hazelnut yoghurt made by the same producer had been eaten by at least 25 of the patients. The UK Department of Health was informed of the connection between the yogurt and the cases of botulism, and on 11 June manufacture of the yogurt was stopped and the product withdrawn from retail outlets. No more cases of botulism were reported.

Botulinum toxin B was found in yoghurt samples from the homes of the hospitalised patients, in samples of hazelnut conserve (used to flavour the yoghurt) from the manufacturer's premises, and in faeces from the patients.

One patient, a 74-year-old woman, died; the others all recovered within 2 weeks. It was concluded that the botulinum toxin had originated from *C. botulinum*-contaminated hazelnut conserve. The conserve had a pH >4.6, which is ideal for growth of *C. botulinum*, and the preparation of the conserve did not involve temperatures high enough to kill *C. botulinum* spores which might have originated from soil contamination of the hazelnuts at harvest.

Reference to case report: O'Mahony *et al.* (1990) *Epidemiology and Infection*, **104**, 389–395.

Clostridium perfringens

Clostridium perfringens is a bacterium of soil and commonly occurs in faeces; therefore, it is easy to contaminate food if hygiene is not strictly observed. It produces spores that will survive high temperatures (100°C) and it synthesises a series of heat labile enterotoxins that are responsible for the disease the bacterium causes. *C. perfringens* is classified according to the exotoxins (i.e. toxins excreted by cells) produced by different strains, e.g. *C. perfringens* type A produces large amounts of α-toxin (lecithinase) which degrades lecithin (phosphatidylcholine), a phospholipid present in cell membranes.

Some *C. perfringens* strains also produces enterotoxins (from the Greek *enterikos* meaning intestine), i.e. toxins secreted into the host intestines, that are responsible for the disease caused by *C. perfringens*. The exotoxins used to classify the bacterial strains must not be confused with the enterotoxins that are responsible for the disease caused by *C. perfringens* infection.

Clostridium perfringens enterotoxins

C. perfringens types A and B produce a protein enterotoxin (CPE; approx. molecular weight = 35,000 Da). CPE is produced by *C. perfringens* growing in the host's intestine and causes diarrhoea by interfering with water uptake across the intestinal wall because it alters the water permeability of intestinal epithelial cells. It is thought to bind to specific CPE receptors on the surface of intestinal epithelial cells. Interestingly, CPE receptors are also present on other epithelial cells in the body (e.g. liver cells) so they are likely to have biological functions that CPE inhibits only in the intestine because CPE is not absorbed and therefore does not reach epithelial cells in other parts of the body (e.g. the liver). CPE is heat sensitive and does not survive cooking.

 C. perfringens type C behaves rather differently to both types A and B in the intestine; instead of simply growing in the intestine contents it adheres to the intestine wall epithelial cells (e.g. cells of the villi), causing necrosis of the surrounding tissue. It is thought that the intimate association between *C. perfringens* type C and the intestinal epithelium cells allows efficient transfer of the enterotoxin from the bacterial cells to the epithelium leading to a profound local effect resulting in necrosis. The necrotic areas of the intestine then become infected with more *C. perfringens* type C, which sets up a vicious cycle of infection resulting in a massive necrosis which can lead to perforation of the intestine wall.

The *C. perfringens* type C secretes a necrotic enteritis B-like toxin (NetB) which is a barrel protein (i.e. the shape of a barrel with a channel through the middle) which upsets ion and water balance by facilitating ion transport across the epithelial cell membrane (Figure 3.26).

Foods associated with *C. perfringens*

Foods that have been cooked, cooled slowly and reheated are typically associated with *C. perfringens* food-borne illness (Table 3.5). This combination of processes allows bacteria in the original raw food to form endospores which survive the temperature of cooking and germinate when the cooked food is cooled down slowly. When the food is reheated before being eaten the bacteria grow as the temperature rises and infect the unsuspecting consumer.

Symptoms of *C. perfringens* infection

Food-borne *C. perfringens* types A and B result in profuse diarrhoea approximately 8-22 hours after consumption of contaminated food. The time possibly reflects the number of bacteria in the food; higher levels of contamination will probably cause disease sooner because the bacteria are able to synthesise sufficient toxin in the host's gut to cause disease. On the other hand, lower levels of bacterial contamination will require growth time in the gut to allow the bacteria to multiply and accrue a sufficient number to make enough toxin to cause illness. Other symptoms of *C. perfringens* types A and B gastric infection include abdominal pain and nausea, but rarely vomiting.

Infection with *C. perfringens* type C causes a far more serious, but very rare (in the developed world; two cases in the USA between 1984 and 2002) disease known as haemorrhagic or necrotic enteritis (enteritis necroticans) or pigbel. The term 'pigbel' is pigeon English for abdominal pain and was used as a colloquial name for *C. perfringens* type C necrotising enteritis because the first cases were reported from Papua New Guinea where pigeon

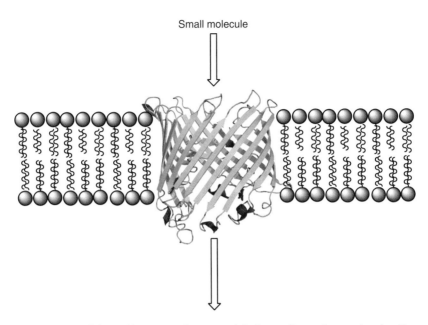

Small molecule

Figure 3.26 A β-barrel transmembrane protein in a cell membrane showing the central channel which allows diffusion of small molecules across the membrane. This protein is a sucrose-specific porin from *Salmonella typhimurium*, but is of the same general structure as NetB from *Clostridium perfringens* type C. (Barrel protein structure from http://upload.wikimedia.org/wikipedia/commons/f/fb/ Sucrose_porin_1aOs.png.)

Table 3.5 The ideal conditions for *C. perfringens* food-borne illness are illustrated well by cases reported in New Zealand. (Data from http://www.nzfsa. govt.nz/science/data-sheets/clostridium-perfringens.pdf.)

Food	Number of cases in outbreak	Situation
Chicken biryani and mutton curry	58	Social function – inadequate cooling and reheating, improper hot holding
Roast meat and gravy	4	Restaurant – inadequate cooling of beef stock, poor hot holding practices
Roast turkey	57	Restaurant – inadequate cooking, cooling and holding procedures, insufficient reheating

English is spoken. The symptoms are severe bloody diarrhoea and severe abdominal pain – it is often fatal.

Clostridium perfringens food-borne illness case examples
C. perfringens *type A/B*

On 8 August 2008 more than 100 inmates in a Wisconsin county jail developed nausea, vomiting and diarrhoea which was noted during an early morning inspection. The inmates had been given a ground turkey and beef casserole for their evening meal on 7 August. Further investigation revealed that some of the inmates developed symptoms only a few hours after eating their evening meal.

The casserole had been made by combining and reheating leftover foods from previous meals. There was limited information about the origins of the leftover foods used in the casserole, but it was suspected that the leftovers had been allowed to cool slowly before being stored in a refrigerator.

Microbiological analysis of both faeces samples from affected inmates and samples of the casserole revealed *C. perfringens* which confirmed that the outbreak was due to *C. perfringens* and that the source of infection was the casserole.

Reference to case report: Communicable Diseases Center (2009) *MMMR Weekly*, 20 February, **58**, 138–141, http://www.cdc.gov/mmwr/preview/mmwrhtml/mm5806a2.htm.

C. perfringens *type C (necrotic enteritis)*

A 66-year-old African-American woman was found by a family member in her home in an unresponsive state. She was taken to the University of Mississippi Medical Center Emergency Room; she was pronounced dead on arrival.

A post mortem examination revealed a highly distended abdomen containing approximately 400 mL of straw-coloured fluid in the peritoneal cavity. The intestine showed segmented mucosal necrosis. The serosa (the outer layer of the colon) showed a thin layer of acute inflammatory exudates. Microscopic examination of the ileum (upper small intestine) confirmed haemorrhagic necrosis and the presence of Gram positive rods thought to be *Clostridium* sp. DNA isolated from the intestinal tissue was found to be consistent with *C. perfringens* type C.

Approximately 65 hours before she died the woman had eaten a breakfast of boiled turkey sausage, toast, egg and corn meal. She had bought the turkey sausage on a trip back to her home in Mississippi from Chicago (approx. 900 miles/1,450 km; approx. 12 hours driving). Fifteen hours after eating the breakfast she complained of stomach pains and developed diarrhoea. Within 6 hours her diarrhoea became bloody.

Microbiological culture of the turkey sausage failed to grow *Clostridia* and therefore even though the sausage was a likely source of infection – remember it had been transported unrefrigerated for a long distance which would have allowed bacteria to grow – this could not be confirmed. However, the presence of *C. perfringenc* type C in the woman's intestine combined with the symptoms and post mortem findings confirmed *C. perfringens* type C necrotic enteritis as the cause of death.

Reference to case report: Gui *et al.* (2002) *Modern Pathology*, **15**, 66–70.

Escherichia

Escherichia is a genus of Gram negative, non-spore forming, motile (flagellate), rod-shaped bacteria that naturally inhabit the intestine of mammals including humans. *Escherichia coli* is by far the most common bacterial species and is an important member of the normal gut microflora of humans and plays an important role in nutrition, e.g. it synthesises vitamin K which is absorbed by its human host. The magnitude of *E. coli*'s prevalence in the human gut is well illustrated when you consider that it accounts for about 50% of the dry weight of faeces.

Native *E. coli* is not pathogenic; however, *E. coli* is promiscuous and conjugates with other bacteria and therefore is able to acquire toxin-coding plasmids from them, thus making the new *E. coli* strain able to synthesise toxins. The new strain is therefore pathogenic; its pathogenicity depends upon the level of toxicity of the toxin coded for by the acquired plasmid.

One of the most important food-borne bacterial pathogens is *E. coli* O157:H7, a shiga-like toxin-producing *E. coli* (STEC) resulting from a conjugation between *E. coli* and *Shigella* sp. (e.g. *Shigella dysenteriae*, the causative organism of dysentery).

E. coli O157:H7

E. coli O157:H7 was first isolated in 1975 from a sick patient, but it was not until 1982 that it was recognised as a food-borne pathogen when it was isolated from a hamburger that had caused an outbreak of gastroenteritis in 47 people in Oregon and Michigan, USA. The serotype (O157:H7) refers to specific antigens on the bacterial cell surface (O-antigens) or the flagella (H-antigens). The numbers refer to the specific glycoprotein antigens of either the flagellum or cell surface. *E. coli* O157:H7 produces a shiga-like toxin which is responsible for its pathogenicity.

Shiga-like toxin

Shiga toxin (Stx) was isolated from *Shigella dysenteriae, the* organism that causes dysentery. It was named after Professor Shiga Kiyoshi (1871-1957) (Figure 3.27) who first isolated *S. dysenteriae* from a case of dysentery in Japan in 1897.

The toxin is a protein (molecular weight=68,000 Da) comprising two subunits - A and B. Subunit B binds to the outside of the target cell (e.g. an intestinal epithelial cell of an infected person) and subunit A enters the cell and inhibits protein synthesis by interfering with ribosomal function. Inhibition of protein synthesis prevents cellular activity and kills the cell.

The shiga-like toxin (SLT) from *E. coli* O157:H7 differs from Stx by only one amino acid residue and has exactly the same mechanism of action as Stx. Both Stx and SLT are heat stable and therefore are not destroyed by cooking. The mechanism of pathogenicity of *E. coli* O157:H7 involves the bacteria growing in the host gut and synthesising STL *in situ* rather than the bacteria growing on food and the SLT-contaminated food being ingested.

Foods associated with *E. coli* O157:H7

E. coli O157:H7 lives in the intestines of farm animals (e.g. cattle) and therefore when animals are slaughtered there is the possibility that their meat will be contaminated with the bacterium. Providing the meat is cooked thoroughly the bacteria will be killed and therefore do not present a risk to the consumer. Beef is the meat most commonly associated with *E. coli* O157:H7, but lamb, venison and pork have also been implicated.

When meat is contaminated by *E. coli* O157:H7 the bacteria are only present on the outer surface of the meat and therefore providing it is cooked well on both sides the bacteria will be killed. This allows internally rare meat

Figure 3.27 Professor Shiga Kiyoshi (1871–1957) who first isolated *S. dysenteriae* in Japan in 1897. (From http://en.wikipedia.org/wiki/Kiyoshi_Shiga.)

(e.g. a rare steak) to be eaten safely. However, if minced meat is contaminated, the *E. coli* O157:H7 that was on the outside of the original piece of meat is distributed onto the multiple surfaces of the minced meat. If the meat is used to make hamburgers the risk of consuming a rare hamburger is great because the *E. coli* O157:H7 will not be killed in the 'pink' middle of the hamburger during cooking (see *Case example*).

Any food that comes into contact with *E. coli* O157:H7-contaminated faeces is a potential cause of *E. coli* O157:H7 food-borne illness. Unpasteurised milk has been associated with *E. coli* O157:H7 outbreaks probably due to its contamination during milking – clearly without pasteurisation the bacteria would remain alive in the milk and infect the consumer. Some vegetables (e.g. lettuce) have also been associated with *E. coli* O157:H7 outbreaks possibly due to contamination by cattle manure-based fertilisers from contaminated animals.

Incidence of *E. coli* O157:H7 food-borne illness
E. coli O157:H7 is a relatively new food-borne pathogen and consequently its incidence is likely to increase as contamination of food animals and their environments progresses. In 1982 human *E. coli* O157:H7 infection was very rare, but by 1998 in the USA the incidence of STEC food-borne illness was 2.8/100,000 population (i.e. 0.002% of the US population suffered from STEC in 1998); the incidence was similar – 2/100,000 population – for New Zealand in 2001.

Symptoms of *E. coli* O157:H7 infection

The symptoms of *E. coli* O157:H7 infection are similar to other food-borne bacterial illnesses; patients present with bloody diarrhoea, severe abdominal pain and tenderness, but with no fever. The symptoms begin 1-10 days after ingestion of the *E. coli* O157:H7-contaminated food. It is likely that the time to onset of symptoms is related to the bacterial contamination level – higher levels of contamination will result in a faster onset of the disease because there are sufficient toxin-producing bacteria to affect the intestinal epithelium almost immediately. On the other hand, if the bacterial contamination level is low the bacteria must multiply in the host's gut before they can synthesise sufficient SLT to affect the intestinal epithelium and cause the disease symptoms.

The disease usually clears up within a week and leaves no lasting effects. However, on rare occasions a very much more serious disease can develop due to the cell-toxic effects of SLT causing larger areas of intestinal cell death leading to necrosis (a significant region of tissue death). This might result in perforation of the intestine which in turn leads to infection of the abdominal cavity (peritonitis – similar to untreated appendicitis when the infected appendix perforates). This is serious and can be fatal.

Damage to the intestinal epithelium by SLT can also allow access of SLT to the circulatory system via the intestinal lymphatic and capillary system – the intestines are very well supplied with blood vessels to take the absorbed food digestion products (e.g. glucose) to the liver for storage, metabolism or onward passage to other tissues in the body. The systemic SLT binds to white blood cell surface receptors and is transferred to the kidney where the SLT binds more strongly (via the SLT B subunit) to kidney cell surface receptors and gains entry to kidney cells where it inhibits protein synthesis and impairs kidney function. The resulting disease is haemolytic uraemic syndrome (HUS) and is serious, often fatal (5-10% mortality), particularly in children under 5 years old. About 2-7% of *E. coli* O157:H7 infections result in HUS.

E. coli O157:H7 food-borne illness case example

In September 1984 34 (median age=89 years) of the 101 residents of a nursing home in Nebraska, USA, developed diarrhoea; 14 of the sufferers were hospitalised because of the severity of their symptoms which included abdominal cramps, marked abdominal distension and grossly bloody diarrhoea. *E. coli* O157 was found in faeces from 11 (32%) of the patients.

On 13 September hamburgers had been given to residents of the nursing home who had special requirements for ground or pureed diets (other residents received ham). The first case of diarrhoea was noted on 18 September. Forty-three percent of the patients who ate hamburgers became ill, while only 11% of those who ate ham developed diarrhoea. It is possible to transmit *E. coli* O157:H7 from person to person and by cross contamination in the kitchen – this probably accounts for why some of the people who ate ham became ill. The cook responsible for preparing the hamburgers reported that she had cooked them in an oven for 85 minutes; on investigation the oven temperature was 5-25°C lower than that indicated on the oven temperature setting. Despite this, studies showed that heating similar hamburgers in the same oven for the time the cook said she had cooked the hamburgers for revealed that the internal temperature of the hamburgers was 71°C, i.e. high enough to kill *E. coli*.

Of the 34 patients, four (12%) died, one (3%) developed HUS and the others recovered. The high death rate is probably because the patients were old. Similarly the HUS case is likely to be related to the patient's age since it generally only occurs following *E. coli* O157:H7 infection in the very young or old.

The most likely cause of this *E. coli* O157:H7 outbreak was contaminated ground beef used for the hamburgers and that the cooking time/temperature was not sufficient to kill the bacteria in the centre of the hamburger.

Reference to case report: Ryan *et al.* (1986) *Journal of Infectious Diseases*, **154**, 631–638.

Listeria

Listeria is a genus of only six species of flagellate, rod-shaped, Gram positive bacteria (Figure 3.18). The genus was named in 1940 after Lord Joseph Lister (1827–1912), the pioneer of antiseptic surgery. Only one species causes human food-borne illness, namely *Listeria monocytogenes* – it causes listeriosis.

Listeria monocytogenes

L. monocytogenes is found in soil and waterways and is sometimes present in the intestines of animals and therefore can contaminate their milk. It has a very broad spectrum of temperatures (4–37°C) that it will grow and reproduce in; the lowest of these is the temperature of the domestic refrigerator which means that even if food is kept in the fridge, *L. monocytogenes* can still grow.

L. monocytogenes has a very interesting mechanism of pathogenesis that does not rely on the synthesis of toxins. When the bacteria enters the warm temperatures of the host's intestine its flagellum becomes inactive and new flagella proteins are not formed; instead it commandeers actin filaments from the cytoskeleton of a host cell and pulls itself along the surface of the cell aided by the actin filaments' contractile properties. The host cell internalises the bacterium by phagocytosis (engulfing) and the *L. monocytogenes* becomes an intracellular parasite that grows and reproduces in the cytoplasm.

Listeriosis

There are two forms of listeriosis – non-invasive and invasive. The former involves infection of intestinal cells and while serious it is far less serious than the invasive disease which results in infection of, for example, the central nervous system. Invasive listeriosis is often fatal.

The symptoms of non-invasive listeriosis include fever, muscle aches and vomiting; nausea and diarrhoea are less common. Symptoms of the invasive disease are much more serious, including serious neurological effects.

The symptoms of listeriosis usually last from 7–10 days and their onset is about 21 days after exposure to contaminated food – this is a long incubation period because the bacterial contaminant of the food has to enter the gut, infiltrate the gut epithelial cells and reproduce before there are enough bacteria in the right place to cause symptoms.

As discussed above, *L. monocytogenes* enters the host's cells where it reproduces. This cellular internalisation gives the bacterium access to the transport mechanisms of the body (e.g. the circulatory system) via capillaries and the lymphatic system which means that it can get to other, distant, tissues. For example, a very small proportion of listeriosis cases result in *L. monocytogenes* infecting the brain or spinal cord; this leads to meningitis (inflammation of the membrane covering of the brain (the meninges)) which is very serious and often fatal.

L. monocytogenenes appears to be particularly pathogenic in pregnant women – 30% of listeriosis cases in the USA are in pregnant women and since 30% of the population are not pregnant at any one time this means that the disease incidence is skewed towards this population cohort. It has been suggested that the reason for the high susceptibility of pregnant women to listeriosis is because their immune system is challenged and therefore they are not as effective at 'fighting' the disease as others. Listeriosis can lead to serious complications in pregnancy, including abortion and infection and death of the baby.

Listeriosis occurs more commonly in the very young (including *in utero*) and the very old (i.e. ≥70 years) or people with diseases that might affect their immune system (e.g. AIDS).

Foods associated with *L. monocytogenes*
L. monocytogenes outbreaks have been associated with many foods, including meat (particularly cold cut cooked meats, e.g. boiled ham), dairy products (particularly soft cheeses, e.g. Brie), seafood, milk (usually unpasteurised), pâté and vegetables (e.g. salads stored in a refrigerator). Foods involved in outbreaks have often been stored in a refrigerator – remember *L. monocytogenes* grows well at +4°C.

Incidence of listeriosis
Listeriosis (invasive and non-invasive) is a serious but rare disease. Approximately 2,500 people (i.e. 0.0008% of the population) contract listeriosis a year in the USA, and of these approximately 500 die (i.e. 20%); this is a very high death rate for a food-borne illness (data from Centers for Disease Control & Prevention (CDC), USA, http://www.cdc.gov/pulsenet/pathogens_pages/listeria_monocytogenes.htm). In 2008 in New Zealand there were 27 cases of listeriosis (i.e. 0.0007% of the New Zealand population), and of these three died (11% of cases) (data from EpiSurv NZ, http://www.surv.esr.cri.nz/PDF_surveillance/MthSurvRpt/2008/200812DecRpt.pdf).

There is evidence that the incidence of listeriosis is increasing. For example, in the UK the incidence increased from 145 cases (0.002% of the population) in 2001 to 213 cases (0.003% of the population) in 2004 (data from Gillespie *et al.* (2006) – see *Further reading*). This might reflect the increasing age profile of the population – remember older people are more susceptible to *L. monocytogenes* infection.

Listeriosis case example
In February 2000 two people were ill after eating corned (salted) beef in Christchurch, New Zealand. They presented with febrile (i.e. fever) gastroenteritis 17 and 32 hours after eating the corned beef (NB: this is a very short incubation period). Shortly afterwards another two cases presented after eating boiled ham. A few days later seven more people in the same geographical area (i.e. South Island) became ill after eating a meal that

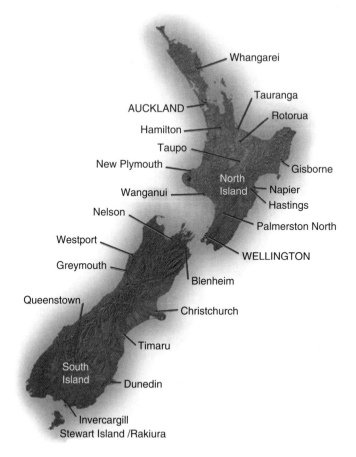

Figure 3.28 New Zealand. (From http://en.wikipedia.org/wiki/File:New_Zealand_ towns_and_cities_copy.jpg.)

included corned beef. And shortly after this 21/24 people attending a lunch including boiled ham on the North Island (i.e. geographically separated from the above cases – Figure 3.28) became ill with the same symptoms – this is a total of 32 cases of gastroenteritis in less than a week.

The corned beef and boiled ham involved in all of these cases were from the same South Island manufacturer. The manufacturer ceased operation while a public health investigation was carried out and meat manufactured by the company was recalled from the retail outlets to prevent further cases. Unfortunately, the supermarket from which the ham was purchased for the North Island lunch did not withdraw the meat from sale.

L. monocytogenes was found in faeces from the patients and in samples of both corned beef and ham from the meals. Serotyping demonstrated that the same bacterial strains were present in the food and the patients.

The source of contamination at the manufacturer was not found. However, the products were packed and had long shelf lives (3 months) which would make finding the source of infection difficult. On the other hand, the long shelf life explains the short incubation period in the infected people. A large

Figure 3.29   Dr Daniel Elmer Salmon (1850–1914) after whom the genus *Salmonella* is named. (From http://en.wikipedia.org/wiki/File:Daniel_Salmon.jpg.)

number of *L. monocytogenes* are needed to cause listeriosis; this usually results from a small number of bacteria in the food which, following consumption, multiply in the host's gut until the bacteria reach numbers that can cause disease. In this case the contaminated meat had been stored for a long time in a refrigerator in the shop which allowed the *L. monocytogenes* to multiply sufficiently to give the consumers a dose of bacteria sufficient to cause disease immediately.

All of the affected people were healthy before they ate the contaminated food and recovered completely from their ordeal.

Reference to case report: Sim *et al.* (2002) *Letters in Applied Microbiology*, **35**, 409–413.

Salmonella

Salmonellae are rod-shaped, Gram negative, flagellate bacteria that do not form spores. They were named in honour of an American veterinary surgeon, Dr Daniel Elmer Salmon (1850-1914) (Figure 3.29) by his assistant Theobald Smith who first described S*almonella choleraesuis* (the former name for *S. enterica*). *S. enterica* is present in the intestines of cattle and poultry and is the cause of *Salmonella* food-borne illness.

S. enterica can infect egg whites and as the egg ages the yolk membrane breaks down which allows the *S. enterica* to infect the egg yolk. This is important because egg yolks are used raw in some foods (e.g. mayonnaise) and often eggs are consumed with 'runny' (i.e. uncooked) yolks. These are important sources of food-borne *S. enterica*.

S. enterica is not killed by freezing and therefore frozen raw meat (particularly poultry) and uncooked egg-based foods are still potential causes of

Table 3.6 *Salmonella enterica enterica* serotypes responsible for food-borne illness.

Salmonella enterica enterica serotype	Disease caused
Salmonella Typhi	Typhoid fever
Salmonella Paratyphi	Gastroenteritis Enteric fever
Salmonella Typhimurium	Gastroenteritis
Salmonella Choleraesuis	Septicaemia

salmonellosis. *S. enterica* is destroyed by cooking (≥60°C for 2-6 minutes) and therefore even contaminated food is safe to eat after being cooked thoroughly (e.g. hard boiled eggs).

Sub-species and serotypes of *Salmonella enterica*
There are six sub-species of *S. enterica*. The most important cause of *Salmonella* food-borne illness is *S. enterica enterica*; this sub-species can be further divided into five serotypes – the most important serotype from the perspective of food-borne illness is *S. enterica enterica* Enteritidis – usually abbreviated to *Salmonella* Enteritidus or *S.* Enteritidis (note: the serotype is not written in italics and is capitalised because it is not a part of the species name). *S.* Enteritidis causes salmonellosis.

The other *S. enterica enterica* serotypes that cause food-borne disease are listed in Table 3.6.

S. Typhi is usually water borne, but in theory could be transmitted via food (e.g. vegetables washed in contaminated water and eaten raw). *S.* Paratyphi and *S.* Typhimurium are both food-borne pathogens, but *S.* Choleraesuis is not transmitted via food.

S. enterica, its sub-species and serotypes have a common, and ingenious, method of infecting cells. A bacterium swims (using its flagellae) to the surface of an intestinal epithelial cell, locates the cell surface and injects a specific protein into the cell. This protein induces the cell to modify its cell membrane, causing surface ruffles. The bacterium moves into the grooves of one of the ruffles and the cell engulfs and internalises it. Once inside the cell the *Salmonella* multiplies.

Salmonella food-borne illness came to the fore in the UK in 1988 when the Junior Health Minister, Edwina Curry, issued a press statement saying that most of the British egg production was contaminated with *Salmonella*. She was wrong and resigned later the same year; despite this the UK egg industry collapsed because of public fears of contracting salmonellosis from eggs. There is a very great difference between egg production (i.e. poultry) being contaminated – *Salmonellae* are natural poultry gut microflora – and eggs being contaminated!

Salmonellosis
The symptoms of salmonellosis are diarrhoea, vomiting and fever. They begin 8-72 hours after infection. The time from infection to symptoms depends, in

part, on the number of bacteria contaminating the food. A larger number of bacteria will result in a faster onset of symptoms because they will not have to multiply as many times to achieve pathogenic numbers in the intestinal epithelial cells. There is, however, always about 8 hours delay to allow the *Salmonella* to invade the epithelial cells in sufficient numbers.

The illness usually lasts for 4–7 days and most people recover completely. In rare cases the *Salmonellae* can spread from the intestinal epithelial cells to the blood stream resulting in a severe septicaemia which can be fatal – this is more common (but still rare) in immuno-compromised people (e.g. AIDS sufferers). Another rare outcome of salmonellosis is reactive arthritis caused by *Salmonella* eliciting an immune response with the production of antibodies that cross-react with joint proteins so causing inflammation in the joints (i.e. arthritis).

Foods associated with *Salmonella* Enteritidus and incidence of salmonellosis

Salmonella enterica is a common component of the gut microflora of most warm-blooded animals, including farm animals. Similarly the serotype S. Enteritidus can be found naturally in the same animal species and therefore can contaminate the meat derived from these animals. Thus most meats can be contaminated, but some meats are more susceptible to contamination than others – this could be related to slaughter methods or differences in the infection rates of different species. Chicken is by far the highest risk meat with respect to outbreaks of salmonellosis. Between 1997 and 1999 in New Zealand there were 24 outbreaks of salmonellosis of which 12 (50%) were associated with poultry and eggs, eight (34%) with red meat, two (8%) with seafood, one (4%) with raw milk, and one (4%) with vegetables (data from Lake *et al.* (2002), http://www.foodsafety.govt.nz/elibrary/industry/Risk_Profile_Salmonella-Science_Research.pdf). The fact that chicken is the commonest vector of salmonellosis is the case worldwide; however, the incidence of the disease varies from country to country (Table 3.7).

Table 3.7 Incidence of salmonellosis around the world showing that the incidence varies from country to country and that in some it is stable (e.g. Australia) whereas in others (e.g. Denmark) it is declining. (Data from http://www.nzfsa.govt.nz/science/risk-profiles/salmonella-in-poultry-meat.pdf.)

Country	Year	Incidence (cases/100,000 population)
Australia	1996	31.6
	2000	32.1
Canada	1998	24.4
Denmark	1998	73.3
	2000	43.3
England & Wales	2000	25

Salmonellosis case examples
Case 1
On Friday 3 October 2008 the US Department of Agriculture (USDA) became aware of 32 people from 12 US states having contracted salmonellosis. Since this was a very high incidence of infection over a short time period they investigated further. Laboratory studies on the *Salmonellae* isolated from the infected people's faeces showed that they all had the same DNA fingerprint which suggested they were all from the same source (e.g. the same manufacturer). After some detective work the USDA found that all of the cases were associated with consumption of raw chicken meals (e.g. chicken Kiev and chicken cordon bleu) from the same manufacturer. The meals had been cooked in the microwave by their unsuspecting purchasers. The problem was that they had not read the instructions on the packets properly and had assumed that the chicken meals were cooked and only needed warming up in the microwave. They were wrong! The chicken in the meals was raw and should have been cooked thoroughly in a conventional oven (to an internal temperature of 74°C) and not just warmed up in a microwave oven. It turned out that the reason for so many people making the same mistake was that the packaged meals had been pre-browned at the manufacturers (presumably to make them look appetising) and they looked 'ready to eat' – clearly they should have read the label more carefully! All of the infected people recovered within a week or so.

This is a classic case of undercooked chicken-associated salmonellosis.

Case 2
A second report illustrates that *Salmonella* infection can result in serious septicaemia. In this case a 58-year-old man was admitted to a hospital in Marseilles, France, in 1999 suffering from jaundice and extensive gangrene; blood cultures were positive for *Salmonella* Enteritidis and the doctors concluded that the gangrene was due to *Salmonella* septicaemia which led to multi-organ dysfunction syndrome (MODS). The patient's jaundice was caused by his high chronic alcohol intake; he had a history of alcohol consumption of 300 g/day which is equivalent to 34 pints of beer a day which probably resulted in his immune system not being able to combat the *Salmonella* infection. This resulted in a generalised infection and septicaemia rather than the usual gastric infection leading to gastroenteritis.

The patient's gangrenous foot was amputated and he recovered from the *Salmonella* septicaemia and was well a year later.

References to case reports:

Case 1 – Paddock (2008) 32 cases of *Salmonella* linked to microwaving raw chicken. *Medical News Today*, http://www.medicalnewstoday.com/articles/124286.php.

Case 2 – Retornaz *et al.* (1999) *European Journal of Clinical Microbiology and Infectious Diseases*, **18**, 830–841.

Shigella

Shigellae are Gram negative, rod-shaped, non-spore-forming bacteria that naturally occur only in the intestines of primates (including humans). They were first described by Professor Kiyoshi Shiga (see *Shiga-like toxin*) and their very potent exotoxin – shiga toxin – is named after him.

The *Shigellae* are divided into four serogroups, which are subdivided into serotypes as follows:

Serogroup A→*Shigella dysenteriae* →12 serotypes
Serogroup B→*Shigella flexneri* →6 serotypes
Serogroup C→*Shigella boydii* →23 serotypes
Serogroup D→*Shigella sonnei* →1 serotype

Groups A, B and D are important human pathogens and of these *S. dysenteriae* is the most important – it causes dysentery (also called shigellosis). *Shigellae* are incredibly potent; only 100 bacteria are needed to cause disease.

Shigellae are transmitted by human faecal contamination of both water and food, but arguably water transmission is the most important route worldwide. In places where a large number of people are confined (e.g. in refugee camps in developing countries) without a proper water supply and toilet facilities, faecal contamination of water often results in epidemic dysentery which, due to the dehydration it causes, kills a large number of people. There are about 165,000,000 cases of shigellosis worldwide each year and, of these, 1,000,000 die, most in developing countries. Clearly *Shigellae* are very important pathogens in a world context; however, they are far less important in the developed world than in the developing world. In addition, they are primarily transmitted via contaminated water rather than food. In this section I will concentrate on *Shigellae* food-borne illness in the developed world.

Shiga toxins

Shiga toxins or Stx (see *Shiga-like toxin*) inhibit protein synthesis in the ribosome and thus stop cells functioning and kills them. Stx has two protein subunits: one binds to the outside of intestinal epithelial cells of the host (i.e. the infected person) and chaperones the cell's uptake of the second subunit. The second subunit has N-glycosidase activity and cleaves bases from ribosomal RNA (rRNA) so inhibiting protein synthesis (Figure 3.30).

Stx is extremely toxic; its LD_{50} [i.p. mouse]=20 µg/kg body weight; this means that about 1.2 mg would kill a human which is why it doesn't take many *Shigellae* to cause disease – just a few Stx molecules per epithelial cell would be enough to debilitate the cell.

Shigellosis (dysentery)

Infection by members of the genus *Shigella* causes shigellosis. As mentioned above, shigellosis is also called dysentery, but strictly speaking only infection with *S. dysenteriae* causes dysentery, but the symptoms of infection by other *Shigellae* are so similar that it is difficult to distinguish between them without laboratory studies.

The symptoms of shigellosis can range from mild abdominal discomfort to severe cramps, diarrhoea, fever, vomiting, bloody faeces and tenesmus (excessive straining to defecate). The death rate from shigellosis is very high (10–15% of cases). *Shigella* infections result in drastic dehydration which is often the cause of death.

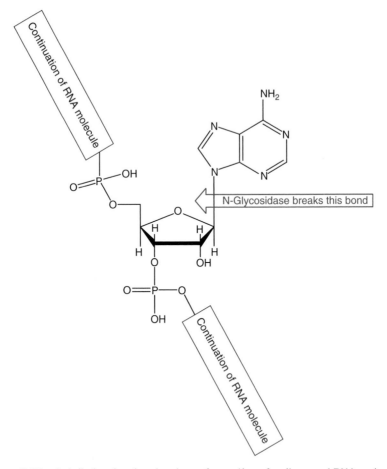

NH₂

N-Glycosidase breaks this bond

Figure 3.30 Detailed molecular structure of a section of a ribosomal RNA molecule showing N-glycosidase catalysed cleavage of the N-glycoside bond between the base (in this case adenine) and ribose of the RNA sugar phosphate backbone.

The incubation period for shigellosis is 12–50 hours and severe infections can last as long as 6 weeks. Antibiotics can be given to old, young or infirmed people with shigellosis as a means of reducing the severity of the infection, but, surprisingly anti-diarrhoeal drugs are not given. This might seem strange because dehydration due to diarrhoea is often the cause of death; however, diarrhoea is also the means by which the *Shigellae* are expelled from the body, so preventing diarrhoea also stops the body getting rid of the pathogen.

As discussed above, shigellosis is a very important disease of developing countries, but it does occur in developed countries and in travellers who have visited developing countries. There are about 18,000 cases of shigellosis in the USA (population 307 million) each year – this represents about 6 cases/100,000 population/year or about 0.006% of the US population contract the disease annually (data from Mead *et al.* (1999), see *Further reading*); shigellosis is therefore rare in the USA. It is difficult to compare the incidence of shigellosis in developing countries, but suffice to say that there are about 165 million cases in the developing world each year – this is equivalent to more than half the population of the USA.

Food associated with shigellosis

Shigellae contaminate the faeces of infected people and so unhygienic food handling can lead to food contamination. Clearly it does not matter which food an infected handler contaminates; if the contaminated food is eaten the consumer might contract shigellosis. Cooking kills *Shigellae* so many cases are traced back to foods that are eaten raw (e.g. salad vegetables). In addition, where sewerage systems are absent or ineffective, *Shigellae* from infected people might contaminate the water supply. If food that is eaten raw (e.g. lettuce) is washed in contaminated water the consumer is at high risk of contracting shigellosis – this is not a common route of infection in the developed world because domestic water sources are reliably clean.

Shigellosis case examples
Case 1

On 11 July 2006 a day care centre in Harvey County, Kansas, USA, reported that 11 of their 135 children and one of their 30 staff members had diarrhoea. Faecal samples were taken for microbiological examination. *S. sonnei* was isolated from the samples. A detailed epidemiological investigation suggested that two children in the day care centre had suffered from diarrhoea in May of the same year – they were not investigated at the time, but it was thought that they represented the first cases in the outbreak.

The bacterial isolates from the faeces samples were investigated by pulsed field gel electrophoresis (PFGE). This technique separates proteins present in the bacteria to produce patterns; if the patterns are identical this means that the bacteria are identical. In this case, four very similar patterns were obtained which suggested that the *S. sonnei* were likely to be the same strain – the small differences in protein pattern were likely to reflect minor differences in proteins produced by the individual isolates, possibly because of mutations in the clones growing in different patients (remember bacteria divide often and so mutations occur often).

It is likely that the two children showing signs of infection in May brought the *S. sonnei* to the day care centre and it slowly moved from person to person via the oral faecal route with food as the vector. The median age of the children in the centre was 4 years – it is very difficult to stop a 4-year-old touching food, and equally difficult to make sure they wash their hands properly after going to the toilet. This is why shigellosis is more common in child day care facilities than other places in the developed world.

Case 2

In January/February 2001 there were two separate outbreaks of shigellosis (*S. sonnei*) in Auckland, New Zealand. The first was at a camp for socially deprived children and the second was at an elderly care facility 40 km from the first outbreak.

- *Children's camp cases* – there were 43 children in the camp (41 were studied) and of these, 15 (37%) contracted shigellosis. Fifteen of the 53

(28%) camp staff also contracted shigellosis, giving an overall infection rate of 33%. Exposure to *Shigella* was thought to be via food at the camp and the ultimate source of the bacteria was thought to be a faulty waste-water discharge sluice. All of the ill children and staff recovered within 2 weeks.

● *Elderly care facility cases* – four of 17 staff (24%) and four of 17 residents (24%) from the home developed shigellosis over 11 days. All except the initial infection were attributed to person-to-person infection; the question is, how did the first case get exposed to *Shigella*? All of the patients recovered within 10 days.

The first case in the children's camp was a Pacific Island boy who had recently returned from a trip to his parent's home near Auckland; two other members of his family also developed gastroenteritis. The boy's mother regularly bought apples from a store near to her home.

The first case in the elderly care home was cared for by one of the home's care staff who also bought apples from the same store as the boy's mother.

Interestingly there were a number other shigellosis cases in and around Auckland at about the same time as the children's camp and care home cases. Several were in air crew from a flight returning from Samoa (a South Pacific island) all of whom had stayed in the same hotel in Samoa. All of the cases had exactly the same strain of *S. sonnei* – interesting! It is likely that the *S. sonnei* originated in Samoa and was contracted by people staying in a particular hotel. When they returned to New Zealand they brought the bacterium with them and infected someone in the shop that both the boy from the camp and the caregiver from the care facility bought their apples from. The apples were therefore the *Shigella* vector by which means the camp and care facility received their infection.

References to case reports:

Case 1 – http://www.kdheks.gov/epi/download/HV_shigella_May06_report.pdf.
Case 2 – Hill *et al.* (2002) *New Zealand Medical Journal*, **115**, http://www.nzma.org.nz/journal/115-1156/62/.

Staphylococcus

Staphylococcus is a genus of Gram positive, spherical bacteria. They normally grow in small groups that look rather like a bunch of grapes (Figure 3.31), hence their name which is derived from the Greek, σταφυλή [staphyle] meaning bunch of grapes and κόκκος [kokkos] meaning granule. There are 32 species of *Staphyloccocus*; most are normal flora of human skin and mucous membranes and are harmless. Only *S. aureus* is a food-borne pathogen; it is not the bacteria *per se* that are the problem, but rather the heat-stable protein toxins they produce.

Since *S. aureus* are common skin bacteria in humans, food can easily get contaminated by handling. If the food is a good growth medium for the *S. aureus* the bacteria multiply, particularly if the food is kept in a warm place or allowed to cool down slowly after cooking. While they are growing the bacteria

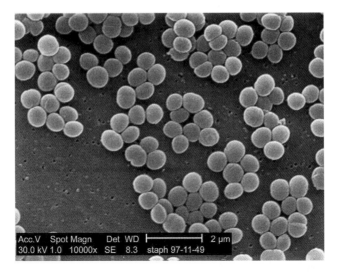

Figure 3.31 Electron micrograph (×10,000) of *Staphylococcus aureus* showing their 'bunch of grapes' clusters. (From http://en.wikipedia.org/wiki/File:Staphylococcus_aureus_01.jpg.)

excrete their toxins into the food. When the food is eaten the consumer succumbs to the effects of the toxin very quickly indeed – it is the toxin that causes the symptoms not the consumer's infection by *S. aureus*.

Staphylococcus toxins
The *Staphylococcus* toxins are heat, acid and protease stable endotoxins; therefore they are not destroyed by cooking or the acids and proteases in the stomach. As little as 0.1 μg of the toxins can cause symptoms in humans.

An example of a staphylococcal toxin is *S. aureus* α-toxin. This is a pore-forming β-barrel toxin. It has a most ingenious mechanism of toxicity (see Clostridium perfringens *enterotoxins* and Figure 3.26) which involves the barrel-shape protein inserting across the cell membrane, forming pores due to the channel in the middle of the barrel. These pores allow molecules and ions to escape from inside the cell; this upsets the biochemistry of the cell, resulting in dysfunction and cell death.

Staphylococcus food poisoning
Staphyloccocus genuinely causes food *poisoning* (often called Staph food poisoning) because it is the toxins that cause the disease not the bacterium itself. *S. aureus* grows on food that has been infected by poor hygiene techniques if the food is not kept hot (i.e. ≥60°C) or cold (i.e. ≤4°C) enough. For example, if a hot food product (e.g. a meat pie) sold by a corner shop is kept warm in a display oven, the oven temperature is not at or above 60°C and the food was contaminated with *S. aureus*, the bacteria will multiply on the food and synthesise α-toxin. The consumer of the pie will receive more than they expected!

The symptoms of *Staphylococcus* food poisoning begin soon (30 minutes to 7 hours) after consumption of contaminated food. There is no incubation period because the bacterial toxin is already in the food. The symptoms are nausea, vomiting, retching, stomach cramps and diarrhoea. The severity of the symptoms depends on the amount of toxin (i.e. the dose) in the food. In very severe cases headache and fever occur. The disease usually clears up within a week.

Foods associated with *Staphylococcus* food poisoning
Any food that is handled during its preparation and is a good culture medium for *S. aureus* can be the vector for *Staphyloccocus* food poisoning. There is a good relationship between foods that are handled a great deal during their preparation and *Staphylococcus* food poisoning. Typical foods include:

- Cooked meats, poultry and egg products (e.g. mayonnaise)
- Salads – egg, tuna, chicken, potato, macaroni
- Cream-filled pastries, chocolate éclairs
- Sandwich fillings
- Milk and dairy products

Incidence of *Staphylococcus* food poisoning
It is difficult to assess the incidence of *Staphylococcus* food poisoning because it is a reasonably mild disease and therefore many people do not visit their doctor. In addition, the symptoms are very similar to those of other food-borne illnesses and so even if sufferers do go to their doctor it is unlikely that the doctor would take samples to allow the toxin to be identified, thus confirming *Staphylococcus* food poisoning. Despite their being few reliable data on the incidence of *Staphylococcus* food poisoning it is thought to be reasonably common – largely because the bacteria are common, handling food is common, and many foods provide excellent culture conditions.

***Staphylococcus* food-borne illness case example**
In early 1984 an outbreak of *Staphylococcus* food poisoning occurred in a family in Florence, Italy. Just before they became ill they had eaten lasagne – sheets of pasta often served as a dish of layered lasagne with a meat and tomato sauce between them and cheese on top ... delicious!

On 4 and 5 February 1984 in Enfield, England, three people suffered severe gastroenteritis and were admitted to hospital. *Staphylococcus* food poisoning was diagnosed. The sufferers had all eaten lasagne shortly before they became ill.

On 4 February in Swansea, Wales, five people contracted *Staphylococcus* food poisoning after eating lasagne in the same restaurant.

On 8 February 30 girls from a school in Winchester, England, became ill after eating lasagne. Their diagnosis was also *Staphylococcus* food poisoning.

There is a clear link between these four *Staphylococcus* food poisoning outbreaks – LASAGNE. The food they had eaten was the same, but they had purchased and eaten it in different parts of the same country (i.e. Wales and England) and a geographically distant country (i.e. Italy), so how could the cases be connected?

Further investigation revealed more *Staphylococcus* food poisoning cases (e.g. in Luxembourg and Reading, England) all associated with the consumption of pasta. Studies on the lasagne from the Luxembourg cases, which occurred in early February 1984, showed that the pasta associated with the disease outbreak was heavily contaminated with *S. aureus* and moreover it had been manufactured by a company in Parma, Italy, and that the company exported their pasta products all over Europe. But the most interesting fact was that the lasagne product batch code associated with the Luxembourg case was the same as that associated with the initial outbreak in Florence. Could all of the disease outbreaks be associated with *S. aureus*-contaminated pasta manufactured by the same company?

Microbiological studies in the UK on similar batches of lasagne from the Italian company revealed heavy contamination with *S. aureus* of the same strain as that associated with all of the cases discussed above – this showed that the manufacturing process resulted in contamination and that it was very likely indeed that the source of contamination for all of the cases of *Staphylococcus* food poisoning was the factory in Italy.

This is an excellent example of a point source contamination leading to a complex array of geographically separated outbreaks and illustrates the importance of manufacturing food hygiene especially in these days of mass production and wide distribution.

Reference to case report: Woolaway *et al.* (1986) *Journal of Hygiene*, **96**, 67–73.

Streptococcus/Enterococcus

Streptococci are spherical, Gram positive bacteria. They tend to grow in chains, hence their name which is derived from the Greek στρεπτος [streptos] meaning twisted chain. They are found naturally on and in humans. For example, different species live in specific parts of the mouth, e.g. *Streptococcus salivarius* lives on the dorsal side of the tongue and usually causes no harm, while *S. mutans* is one of the acid-producing bacteria responsible for dental decay, *S. pneumoniae* causes bacterial pneumonia which is a serious disease particularly in old people, and *S. pyogenes* causes strep throat (streptococcal pharyngitis). *Streptococci* are not all bad; *S. thermophilus* has been used in yoghurt manufacture for thousands of years and is an excellent means of preserving milk for winter consumption and providing dietary proteins at the same time.

Streptococci are subdivided into five groups – Groups A, B, C, D and G. The subdivision is based on their haemolytic properties. Group D is the group most commonly associated with food-borne illness. It comprises the following species:

- *S. faecalis*
- *S. faecium*
- *S. durans*
- *S. avium*
- *S. bovis*

Most of the *Streptococci* listed above (except *S. bovis*) have recently (1990) been reclassified to the genus *Enterococcus* and so *Streptococcus faecalis* is now *Enterococcus faecalis*, etc. Since the old classification is still often used I have included both in this section – it is all rather confusing at the moment!

Because *Streptococci* are on and in humans, it is easy for food handlers to contaminate food, especially if their hygiene standards are not what they should be. Little is known about the mechanisms of pathogenicity of the *Streptococci/Enterococci* associated with food-borne illness; far more is known about the *Streptococci* associated with more serious diseases (e.g. *S. pyogenes* which causes strep throat and necrotising fasciitis which is often fatal). Many of them produce toxins that cause haemolysis and have specific cell surface proteins that allow them to adhere to epithelial cells which begins the infection process. It is possible that the food-borne *Streptococci/ Enterococci* have similar means of causing disease.

If a food handler infected with *S. pyogenes* (e.g. suffering from strep throat) coughs onto the food being prepared, the food might become the vector of *S. pyogenes* and the food's consumers might develop strep throat. There are numerous cases where strep throat has been shown to be associated with consumption of a particular food at a social event (see *Streptococcal/ enterococcal food-borne illness case examples*).

Foods associated with streptococcal/enterococcal food-borne illness

The foods associated with *Streptococcus/Enterococcus* are often highly processed or handled a lot during manufacture or preparation. Often the reason for infection of highly processed food is because some aspect of the processing (e.g. heat treatment) was not sufficient to kill the bacteria. The reason for contamination of food that is handled a great deal during its preparation is obvious; hands can be contaminated with *Streptococci/Enterococci* and easily transfer the bacteria to handled foods. A large number of bacteria (about 10^7) are needed to cause infection in a consumer of contaminated food.

Typical foods associated with *Streptococci/Enterococci* are:

- Sausages
- Evaporated milk
- Cheese
- Meatballs
- Meat pies
- Milk
- Custard

Streptococcal/enterococcal food-borne illness

The symptoms of streptococcal/enterococcal food-borne disease are diarrhoea, abdominal cramps, nausea, vomiting, fever, chills and dizziness; onset of symptoms is usually 2–36 hours after eating contaminated food. The symptoms are very similar to staphylococcal food-borne disease and are relatively mild; this means that most people do not go to their doctors. This, in turn, means that there are no reliable statistics on the incidence of streptococcal/ enterococcal food-borne illness. However, it has been estimated that there are about 50,920 cases of food-borne *Streptococcus* illness per year (i.e. an incidence of 0.16 cases/100,000 population/year) in the USA with no fatalities (data from www.cdc.gov/foodborneburden/2011-methods-tables.html).

The symptoms of *S. pyogenes or* Group G *Streptococci* food-borne illness are sore throat, pain when swallowing and other symptoms associated with strep throat.

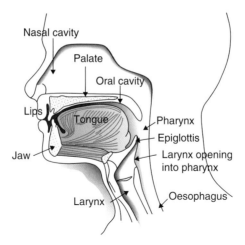

Figure 3.32 Cross section of the head and neck showing that food en route to the stomach via the oesophagus passes over the pharynx. If the food is contaminated with *Streptococci* it can infect the pharynx, resulting in pharyngitis (strep throat). (From http://en.wikipedia.org/wiki/File:Illu01_head_neck.jpg.)

Streptococcal/enterococcal food-borne illness case example

Seventy-two of 231 people (31%) who attended a convention in Florida, USA, in June 1979 developed pharyngitis. Group G *Streptococci* were isolated from their throats. Of the 231 people at the convention, 111 had attended the convention lunch and 57 (51%) of them developed pharyngitis, compared to 12 (10%) of the 117 people who did not go to the luncheon. This suggested that there might be a link between contracting pharyngitis and taking lunch. All of the people who attended the lunch and developed pharyngitis had eaten chicken salad. Shortly after the convention the cook who had prepared the chicken salad developed pharyngitis and therefore it was likely that she was incubating the bacteria in her throat and contaminated the chicken salad when she prepared it. The poor conference delegates who ate the chicken infected their throats with *Streptococcus* Group G as the food passed the pharynx en route to their stomachs via the oesophagus (Figure 3.32).

Reference to case report: Stryker *et al.* (1982) *American Journal of Epidemiology*, **116**, 533–540.

Vibrio

Vibrio is a genus of non-spore-forming, curved rod-shaped, Gram-negative bacteria that are found in marine environments and so are associated with eating undercooked seafood (especially shellfish, e.g. oysters). They have a flagellum at each end of their cells which makes them very motile. They are named after 'vibrions', a term coined in 1854 by the Italian anatomist Filippo Pacini (1812–1883) to describe the microorganisms he isolated from cholera patients – cholera is caused by *Vibrio cholerae*.

Vibrio spp are unusual among food-borne bacterial pathogens because they cause disease in other species as well as humans (i.e. they are zoonotic).

They are a common cause of mortality in marine life, for example *V. harveyi* causes luminous vibriosis in shrimps.

Vibrio food-borne illness is subdivided into cholera *Vibrio* and non-cholera *Vibrio* (NCV) infections. The former is caused by a single species, *V. cholera*, and the latter by many species including:

- *V. parahaemolyticus*
- *V. fluvialis*
- *V. mimicus*
- *V. furnissi*
- *V. hollisae*
- *V. alginolyticus*
- *V. vulnificus*

V. parahaemoliticus is by far the commonest species of *Vibrio* associated with food-borne illness; indeed in the USA it is the leading cause of shellfish-associated gastroenteritis.

Vibrio spp excrete complex protein toxins which are responsible for the effects they cause in the animals (including humans) they infect.

Vibrio cholerae and cholera

V. cholera causes the very serious disease cholera. It is a bacterium associated with faeces-contaminated water and is more common in developing countries where sewage systems are not in place. It also occurs after disasters, such as hurricanes and earthquakes that result in people losing their homes and having to live in unsanitary conditions that result in faecal contamination of drinking water.

Cholera is therefore not an exclusively food-borne disease – it is, perhaps, more commonly associated with drinking contaminated water. However, cholera can be food-associated if food is washed in *V. cholera* contaminated water or poor personal hygiene results in faecal contamination of food.

V. cholerae cannot withstand the acid conditions of the stomach and therefore most are killed during their passage through the gastrointestinal tract – for this reason a huge *V. cholerae* dose (in excess of 10^6 bacteria) is needed to cause cholera. The few bacteria that survive the stomach acid grow flagella when they reach the intestine and swim through the mucous lining of the small intestine and when they reach the intestinal epithelium they begin to secrete cholera toxin (CTX). CTX is a complex protein consisting of multiple subunits. One of the subunits aids its adhesion to the epithelial cells, another results in its uptake by the cells, and yet another causes the cell to excrete water, K^+, Na^+ and HCO_3^- (hydrogen carbonate or bicarbonate) which causes the characteristic watery diarrhoea and results in dehydration. It is the latter that is often the cause of death. Treatment, of course, includes drinking plenty of water, but in a situation where the water supply is contaminated with *V. cholera* this exacerbates the situation.

Cholera is reasonably common. In 2009 the World Health Organisation recorded 221,226 cases; 98% of these were in Africa, i.e. incidence approx. 21 cases/100,000 population/year in Africa (data from http://www.who.int/gho/epidemic_diseases/cholera/cases_text/en/index.html). Cholera is one of the world's major causes of death.

Table 3.8 Cases of *Vibrio* food-borne illness in Georgia, USA, in 2008. (Data from http://health.state.ga.us/pdfs/epi/fbd/2008/Updated%20-%202008%20Vibrio%20YE%20Data%20Summary.pdf.)

Species	Number of cases	Percentage of total cases
V. parahaemolyticus	5	26
V. vulnificus	4	21
V. alginolyticus	3	16
V. mimicus	1	5
V. fluvalis	1	5
Non-cholera *Vibrio* species not identified	4	21
Total non-cholera *Vibrio* cases	18	95
V. cholerae	1	5

Non-cholera *Vibrio* food-borne infections

As discussed above, *Vibrio* spp are inhabitants of marine ecosystems. The food-borne illnesses they cause in people are particularly associated with eating uncooked or undercooked shellfish. Shellfish are important as vectors of *Vibrio* spp because they are filter feeders and filter the bacteria out of their marine environment, harbour them in their filtration apparatus and gastrointestinal organs, and pass them on to their unsuspecting human consumers.

V. parahaemolyticus is the most common food-borne *Vibrio* (59% of *Vibrio* cases in the USA), but *V. vulnificus* (5% of US cases) is associated with 94% of *Vibrio*-associated deaths (data from Ho *et al.* (2009), http://emedicine. medscape.com/article/232038-overview). As for *V. cholera* infection, non-cholera *Vibrios* are also associated with natural disasters which result in faecal contamination of drinking water and unsanitary food preparation. For example, following Hurricane Katrina in southern USA in August 2005 there were 22 cases of *Vibrio* infection reported in Louisiana and Mississippi (Table 3.8); most were wound infections caused by cuts being exposed to *Vibrio* spp from the flood waters, but this makes the point that natural disasters result in water contamination and therefore increase the risk of infection.

Vibrio food-borne illness is increasing worldwide. In the USA food-borne illness generally is decreasing, but food-associated *Vibrio* infections increased by 80% between 1996 and 2001.

The symptoms of non-cholera *Vibrio* gastroenteritis are similar to, but less severe than, cholera; symptoms include diarrhoea (25% of sufferers have blood in faeces), vomiting and fever. The incubation period is about 48 hours from ingestion of contaminated food and symptoms last for up to 10 days (usually 6-7 days). The non-cholera *Vibrios* all secrete toxins similar to, but less virulent than, cholera toxin.

Vibrio food-borne illness case example

A 69-year-old man developed continuous watery diarrhoea on 14 February 1995, 3 days after returning home from a cruise around South-East Asia. His condition got worse over the next few days and he was admitted to hospital.

On day 6 of his hospitalisation blood cultures revealed *V. cholera* O139 Bengal (a cholera serotype). He was treated and recovered.

There were 630 passengers on the cruise; 490 were contacted and 62 (13%) reported having watery diarrhoea onset between 8 and 28 February. Serum samples from these passengers showed antibodies against cholera toxin (i.e. they had been infected with *V. cholerae*) approximately 80 days after returning from the cruise – of these, six passengers met the antibody titre criteria for *V. cholera* infection and faeces from two cases cultured *V. cholerae* O139 Bengal.

The cruise ended in Thailand on 10 February 1995 with a lunch at a buffet restaurant in Bangkok. All of the passengers infected had eaten at this restaurant, and most had eaten yellow rice and therefore it is likely that the yellow rice was the cholera vector.

The epidemiological implications of this case are significant because, of the six confirmed *V. cholerae* infections, five returned home to the USA and one to the UK ... and they all took *V. cholerae* O139 Bengal with them!

Reference to case report: Boyce *et al.* (1995) *Journal of Infectious Diseases*, **172**, 1401-1404.

Yersinia

Yersinia is a genus of 11 species of Gram-negative, rod-shaped bacteria which includes the causative agent of bubonic plague – *Yersinia pestis*. There are two species associated with food-borne illness, *Y. enterolitica* and *Y. pseudotuberculosis*; the former is by far the most common. The disease caused by *Y. enterolitica* or *Y. pseudotuberculosis* is called yersiniosis.

Yersinia enterolitica
Y. enterolitica has relatively recently been identified as a food-borne pathogen and consequently far less is known about it than other pathogens that have been studied for longer. Recent studies have shown that the bacterium may possess a plasmid that codes for proteins involved in cell invasion and infectivity. This might explain the differences in infectivity between different *Y. enterolitica* strains.

There is no doubt that the frequency of *Y. enterolitica* infection (yersiniosis) is increasing around the world – food safety experts are concerned about its future role, especially as it is able to grow at refrigerator temperatures (+4°C).

Yersiniosis
Y. enterolitica infection is associated with pork and milk and is uncommon in adults, but more common in infants – the annual incidence in the USA is about 9.6 per 100,000 population for infants, 1.4/100,000 for children and 0.2/100,000 for adults (data from Foodborne Disease Active Surveillance Network (FoodNet), USA, http://www/diet.com/g/yersinia).

The symptoms are fever, abdominal pain and diarrhoea (often bloody); they develop 4-7 days after infection and can last up to 3 weeks, but usually clear within 2-3 days. The abdominal pain is often localised and can lead to doctors misdiagnosing as appendicitis. Most patients recover completely from yersiniosis, but in rare cases *Y. enterolitica* can lead to systemic infection which causes joint pains and skin rashes (erythema nodosum) or more serious septicaemia. It is likely that the arthritis-like joint pains and skin rashes are due to an autoimmune response caused by the body's immune system producing antibodies against *Yersinia* that cross-react with the body's own proteins (e.g. proteins in the synovial fluid of the joints).

Y. enterolitica can be contracted from undercooked pork or unpasteurised milk, or directly from pigs (e.g. on farms or their organs in abattoirs or butchers). Preparation of chitterlings (pigs' intestines often cooked in broth) is particularly associated with yersiniosis – providing the chitterlings are well cooked the risk from eating the dish is low; however, during the cleaning of the intestines before they are cooked there is a very great risk of *Y. enterolitica* contaminating the cook's hands and then being transferred to other foods.

The incidence of yersiniosis varies greatly from country to country. In Georgia, USA, in 2008 the incidence was 0.47/100,000 population, whereas the incidence in New Zealand in 2009 was 10/100,000 population (data from 2008 Georgia Data Summary – Yersinia, http://health.state.ga.us/epi/food-borne). There are known to be many different *Y. enterolitica* strains (sero-types) and their prevalence varies from country to country which might, at least in part, explain the different incidences around the world, and some of them are not human pathogens.

Yersiniosis case example

This is an interesting case because it is one of the early reports of yersiniosis.

On 25 February 1981 a 2-year-old boy from Saskatchewan, Canada, developed fever (40°C), diarrhoea and vomiting. He was admitted to a local hospital. Initially there was no localised abdominal pain, but 3 days later appendicitis was suspected because of the changed abdominal pain profile. The boy was transferred to University Hospital, Saskatoon, in case he needed surgery. On 3 March a faecal culture showed *Y. enterolitica* serotype O:21.

On 26 March 1981 the boy's sister was admitted to hospital with very similar, but more severe, symptoms. On 1 April a laparotomy (surgical abdominal investigation) was conducted because peritonitis was suspected. The appendix was removed. Two days after surgery rectal washing cultures were positive for *Y. enterolytica* serotype O:21. The girl developed severe multi-system failure and died on 13 May.

On 3 April the mother of the two children went to the hospital to stay with her very ill daughter. She told staff that she did not feel unwell, but that she had diarrhoea. A faeces sample was taken and cultures were positive for *Y. enterolitica* serotype O:21.

The boy had drunk milk (unpasteurised) from a cow kept by his grandparents and it is possible that this was the source of infection and that spread to other members of the family was person to person.

Reference to case report: Martin *et al.* (1982) *Journal of Clinical Microbiology*, **16**, 622–626.

Take home messages

- Bacteria are responsible for a high proportion of food-borne illnesses.
- There is a large number of different bacteria that cause food-borne illness, ranging from mild gastroenteritis to life-threatening diseases.

- Some bacteria produce toxins that are responsible for their clinical effects. Some of these toxins are not destroyed by cooking.
- The plasmids that code for bacterial toxins can be transferred from one bacteria to another during conjugation – this means that toxin-mediated virulence can be transferred from one species to another (e.g. *E. coli* O157:H7).
- Some bacteria can grow at refrigeration temperatures (e.g. *Listeria mono-cytogenes*).
- Mass produced foods not cooled quickly and stored or transported at too high temperatures are often responsible for bacterial food-borne illness outbreaks.

Further reading

BBC On This Day. http://news.bbc.co.uk/today/hi/today/newsid_9415000/9415640. stm. This is an interview with Edwina Curry who resigned as a result of the debacle over *Salmonella* in eggs.

Forsythe SJ (2010) *The Microbiology of Safe Food*, 2nd Edn. Wiley-Blackwell, Oxford. This is an excellent in-depth book about food microbiology if you want to learn more about the subject as a whole.

Gillespie IA, McLauchlin J, Grant KA, *et al*. (2006) Changing pattern of human listeriosis, England and Wales, 2001-2004. *Emerging Infectious Diseases*, **12**, 1361-1366.

Hudson JA, Whyte R, Armstrong B & Magee L (2003) Cross contamination by *Campylobacter* and *Salmonella* via tongs during the cooking of chicken. *New Zealand Journal of Environmental Health*, **26**, 13-14.

Mead PS, Slutsker L, Dietz V, *et al*. (1999). Food-related illness and death in the United States. *Emerging Infectious Diseases*, **5**, 607-625.

Chapter 4

Viruses

Introduction

Viruses, like bacteria, are everywhere, but unlike bacteria they cannot survive alone - they can only live and multiply in specific host cells. Most viruses don't survive for long outside a cell. They need the host cell's biochemistry to latch onto and turn it around to their own reproductive mission; when they do this a single virus can form many tens of thousands of new viruses by this hijack approach to reproduction.

Arguably viruses, in fact one virus in particular (Norovirus), are the commonest cause of food-borne illness worldwide. Most viruses are harmful to their host cells, but there is one group, the phages, that prey on bacterial cells, killing them. Phages that kill food-borne pathogens might play a part in preventing food contamination and therefore food-borne illness. It might even be possible to use such phages in the future to reduce food bacterial pathogen contamination - so viruses might not be all bad after all!

This chapter is about the very few viruses that use food as their means of infecting people and the diseases they cause; even though most of these diseases might not be particularly severe they are very common and result in a huge number of work hours lost per year and so have a significant economic impact worldwide.

The discovery of viruses

While studying rabies in the late 1800s, Louis Pasteur (1822–1895) was perplexed; he could not see the causative agent under his microscope. He postulated that the rabies pathogen must be too small to see using a microscope - this might seem obvious now, but this was revolutionary thinking in the 1800s; to be smaller than a bacterium was unthinkable!

Around about the same time, Charles Chamberlain (1851–1908), with whom Pasteur worked, invented a minute pore size filter (the Chamberlain-Pasteur filter) capable of filtering out bacteria. The Russian biologist Dmitry Ivanovsky (1864–1920) was studying diseases of tobacco plants at the University of St.

Food Safety: The Science of Keeping Food Safe, First Edition. Ian C. Shaw.
© 2013 John Wiley & Sons, Ltd. Published 2013 by John Wiley & Sons, Ltd.

Figure 4.1 Martius Beijerink (1851–1931), the discoverer of viruses. (From http://en.wikipedia.org/wiki/File:Mwb_in_lab.JPG.)

Petersburg in 1892 and showed that the diseases were caused by something that passed through the Chamberlain–Pasteur filter; therefore it was smaller than a bacterium - unbelievable! Ivanovsky had discovered viruses; indeed he had discovered what is now called the tobacco mosaic virus. Ivanovsky, however, did not recognise what he had discovered, concluding that the agent of disease must be a toxin produced by bacteria. It was not until the Dutch microbiologist Martius Beijerink (1851–1931) (Figure 4.1) took a closer look at Ivanovsky's experiments that the identity of the pathogen was revealed. In 1898 Beijerink postulated that the Chamberlain–Pasteur filtrate contained a pathogen smaller than bacteria rather than a toxin, he called it *contagium vivum fluidum* (soluble living germ). Beijerink later used the term 'virus' to describe his soluble living germ; virus is from the Latin for poison, which is intriguing in the context of Ivanovsky's thoughts that toxins were the causative agents of filterable diseases.

Then in 1899 the American biochemist Wendell Stanley (1904–1971) postulated that viruses are particles; Beijerink was convinced they were liquids (or solutions). Arguably Stanley was the first virologist. Also in 1899 the German bacteriologists Friedrich Loeffler (1852–1915) and Paul Frosch (1860–1928) were working on foot-and-mouth disease (now known to be caused by *Aphthovirus*) and showed that when filtrates of the virus were diluted they were still infectious; therefore, they postulated, viruses must replicate in order to produce infectious concentrations. The science of virology was born.

By the late 1800s viruses were known to be very small particles that could not be seen under the microscopes of the time and, very importantly, that they could not survive for long, or replicate outside living cells. In fact it was not until the 1930s when the electron microscope was invented that the first viruses were seen. It was only then possible to explore their structure. It must have been an amazing moment to see the strange symmetry of a viral particle. Nothing like this had ever been seen before.

Stanley continued to study the tobacco mosaic virus and showed that it was mainly protein. Later nucleic acid was shown to be a component. Then in 1941 F.C. Bernal and I. Fankuchen produced tobacco mosaic virus crystals (yes, viruses can be crystallised) and studied their X-ray diffraction. In 1955 Rosalind Franklin's (1920–1958) work in Cambridge, UK, showed the importance of the nucleic acid component following the X-ray diffraction studies carried out by Heinz Fraenkel-Conrat (1910–1999) and Robley Williams (1908–1995) at the University of California at Berkeley, USA.

So, from Pasteur's suspicions that a pathogen smaller than a bacterium existed, it took about 75 years for scientists to develop a pretty well complete understanding of the structure of viruses. But how does this viral structure translate into infectivity? After all, viruses are not alive.

The biology of viruses

Viruses are fantastic (in the literal sense of the word); they are simply a collection of macromolecules (nucleic acid, proteins and sometimes lipids) that can alter the host cell's biochemistry to make it possible for them to replicate viral particles – viruses cannot replicate without a host cell. They are not alive, just pure chemistry. Their evolution is obscure since they cannot be classified as life and they cannot be the chemical precursors of life because they cannot survive without living cells and so could not have evolved before living cells – a conundrum that scientists have not yet solved.

A virus (sometimes called a viral particle or virion) consists of a protein coat which sometimes (depending on the viral species) has an outer lipid membrane-like envelope surrounding a single nucleic acid molecule – sometimes DNA, sometimes RNA depending on the species of virus (Figure 4.2). Viruses are incredibly small; for example, norovirus is about 28 nm in 'diameter' which means that about 500 noroviruses would fit end to end on a pinhead.

Viruses infect only cells of specific species, and often specific cell types (e.g. liver cells) within the species. Their specificity is determined by proteins protruding from the outside of the capsule or viral particle (in viruses that do not have capsules). These proteins recognise and bind to specific proteins on

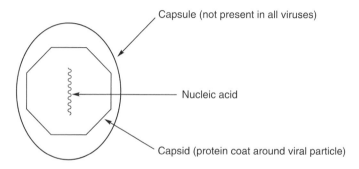

Figure 4.2 A stylised virus showing its component parts.

the outer surface of host cells. This docking process attaches the virus to the outside of its host cell. The docked virus then enters the host cell and enzymes break down the capsule and/or viral particle wall to release the viral nucleic acid. The viral nucleic acid is them replicated by the host cell's replication apparatus and viral components are synthesised from the nucleic acid codes. The host cell-synthesised viral components and nucleic acid are assembled into new virions; the host cell then lyses to release thousands of new viruses which can infect neighbouring cells or be expelled from the host's body (e.g. by coughing) to infect another unsuspecting host (Figure 4.3).

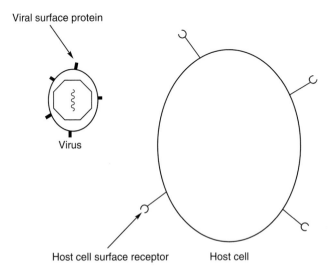

Viral surface protein

Virus

Host cell surface receptor   Host cell

Figure 4.3a   A virus in the vicinity of its host cell showing the viral surface proteins and host cell receptors.

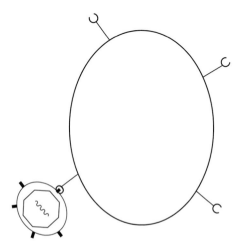

Figure 4.3b   The virus docks onto the host cell surface via its surface proteins and their specific interactions with the host cell receptors.

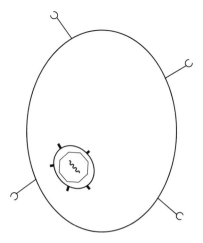

Figure 4.3c The host cell internalises the virus.

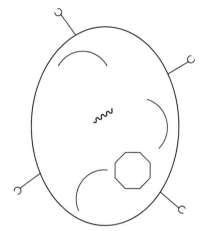

Figure 4.3d Enzymes break down the viral structure releasing its nucleic acid.

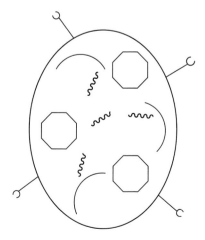

Figure 4.3e The viral nucleic acid hijacks the host cell synthetic apparatus and makes more viral components.

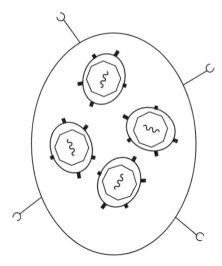

Figure 4.3f   Viruses are assembled within the host cell.

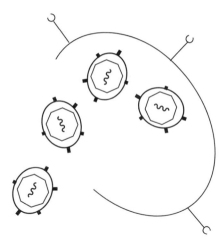

Figure 4.3g   The host cell lyses to release the new viruses.

Diseases caused by viruses and mechanisms of viral transmission

Viruses cause a wide range of diseases, from the very serious ebola and AIDS right down to the common cold. Until very recently there were no treatments for viral diseases, but there are now pharmaceuticals which interfere with viral replication, so slowing down their reproduction and giving the host's immune system a chance to fight them. These are used for treatment of the more serious viral diseases (e.g. AIDS) because their side effects can be quite serious too.

Viruses cannot survive for long outside their host and therefore transmission routes generally involve direct contact or droplet infection. In the latter the virus is contained in a droplet, perhaps exhaled by an infected host (this is how the common cold virus is spread, i.e. by sneezing and coughing). In the case of food-borne viruses someone infected with or carrying (i.e. harbouring the virus without showing any symptoms) a virus transfers the virus in droplets which land on food or by touching the food with virus-contaminated hands. The consumer of the food might get infected if they eat the contaminated food – viruses are remarkably resilient to human digestive acids and enzymes. Droplet infection might result if someone infected with a virus vomits, so forming vomit droplets which might land on food. Some food-borne viruses (e.g. norovirus) cause vomiting which promotes their transmission. Transfer by infected hands generally involves poor hygiene practices following defecation – the viral particle count in faeces from people infected with gastrointestinal viruses (e.g. norovirus) can be very high indeed.

There are only two major food-borne viruses, norovirus and hepatitis A.

Norovirus

Norovirus used to be called Norwalk-like virus (renamed in 2002). Norwalk is a town in Ohio, USA, where the first outbreak occurred in 1968. Norovirus is a member of the family of viruses called Caliciviruses which have their genetic material on a single strand of RNA, do not have a capsule and have an hexagonal capsid. Norovirus is approximately 28 nm 'diameter'.

Norovirus is likely to be the most common form of food-borne gastroenteritis worldwide. I use the word 'likely' because the disease is relatively mild and short-lived and so most sufferers do not visit their doctors and so the disease statistics are not complete because not all cases are recorded. For example, there are about 180,000 confirmed norovirus cases per year in the USA, i.e. 58 cases/100,000 population/year (data from CDC, USA, http://www.cdc.gov/ncidod/dvrd/revb/gastro/norovirus. htm). However, using immunological techniques to look for norovirus antibodies in a sample of the US population shows that approximately 60% of the population has been exposed to the virus by age 50 years which suggests that in reality there are about 9,200,000 cases/year, i.e. 3,150/100,000 population/year. This makes norovirus gastroenteritis very common indeed, and much more common than the confirmed (i.e. cases diagnosed by doctors) cases would suggest. This is supported by data from New Zealand which suggest that only one case of norovirus is reported (and therefore included in national statistics) for every 226 people suffering from norovirus gastroenteritis.

Like bacteria, viruses of a particular species often have many different serotypes, i.e. they have different genotypes which might code, for example, for different viral surface proteins that lead to different antibodies being produced by the human host. Norovirus serotypes are very useful in determining the source of a particular outbreak. For example, if a group of people who ate lunch together at a restaurant develop norovirus gastroenteritis and the norovirus serotype is the same as that isolated from the chef at the restaurant, it is clear where the infection originated.

Norovirus is an unusual virus in that it survives for long periods outside its host. Norovirus-infected food stored refrigerated or frozen can remain infectious for several years. Indeed, work with original norovirus extracts used in the 1970s showed that they were still active 30 years later. Norovirus will survive reasonably high temperatures (e.g. 60°C for 30 minutes) and is not deactivated by pasteurisation.

Cooking almost always destroys the virus providing the temperature reaches at least 90°C for 90 seconds. Foods that are not thoroughly cooked, eaten raw or contaminated after cooking are those most commonly associated with transmission.

Foods associated with norovirus

Almost any food can result in norovirus food-borne illness if it is contaminated after cooking (or is eaten raw) by a food handler infected with norovirus and practising poor personal hygiene. It can take as little as 10 viruses to result in infection and a human sufferer can shed 10,000,000 viruses/g faeces at the peak of infection. Clearly a very small faecal/hand/food transfer would be sufficient to give the consumer a pathogenic norovirus dose. This is one reason why norovirus infections spread very quickly.

Some foods, however, are particularly associated with norovirus. Filter-feeding shellfish (e.g. mussels and oysters) are by far the most commonly encountered foods in norovirus outbreaks. This is because they live in marine aquatic environments that can be contaminated with human sewage – and therefore norovirus. Their filter feeding results in norovirus being filtered out and contaminating their tissues. Then if a norovirus-contaminated oyster is eaten raw its consumer is very likely to be infected. Mussels are usually cooked before eating, but it is conventional to cook them very lightly, usually only until the shells open, and this might not be enough to reach the all important internal temperature of 90°C necessary to inactivate norovirus.

Norovirus gastroenteritis

The symptoms of norovirus infection are very easily recognised – severe nausea, projectile vomiting and violent diarrhoea. The incubation period is 12-48 hours and most patients recover fully within 3 days, but on rare occasions the dehydration resulting from diarrhoea can be fatal in the very old or very young.

Norovirus outbreaks

Norovirus outbreaks are associated particularly with institutions (e.g. retirement homes and hospitals), cruise ships, restaurants and catered events. This is because a point source of viral infection (e.g. an infected kitchen staff member) can result in contamination of food that is eaten by a large number of people. The situation in retirement homes, cruise ships and hospitals is even more extreme because here the sufferers are living in close proximity and therefore person to person transfer might also occur, thus increasing the size of the outbreak.

Norovirus food-borne illness case example

The game of rugby is of national importance to New Zealand so imagine how embarrassing it was to have a norovirus outbreak following a corporate hospitality event at an international rugby match between the New Zealand

All Blacks and Ireland at Eden Park, Auckland. And what's more, this was the largest food-borne norovirus outbreak in New Zealand.

On 20 June 2006 Auckland Regional Public Health Services was notified of an outbreak of gastroenteritis among a group of people who had been at a corporate hospitality event at Eden Park on 17 June during the All Blacks/Ireland game. Faeces samples were collected from four ill patrons and 55 food handlers and samples of foods served at the event were taken. The samples were subjected to microbiological analysis.

Three hundred and eighty-seven people who had attended the hospitality event were interviewed and 115 reported symptoms of gastroenteritis (i.e. 30% attack rate). Consumption of oysters was associated with a 65% gastroenteritis attack rate.

Norovirus Genogroup I (GI) and Genogroup II (GII) were found in the faeces samples from four of the sick patrons. One food handler's faeces sample was also positive for norovirus GII, but the food handler had also consumed oysters which explained the finding (i.e. the food handler was not the source of the infection).

Four brands of oysters were served at the event. Two brands were locally produced and two were imported frozen Korean oysters. Norovirus GI and GII were detected in both batches of Korean oysters; the locally produced oysters were negative for norovirus. Clearly the source of the outbreak was the frozen Korean oysters.

The outbreak could have been prevented if the caterers had heeded the label 'COOK PRIOR TO CONSUMPTION' on the frozen Korean oyster packets, but they didn't.

Reference to case report: Simmons *et al.* (2007) *New Zealand Medical Journal*, **120**, No. 1264, http://www.nzma.org.nz/journal/120-1264/2773/.

Hepatitis

Hepatitis (from the Greek *hepatikos* meaning liver) is a group of liver diseases caused by five different viruses: hepatitis A, B, C, D and E. Only hepatitis A is food-borne.

Hepatitis A

The hepatitis A virus (HAV) is a member of the viral family Picornaviridae which includes polio virus and rhinoviruses (the common cold viruses). The HAV has a single RNA molecule surrounded by a capsid; it has no capsule. Unlike some of the other hepatitises, hepatitis A is not a serious disease. HAV not only is food-borne but also can be contracted from water and person to person.

HAV attaches to epithelial cells – particularly in the oropharynx (throat) and the intestine – and enters the blood stream via this route. It is carried in the blood to its target cells (hepatocytes and kupffer (phagocytic) cells) in the liver. The virus attaches to the surface of its target cells and is internalised. Once in the cells the viral RNA hijacks the liver cells' nucleic acid and protein synthetic apparatus and produces tens of thousands of new HAVs. The daughter HAVs are excreted in bile, via the bile duct into the intestine where they are expelled in faeces to infect someone else. Clearly the faecal/oral route (with food as the vector) is the food-borne mechanism of HAV transmission. Poor personal hygiene is responsible for this route of transmission.

The incubation period for hepatitis A is 2-6 weeks. Some people, even though they have an immunological response (i.e. produce antibodies against the virus) to their HAV infection, do not develop any symptoms. Others have fairly mild symptoms including fever, headache and aches - these symptoms are sometimes confused with influenza (flu). In most people the symptoms disappear quite quickly and they recover completely; the mortality rate is very low (<0.5%). In some people the disease might reappear during the following 2-6 months; this time the symptoms can be different to its first manifestation and sometimes include jaundice (yellow appearance of the skin and whites of the eyes due to high blood bilirubin levels because the liver is not efficiently excreting bilirubin in bile) and weight loss. Infection with HAV results in antibody production that confers immunity for the rest of the person's life.

Foods associated with hepatitis A

There are no particular foods associated with hepatitis A because HAV is excreted in faeces and due to poor personal hygiene contaminates the hands which then contaminate any food that is handled. The virus is destroyed by heat and therefore foods eaten raw (e.g. fruit) or cooked foods handled after cooking are the most likely routes of transmission.

Hepatitis A case examples
Case 1

During November 2003 about 500 people presented (three later died) with hepatitis A in Monaca, Pennsylvania, USA; 13 of them worked in a restaurant in Monaca, the others had dined at the restaurant. In addition to the Monaca cases, there were 75 others from other states, but they had all also dined at the Monaca restaurant. Clearly the Monaca restaurant was the source of HAV.

Some of the infected diners ($n = 207$) were interviewed; they had all eaten at the Monaca restaurant between 14 September and 17 October. Discussions with the infected staff from the restaurant revealed that the earliest date that any of them became sick was 26 October. Based on an incubation period of 2-6 weeks this would mean that the earliest a staff member could have been infected was about 14 September (i.e. 6 weeks before 26 October) which suggests that they were unlikely to be the source of infection, but perhaps acquired it from the same source as the sick restaurant diners.

Further investigations involved asking some of the hepatitis A patients ($n = 181$) what they had eaten at the restaurant. There were 133 items on the restaurant's menu and only two, chili con queso and mild salsa, were consistently associated with the disease. Drilling down further revealed that 94% of the hepatitis A patients had eaten the mild salsa and 15% had eaten the chilli. There was one ingredient common to both the salsa and the chilli - fresh green onions.

The green onions were traced back to a grower in Mexico whose produce had been associated with previous hepatitis A outbreaks in Tennessee, Ohio (see below) and Georgia, USA. Green onions require significant handling during picking and packing and therefore are a prime case for faecal/hand contamination if the packers have poor personal hygiene.

It was concluded that contaminated green onions caused the outbreak.

Case 2

In November 1998 the number of cases of hepatitis A in Mansfield, Ohio, USA, rose sharply from one case over the previous 4 months to 26 cases in a 3-week period. Of the 26 cases, 16 (61%) had eaten in a particular restaurant between 14 and 30 October. The Mansfield-Richland County Health Department carried out a full investigation and found that no particular menu item was associated with the disease, but that 95% of the sick diners had eaten food containing one ingredient – green onions. Viral nucleic acid isolated from the blood of 14 of the patients showed that they were all infected with HAV Genotype Ia and therefore likely to have been infected from the same source. The green onions had been imported from Mexico; it was thought very likely that contaminated green onions (as in case 1) caused the outbreak.

References to case reports:

Case 1 – Dato *et al.* (2003) *MMWR Weekly*, 28 November, http://www.cdc.gov/mmwR/preview/mmwrhtml/mm5247a5.htm.
Case 2 – Dentinger *et al.* (2001) *Journal of Infectious Diseases*, **183**, 1273–1276.

Bacteriophages

Bacteriophages (from the Greek φᾰγεῖν [*phagein*] meaning to eat; often called phages) are viruses that infect, and often kill, bacteria. I have included them here not because they cause food-borne illness, but because they might kill food-borne pathogens and might, one day, play a role in the control of bacterial food-borne illness.

The discovery of bacteriophages

It has long been known that river water can 'treat' bacterial diseases (it can also cause bacterial diseases!). The British bacteriologist Ernest Hankin (1865–1939) working in India showed that River Ganges water filtered through a very fine porcelain filter was able to kill *Vibrio cholerae*. In 1915 another British bacteriologist, Frederick Twort (1877–1950), termed these minute filterable 'things' bacteriolytic agents. He speculated that they might be viruses, but unfortunately World War I stopped his research and he never actually showed that his elusive 'agents' were indeed viruses. In 1917 the French-Canadian microbiologist Félix d'Hérelle (1873–1949) working at the Pasteur Institute in Paris identified Twort's bacteriolytic agents as viruses that infect bacteria and called them bacteriophages.

The biology of bacteriophages

The structure and function of bacteriophages can only be described as amazing. They consist of a capsid surrounding either single or double stranded RNA or single or double stranded DNA (according to the species), but the amazing thing is their physical structure and the way this is utilised to inject their genetic material into their bacterial host, thus resulting in bacteriophage replication in the host. Their viral body is attached to a tube with tail fibres. The tail fibres attach the bacteriophage to the host bacterium, and they contract and pull the tube down so that it penetrates the bacterial cell wall and cell membrane; they then inject their nucleic acid into the host through

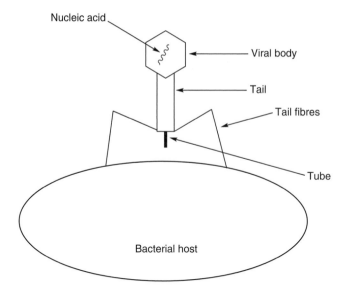

Figure 4.4a A bacteriophage attached to the surface of its bacterial host.

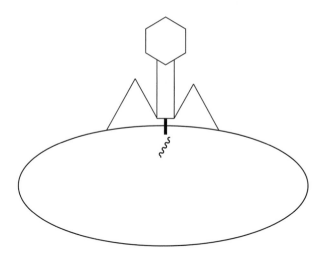

Figure 4.4b The tail fibres contract, pushing the tube through the bacterial cell wall and membrane. The bacteriophage's nucleic acid is injected into the host via the tube.

the tube (Figures 4.4 and 4.5). The process is akin to using a hypodermic syringe and needle to administer a medicine.

Could bacteriophages be used to kill food bacterial pathogens?
The answer to this question is most definitely yes. The problem is that we have to find the best bacteriophages for important food bacterial pathogens and somehow stabilise them because bacteriophages, like other viruses,

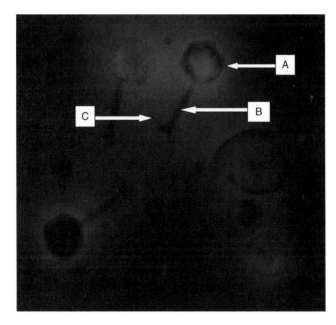

Figure 4.5 Electron micrograph of bacteriophages showing their unique morphology; A – viral body; B – tail; C – tail fibres. (Electron micrograph taken by Manfred Ingerfeld at the University of Canterbury and provided by Gwyneth Carey-Smith.)

mutate rapidly and soon change their characteristics and infectivity profile. Therefore a good killer phage for *Campylobacter jejuni* might be found, but after just a few generations it might have lost its effectiveness, and after another few generations might be useless. So at the moment (2012) there are very few bacteriophages used to reduce food-borne illness. However, in 2006 the USA Food and Drug Administration (FDA) approved the use of a bacteriophage to control *Listeria monocytogenes* in the cheese manufacturing process. Who knows, by 2030 bacteriophages might be a key facet of the armoury of food hygienists. Just imagine the day when farm animals are infected with food pathogen-specific phages to rid them of the pathogens that later infect our food!

Other food-borne viruses

In this chapter I have covered only the two viruses that are primarily transmitted by food. There are many other viruses that if they contaminate food might cause infections in the consumers of the food, but that have 'usual' routes of exposure that are not food based.

Poliovirus is a good example. It is a very simple virus with a 30 nm diameter capsid housing a 7,500 base RNA genome; it causes the very serious, debilitating disease poliomyelitis. Poliovirus is excreted by poliomyelitis patients in faeces and therefore might contaminate waterways and marine environments in which shellfish (e.g. mussels) are farmed or harvested from wild stock. Since shellfish are filter feeders they could filter out poliovirus from their environment. The consumer of a mussel, for example, grown in poliovirus-contaminated waters might get an oral dose of poliovirus and might contract poliomyelitis as a result. This scenario is far from theoretical because poliovirus has been detected in mussels and oysters, but there have been no cases of poliomyelitis traced back to food ... yet!

Take home messages

- There are very few viral species associated with food-borne illness.
- Despite the above, one virus, Norovirus, causes most of the world's food-borne illness.
- Food-borne viral infections are usually mild (e.g. vomiting) and short lived, but are very common and so have a significant economic impact (e.g. days taken off work).

Further reading

Fiore AE (2004) Hepatitis A transmitted by food. *Clinical Infectious Diseases*, **38**, 705-715.

Guttman B, Raya R & Kutter E (2005) Basic phage biology. In: Kutter E & Sulakvelidze A (eds) *Bacteriophages: Biology and Applications*. CRC Press, New York.

Wagner EK & Hewlett MJ (2004) *Basic Virology*. Blackwell Publishing Ltd., Oxford. This is an excellent general text that introduces virology very well.

Widdowson MA, Sulka AC, Mead PS & Glass RI (2005) Norovirus and foodborne disease, United States, 1991-2000. *Emerging Infectious Diseases*, **11**, 95-102.

Chapter 5
Parasites

Introduction

A parasite (from the Greek *parasitos* meaning a person who eats at the table of another) is an animal or plant that lives off a host animal or plant giving the host nothing in return. In this chapter I will only discuss animal parasites that have a human host and are transmitted via food.

Food-borne parasites often have complex lifecycles, part of which occurs in humans and part occurs in another animal; for this reason the food-borne parasites are all zoonoses (i.e. can infect humans and animals and can be contracted from animals). All parasites are killed by cooking at sufficiently high temperatures.

Parasite food-borne illnesses are far more common in developing countries than in the developed world because of improvements in food hygiene over the past 50 years in the latter.

There is a broad array of different parasites associated with food-borne illnesses. Indeed they range from the simplest of single-celled animals (e.g. Protozoa) to complex, highly adapted creatures (e.g. tapeworms). Therefore it is impossible to deal with the parasites as a single taxonomic group; for this reason I can't describe their discovery and history because each creature has an individual story.

In this chapter I will deal with each biological group of food-borne parasites separately.

What are parasites?

Parasites comprise a broad array of different creatures all with one goal, to live for free! They utilise nutrients from another animal (or plant) - the host - and give nothing in return. Parasites in a food safety context have a human host as part of their complex lifecycles and are transmitted in food. They are all destroyed by cooking and so are only a problem in undercooked food (e.g. pork) or food eaten raw (e.g. fish sashimi).

Food Safety: The Science of Keeping Food Safe, First Edition. Ian C. Shaw.
© 2013 John Wiley & Sons, Ltd. Published 2013 by John Wiley & Sons, Ltd.

Food-borne parasites are categorised as parasites because of their lifestyle (i.e. robbing their food from others) and not based on phylogenetics. For this reason parasites range from simple single-celled animals (e.g. *Giardia*) to complex multi-celled animals (e.g. flatworms).

Flatworms – Platyhelminthes

The phylum Platyhelminthes (from the Greek πλατύ (platy) meaning flat; ἕλμινς (helminth) meaning worm) includes four classes; three of them are parasites, namely cestodes (tapeworms), trematodes (flukes) and monogenea (parasitic flatworms). Only cestodes and trematodes are human parasites.

Tapeworms – Cestodes

The tapeworms have complex lifecycles usually involving at least two host species; those important as food-borne pathogens have humans as one of the hosts. I will discuss several important food-transmitted tapeworms and describe their lifecycles in a little more detail below.

The symptoms of tapeworm infection are often hunger and loss of weight. This is because the tapeworm is helping itself to the host's partly digested food as it grows in the host's gut which means that less food is absorbed by the host and so she/he feels hungry and loses weight.

Anatomy of a tapeworm

All of the tapeworms have the same basic structure: they have a 'head' or scolex with hooks to allow the animal to hold firmly onto the intestine wall of its host, and a long, often very long (up to 5 m), segmented body (Figure 5.1). The segments furthest away from the scolex contain eggs and break off to form egg-containing sacks which are excreted in the host's faeces.

Fish tapeworms – *Diphyllobothrium* sp.

The lifecycle of *Diphyllobothrium* sp. is complex, involving aquatic crustacea, fish and humans (Figure 5.2). The *Diphyllobothrium* eggs are eaten by aquatic crustacea, the contaminated crustacea are eaten by small fish, the small fish are eaten by larger fish (e.g. salmon) and the large fish are eaten by people. The *Diphyllobothrium* eggs then develop in the intestine of the human host where they mature to form a huge segmented tapeworm which hangs on to the intestine wall with a specially adapted barbed head (scolex). When mature the tapeworm sheds segments containing eggs which find their way back to waterways via sewage, and begin the lifecycle once again.

Until quite recently fish tapeworm infestation in humans was only associated with countries in which raw fish is a part of their traditional cuisine (e.g. sushi in Japan). However, with the globalisation of cuisine has come

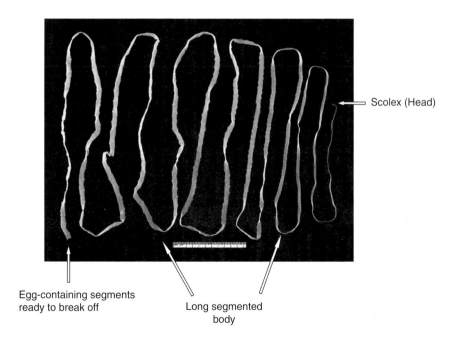

Figure 5.1 Anatomy of a tapeworm. (Image of *Taenia sagitata* taken from http://en.wikipedia.org/wiki/File:Taenia_saginata_adult_5260_lores.jpg, http://en.wikipedia.org/wiki/Tapeworm_infection.)

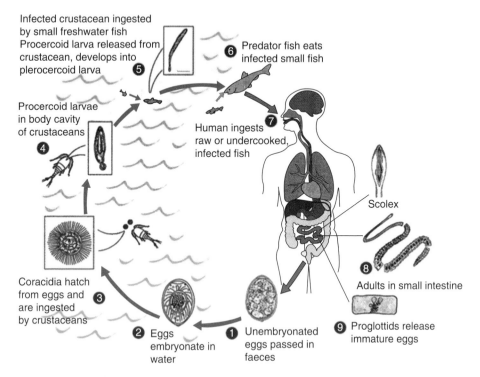

Figure 5.2 The lifecycle of *Diphyllobothrium* showing the importance of fish in infecting humans. (Modified from a diagram produced by the Centers for Disease Control and Prevention (CDC), Atlanta, USA.)

Human eats undercooked beef

Cattle

- Embryos ingested
- Zygotes released by gastric enzymes
- Zygotes penetrate gut wall
- Zygotes enter blood
- Cysts formed in muscles

Human

- Cysts acquired from undercooked beef
- Larvae released by gastric enzymes
- Adult tapeworm lives in intestine
- Eggs excreted in faeces

Human faeces contaminate pasture

Eggs develop into embryos on grass

Figure 5.3 The lifecycle of *Taenia sagitata.*

the globalisation of fish tapeworm infestations in consumers. Despite this, *Diphyllobothrium* infestation is still rare.

Beef tapeworm – *Taenia sagitata*

Taenia sagitata only has two hosts in its lifecycle (Figure 5.3); nevertheless it is a complex lifecycle with several different forms of the parasite developing in each host. In short, the adult tapeworm which is 3–5 m long lives in the small intestine of the human host. It sheds egg-filled segments that are eliminated with the host's faeces. These can find their way to pasture if human faeces is used as a fertiliser or is disposed of near to cattle grazing fields. The eggs hatch on grass to form embryos. The embryos are eaten by cattle as they graze and the cattle's digestive enzymes break down the thick embryo wall to release zygotes (oncospheres). The oncospheres penetrate the mucous layers of the cattle's intestinal tract and find their way into the animal's circulatory system via capillaries in the intestine wall. The oncospheres develop during their time in the blood and eventually come to rest in the cattle's muscles where they form cysts; they can remain in this state for a long time. When the

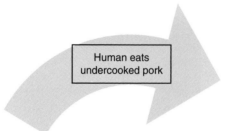

Human eats undercooked pork

Pig

- Embryos ingested
- Zygotes released by gastric enzymes
- Zygotes penetrate gut wall
- Zygotes enter blood
- Cysts formed in muscles

Human

- Cysts acquired from undercooked pork
- Larvae released by gastric enzymes
- Adult tapeworm lives in intestine
- Eggs excreted in faeces

Human faeces contaminate pasture

Eggs develop into embryos on grass

Figure 5.4 The lifecycle of *Taenia solium*.

animal is slaughtered for human consumption and if it is eaten uncooked, or insufficiently cooked, the cysts will survive and infect the human consumer. The cyst's rigid and strong wall is broken down by the human gastric digestive enzymes (e.g. proteases) to release larvae which attach to the intestine wall where they mature, grow and produce eggs to start the cycle over again.

The disease caused by *T. sagitata* in humans is called taeniasis.

Pork tapeworm – *Taenia solium*

The pork tapeworm is very similar to its beef counterpart; it has only two host species – pigs and humans – but nevertheless its lifecycle is complex (Figure 5.4).

Flukes – Trematodes

Flukes are parasitic, most have at least two hosts and they usually inhabit the intestinal tract of one of their hosts. For this reason they have a very resilient outer covering to withstand the digestive enzymes of their host.

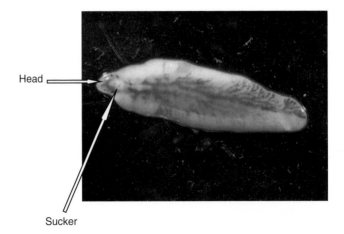

Head

Sucker

Figure 5.5 Anatomy of a fluke. (Image of *Fasciola hepatica* taken from http://en.wikipedia.org/wiki/File:Fasciola_hepatica.JPG.)

Anatomy of flukes

Flukes are usually only a few centimetres long and have two suckers, one near to their 'mouth' and the other in the middle of their body (Figure 5.5). These suckers hold the adult fluke firmly on the gastrointestinal wall of their host and prevent them being washed away by gastric flow.

The most important species in the genus is *Nanophyetus salmincola*; it is responsible for salmon poisoning disease (SPD) which can be contracted from uncooked or cold smoked salmon or trout - particularly in the Pacific Northwest region of the USA. The lifecycle of *N. salmincola* includes three hosts (Figure 5.6), the stream snail (*Oxytrema silicula*), a salmonid fish (e.g. salmon or trout) and a member of the dog family (e.g. domestic dog or wolf). Since humans eat salmon and trout they too can contract SPD. In addition to the three conventional host species there are several reservoir species. Reservoir species are animals that can harbour the tapeworm without showing any symptoms, but can release infectious stages of the tapeworm's lifecycle that then infect the true hosts. The most important *Nanophyetus* reservoir species are mink, skunk and raccoon and they release *Nanophyetus* eggs into the aquatic environment.

Liver fluke – *Fasciola hepatica*

Liver flukes cause fascioliasis in humans - this is a serious disease which often manifests as anaemia because the flukes interfere with liver metabolism and so reduce iron availability for erythrocyte (red blood cell) synthesis (erythropoesis). Liver flukes can be contracted by eating the uncooked or undercooked livers of infected mammals (usually sheep) or vegetation (e.g. water cress) from infected aquatic environments. The aquatic environment is important because the intermediary host of *F. hepatica* is an aquatic snail (usually the pond snail – *Galba truncatula*). The fluke infects and reproduces in the snail; tiny, swimming cercariae are released from the snail and swim until they find suitable vegetation to form cysts on (metacercariae). Humans

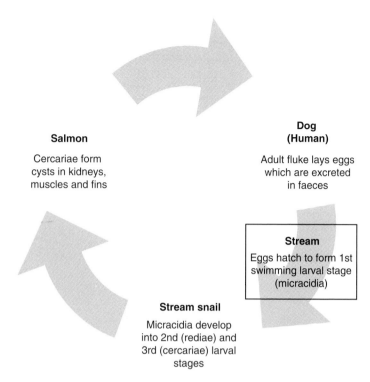

Salmon

Cercariae form
cysts in kidneys,
muscles and fins

**Dog
(Human)**

Adult fluke lays eggs
which are excreted
in faeces

Stream

Eggs hatch to form 1st
swimming larval stage
(micracidia)

Stream snail

Micracidia develop
into 2nd (rediae) and
3rd (cercariae) larval
stages

Figure 5.6   The lifecycle of *Nanophyetus salmincola.*

can contract liver flukes if they eat vegetation (e.g. water cress) with metacercariae. When the metacercariae are ingested they are stimulated by the low pH of the stomach to break open (excyst) to release juvenile liver flukes. The juveniles burrow through the intestine wall into the peritoneal cavity which contains peritoneal fluid through which they swim to the liver. They enter the liver tissue and begin to feed for 5–6 weeks during which time they begin to mature into adult flukes; during this maturation process they migrate to the bile duct where they become fully mature and produce tens of thousands of eggs. The eggs are excreted via the bile into the duodenum, they pass through the small intestine into the colon and eventually out in the faeces, from where they find their way to a stream or pond and infect another pond snail and begin the cycle over again (Figure 5.7).

   The lifecycle described above is a simple pond snail/human cycle; however, often another mammal (usually a sheep) is involved. The sheep eats the vegetation infected with metacercariae and the metacercariae develop and migrate in exactly the same way as described for human infection above. If humans eat undercooked or raw sheep's liver they can become infected by this route. The liver fluke's eggs excreted by the mature fluke into the sheep's bile duct begin to develop into embryos when they reach the acid environment of the human consumer's stomach. The juvenile flukes burrow through the intestine wall and find their way to the liver and continue their development in exactly the same way as when they directly infected humans (Figure 5.7).

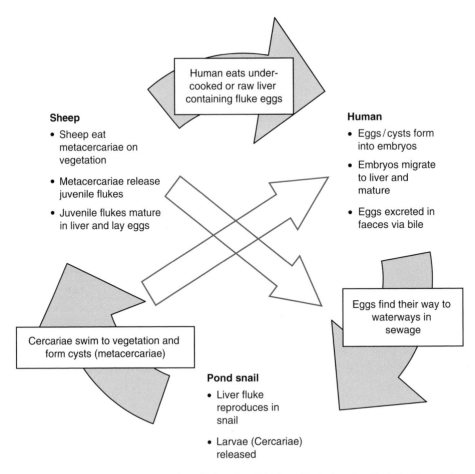

Figure 5.7 The lifecycle of the liver fluke (*Fasciola hepatica*) showing that the 'normal' lifecycle can be shortcut if humans eat infected vegetation (e.g. water cress) or eggs in sheep faeces find their way directly into watercourses without going via humans.

Nematodes

The nematodes (from the Greek *nematos* meaning thread) are a huge, amazingly diverse group of animals – the first to be described were thread-like (hence their name). It has been estimated that there are a million species of nematodes of which over 16,000 species are parasitic. Nematodes were first described in 1808 by the Swedish naturalist Karl Rudolph (1771-1832) who was working in Germany at the time. Nematodes are often called roundworms.

Since there are so many parasitic nematodes it is very likely that many of us have quite a few inhabiting our intestines and they apparently cause us little or no harm. Some, however, are more likely to increase in number and thus present more of a problem because they rob our nutrients while they live inside us. Arguably any nematode able to live in humans could be

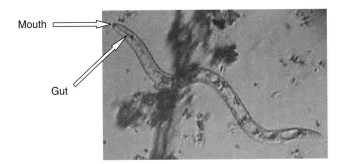

Mouth

Gut

Figure 5.8 Anatomy of a nematode. (Nematode image taken from http://en.wikipedia.org/wiki/File:Roundworm.jpg.)

transmitted by food, but there are several species that 'use' food as their usual vector. I will discuss only this group in this section.

Anatomy of nematodes

As discussed above, nematodes are thread-like, worm-like animals. They are usually less than 2.5 mm long, but some can be as long as 50 mm. Their bodies are cylindrical with a flattened head with central mouth. The mouth has 'teeth' that allow it to fix firmly to its host and leads to a gut which runs down the centre of the nematode's body (Figure 5.8).

Food-borne nematodes that affect humans

Fish nematodes
Anisakis *sp.*

Human infection with *Anisakis* sp. (usually *A. simplex*) is termed anisakiasis and can result from eating raw or undercooked fish. Interestingly, eating *Anisakis*-infected fish can also lead to a severe allergic reaction (anaphylactic shock) because the nematode synthesises a specific protein that can cause a massive immune response in some human consumers. So there are two possible effects of eating *Anisakis*-infected fish: straight intestinal infection with the nematode leading to anisakiasis, and a severe, quite rare, immuno-logical response to a specific *Anisakis* protein that can be fatal.

The lifecycle of *Anisakis* is complex (Figure 5.9); it involves marine mammals (e.g. dolphin), crustacea (e.g. shrimp), fish (e.g. sardine) and squid. The adult *Anisakis* (i.e. 'worms') inhabits the intestine of marine mammals; they shed eggs which are expelled from the marine mammal host in their faeces. The eggs hatch into free-swimming larvae in sea water and the larvae are eaten by crustaceans. Infected crustaceans are eaten by fish or squid. When the fish/squid dies the larvae migrate into the dead creature's muscles which are eaten by other fish, and thus the *Anisakis* larvae are transferred from one fish/squid to another. If an infected fish/squid is eaten by a marine mammal the larvae develop into adult 'worms' in their intestine. The adult *Anisakis* sheds eggs and the cycle starts again. Humans are not a 'real' part of the *Anisakis* lifecycle, but if a human consumer eats an undercooked or raw *Anisakis*-infected fish or squid she/he will get infected too. The *Anisakis* in the human gut are 2 cm long and generally tightly coiled – they can be found

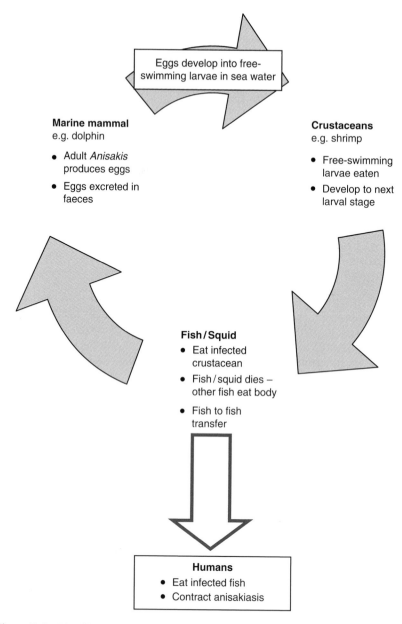

Figure 5.9 The lifecycle of the fish nematode (*Anisakis* sp.) showing how humans can be infected even though they are not part of the lifecycle.

easily and removed using endoscopy (a surgical procedure utilising a camera inserted into the gut).

Eustrongylides *sp.*

Eustrongylides larvae are large, bright red roundworms. They live in marine and brackish (i.e. salty) water fish. The 'normal' *Eustrongylides* lifecycle includes fish where the eggs mature and develop into larvae; the fish are

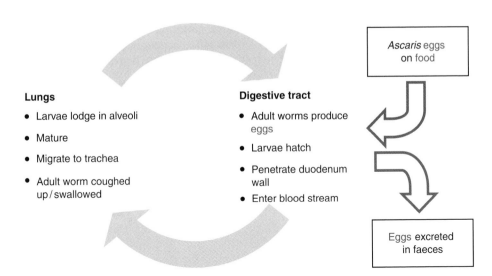

Figure 5.10 The lifecycle of the giant roundworm (*Ascaris lumbricoides*) in humans showing how roundworm eggs on food can lead to infection.

eaten by wading birds such as herons, egrets and flamingos where the larvae mature in the birds' intestines. The mature *Eustrongylides* shed eggs which are excreted in the birds' faeces and begin the lifecycle over again. Human consumers of *Eustrongylides*-containing raw or undercooked fish can contract the nematode.

The disease is very rare indeed – only five cases (contracted from sushi) have ever been reported in the USA. The effects of a single *Eustrongylides* infecting the human gut can be severe because the large nematode (4 cm long) is able to perforate the intestinal wall and result in peritoneal infection (peritonitis) which can be fatal. The infection can be treated surgically by removing the nematode either laparoscopically (key hole surgery) using fibre optics or by conventional surgery.

Giant roundworm – Ascaris lumbricoides

This is the most common 'worm' infection in humans – it is estimated that more than 1 billion (10^9) people are affected worldwide. It is far more common in the developing world where sanitation might be poor.

Ascaris can be food-borne if someone with poor hygiene who is infected contaminates their hands with *Ascaris* egg-containing faeces, then handles food. However, there are many other ways the nematode can be transmitted (e.g. via sewage-contaminated water) (Figure 5.10). Therefore food transmission is just one way that humans can contract *Ascaris*.

Ascaris lumbricoides eggs ingested with food release a larval worm in response to the acid pH of the stomach; the larvae penetrate the wall of the duodenum (upper small intestine) and enter the blood stream. Via the circulatory system they enter the heart and the liver and find their way into the pulmonary circulation, from where they lodge in the alveoli of the lungs. The alveoli are the oxygen exchange centres of the lung and they provide the ideal nutrient- and oxygen-rich environment for the *Ascaris* larvae to mature in. The larvae take about 3 weeks to mature, after which they migrate into the bronchi, then into the trachea and are eventually coughed up and swallowed and thus re-enter the digestive tract where they mature and produce tens of thousands of eggs which are excreted in the faeces, and the cycle begins again.

Trichinella *sp.*

Trichinellae are nematodes that infect pigs and can be contracted by humans if they consume raw or undercooked pork. There are three important species in the human disease context: *T. spiralis*, *T. nativa* and *T. britovi*. *T. spiralis* is by far the most common cause of disease in humans – the disease caused by *Trichinellae* is called trichinosis (sometimes called trichinellosis or trichiniasis). Trichinosis is now very rare in the developed world – between 1997 and 2001 there were an average of 12 cases in the USA (i.e. 4×10^{-6}% of the USA population contract trichinosis *per annum*). However, trichinosis is more common in the developing world where pig feed hygiene is not usually controlled and therefore feed containing raw pig meat might be fed to pigs and so transmit the disease from pig to pig very effectively. Most developed countries have strict rules governing what can be fed to pigs and how pig feed should be treated (e.g. heating to kill parasites) before feeding – one of the key reasons for such regulations is to prevent the spread of *Trichinellae*.

In some places (including the developed world) raw pork is eaten as a delicacy – such consumers are at risk of contracting trichinosis. For example, in Thailand, nam mu sod (raw pork parcels) are eaten (see http://www.khiewchanta.com/archives/snacks/raw-pork-parcels-nam-mu-sod.html for recipe (and warning!)). Clearly this dish carries a significant risk of its consumer contracting trichinosis.

Trichinella *lifecycle*

Trichinellae have complex lifecycles involving pigs, humans and rodents (e.g. rats), the latter acting as an environmental pool of infection that can be transferred to pigs if they eat an infected rodent (see Figure 5.11). Pigs are omnivores and will therefore eat meat scraps in their food, or consume rodents if they get a chance; if the meat (or rodent) is infected with *Trichinella* the pigs become infected too. The larvae of *Trichinella* form cysts in the muscles of their host and if the encysted larvae-containing muscle is eaten by a pig, the acid conditions and proteases in the pig's stomach stimulate the encysted larvae to break out of their cysts and invade the mucosa of the host's small intestine. The larvae develop into adult nematodes (1–2 mm long) in the small intestine. The female *Trichinellae* release larvae during their 4-week life span in the host's intestine; the larvae move through the intestine wall and locate striated muscles (i.e. skeletal and cardiac muscle) where they encyst to start the lifecycle all over again. However, this time the cysts are in pig muscle – pork – which is eaten by people. The human consumer of raw or undercooked *Trichinella* cyst-infected pork swallows the pork plus cysts, the acid and digestive enzymes in their stomach stimulates release of the larvae, which mature in the intestine and release more larvae which burrow into muscles and form cysts.

So why are a few *Trichinella* cysts in your muscles a problem? Often there is little effect; however, sometimes on their migratory path from the intestine to striated muscles the larvae enter the central nervous system (CNS) – this

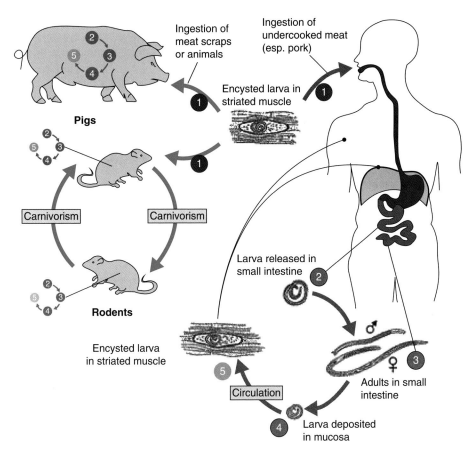

Figure 5.11 The lifecycle of *Trichinella* showing that the entire lifecycle (2–5) can be completed in pigs because pigs might be fed raw pig meat (1); that humans can be infected if they consume (1) raw or undercooked pork and then the entire lifecycle (2–5) can be completed in humans too. Rats are an important *Trichinella* environmental reservoir – they get infected if they eat other infected rats (cannibalism) or pig meat (1), then the entire lifecycle (2–5) can be completed in them too. (Modified from a diagram produced by the Centers for Disease Control and Prevention (CDC), Atlanta, USA.)

can cause significant damage which can be fatal (there is a very rare form of stroke caused by *Trichinella* in the CNS). In addition, if the striated muscle chosen by the larvae for encystment is cardiac (i.e. heart) muscle, myocarditis (inflammation of heart muscle) can result in death. *Trichinella* cysts in skeletal muscle are far less problematic; indeed they may cause no symptoms whatsoever, but often an immune response to the foreign imposters results in muscle inflammation and pain.

Whipworm – Trichuris trichiura

Whipworm infection is very rare and occurs only in tropical regions. Whipworm not only is food-borne, but also can be contracted directly from contaminated water and soil. An infected person excretes whipworm eggs in their faeces which may contaminate food if their hygiene is poor.

Whipworms infect the colon and cause the disease trichuriasis. A female *T. trichiura* living in a human gut can produce 20,000 eggs a day in her 5-year lifespan; the eggs are excreted in faeces and either end up in soil or contaminate food due to poor hygiene practices. After about 2 weeks on food or in soil the eggs develop into embryos and if they are ingested (e.g. by a human) they hatch and develop in the small intestine. The young worms migrate to the caecum and penetrate its mucosa and begin their final development into adult whipworms – the development from embryo to adult takes about 3 months. The adults live in the large intestine and the females lay a huge number of eggs every day. The eggs are excreted in faeces and the cycle begins again.

Interestingly, some people are particularly susceptible to infection by *Trichuris* because of their genetic make-up – they have specific genes on chromosomes 9 and 18 which code for traits that confer susceptibility.

Protozoa

Protozoa (from the Greek *proto* meaning first, and *zoa* meaning animals) are primitive single-celled eukarote (i.e. having a nucleus) animal-like creatures. There is dispute amongst biologists as to whether they are animals or plants; it is generally accepted that they should be referred to as *Protista*, i.e. something between animals and plants. Many protozoa have flagellae and cilia and therefore are motile; this is very important in their ability to move around and therefore infect different environments, including food.

There are some very important pathogenic protozoa that are food-borne; several can result in serious diseases (e.g. *Giardia*). They are a diverse group of creatures with very different mechanisms of infectivity, which makes them fascinating to study.

Amoebae

Most students who study Biology at school learn about amoeba. It is used as the archetypal single-celled *Protista*; it glides along with a flowing movement and has the ability to engulf its bacterial and algael food. Most of us have looked at pond water under the microscope and seen tiny flowing *Amoebae* moving amongst the particulate debris – this was my introduction to microbiology!

The most important *Amoeba* in a food-borne disease context is *Entamoeba histolytica*; it is just like the *Amoebae* you probably looked at under the microscope when you were beginning your studies in Biology. The key difference, of course, is that the pathogenic *E. histolytica* is able to infect its host and cause biochemical changes that manifest as disease. We'll take a closer look at this devious little creature.

Entamoeba histolytica

E. histolytica (*histo-lytic* from the Greek for tissue destroying) causes amoebiasis or amoebic dysentery. It lives in fresh water and therefore can infect humans by several routes – it is not only food-borne. Contact with contaminated water, drinking contaminated water and washing food in contaminated water are all possible routes of infection. When infected the organism lives and multiplies in the gut and might be excreted in faeces where it can only live for a short time. In addition, it can form encapsulated cysts which are also excreted in faeces and can survive for months outside the body. Contamination via the faecal-oral route with food as the vector might occur if personal hygiene practices are poor. So, human infection can result directly from environmental contamination, from environmental contamination of food or from faecal contamination of food.

As mentioned above, oral infection with *E. histolytica* via food can be as either free-swimming organisms (termed trophozoites) or cysts. The trophozoites grow and reproduce in the gut immediately; the cysts rupture to release trophozoites which then reproduce – during this time they feast on the gut microflora. The trophozoites attach to the gut wall and work their way between the cells to the capillaries and once in the blood stream can infect organs. Interestingly, there is evidence that during their journey in the circulatory system the trophozoites eat red blood cells – I have seen lovely photomicrographs of *E. histolytica* engulfing red blood cells which appear to stay in their cytoplasm for a while and then, presumably, they are digested.

The *E. histolytica* trophozoites can, in theory, infect any organ, but the most important, from the human disease point of view, are the brain, spleen, lungs and liver. The most common serious effect of organ infection is liver abscess which can be fatal.

It is estimated that over 50 million people (approx. 0.7% of the world's population) worldwide suffer from amoebic dysentery at any time. The disease is more common in tropical regions, but occurs throughout the world even in countries near to the Arctic Circle – indeed the first case was described in St Petersburg, Russia (latitude 59° North) in 1875.

Cryptosporidium

Cryptosporidium (from the Greek *Kryptos* meaning hidden and 'spore' which refers to the small size of the spore) is a member of the genus *Apicomplexa* which also includes other human pathogens, e.g. the malaria parasite *Plasmodium* which is carried by the Anopheles mosquito. *Cryptosporidium* causes cryptosporidiosis which is characterised by severe diarrhoea – the species most commonly found in human infections are *Cryptosporidium parvum* and *C. hominis*. *C. parvum* is very small (from Latin *parvum* meaning small); its oocyte stage is only 5–6 µm across. Cryptosporidiosis is one of the most common water-borne diseases worldwide and therefore it is rather surprising that despite *Cryptosporidia* being discovered in 1907 the first human infection was not reported until 1976. Since then the incidence has escalated.

Cryptosporidia are transmitted via contaminated water or by the faecal-oral route – *C. hominis* appears to be more commonly transmitted via faeces. *Cryptosporidia* can be food-borne if food is washed in contaminated water or contaminated with faeces due to poor personal hygiene practices.

Cryptosporidia, like the *Amoebae*, have only one host – a mammal (e.g. human). There are three important stages in *Cryptosporidia*'s complex lifecycle (Figure 5.12): sexual and asexual phases and spores (oocysts) which are very resistant to heat and chemicals and facilitate the survival of the organism outside the host. It is possible that farm animals such as cattle, sheep, goats and dogs are an important environmental reservoir from which humans can be infected either directly or via food.

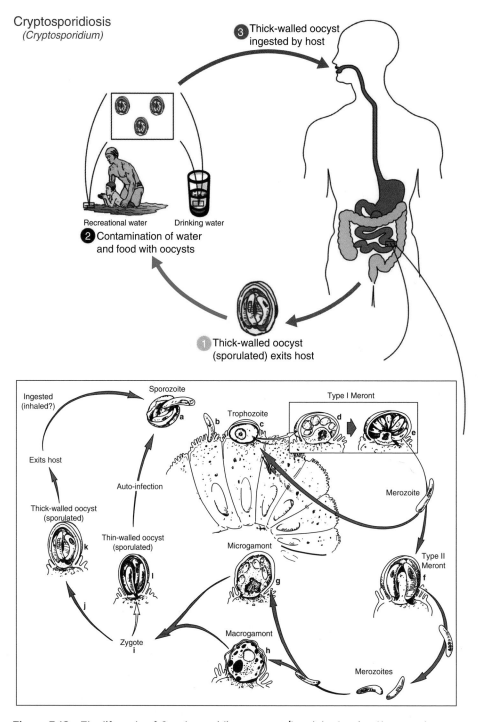

Cryptosporidiosis
(Cryptosporidium)

③ Thick-walled oocyst ingested by host

Recreational water   Drinking water

② Contamination of water and food with oocysts

① Thick-walled oocyst (sporulated) exits host

Ingested (inhaled?)

Sporozoite

Exits host

Auto-infection

Thick-walled oocyst (sporulated)

Thin-walled oocyst (sporulated)

Trophozoite

Type I Meront

Microgamont

Merozoite

Type II Meront

Zygote

Macrogamont

Merozoites

Figure 5.12 The lifecycle of *Cryptosporidium parvum/hominis* showing the complex asexual and sexual reproduction phases in the host's small intestine and the importance of the thick-walled oocyst in its transmission. (Diagram produced by the Centers for Disease Control and Prevention (CDC), Atlanta, USA. *Cryptosporidium* stages reproduced with permission from Current WL & Blagburn BL (1990) *Cryptosporidium*: infections in man and domesticated animals, pp. 155–185. In: Long PL (ed.) *Coccidiosis of Man and Domestic Animals*. CRC Press, Inc., Boca Raton, FL.)

The oocyst is excreted in faeces of infected people and other mammals (e.g. cattle) and can survive either dry or in water (e.g. ponds) for months. Food can become contaminated either via faecal contamination due to poor hygiene practices, or by washing food to be eaten raw in contaminated water. Drinking or bathing in contaminated water is also an important route of infection, but I will not discuss it further here because we are concerned only with food-borne diseases in this book.

Cryptosporidium parvum/hominis lifecycle

This is not an easy lifecycle to follow - I hope Figure 5.12 makes it clearer. In the description below I have referred to the stages of the lifecycle denoted by numbers and letters in the diagram.

The oocysts from contaminated food or water (2) are consumed (3) and pass through the digestive system until they reach the small intestine where they split (a) and release several (up to 4) active sporozoites (b). The sporozoites attach to villi (protuberances from the small intestine to increase surface area and facilitate nutrient absorption) and mature to form trophozoites (c). The trophozoite phase is able to reproduce asexually (d, e). This asexual reproduction involves the formation of many merozoites within the trophozoite. The merozoites are released and can either attach to villi where they form another trophozoite (c), which multiplies asexually to form more merozites, etc. Or they can undergo sexual reproduction; in this case the merozoites attach to villi and mature into a gamont which can differentiate into either a microgamont (male) (g) or macrogamont (female) (h). The microgamont releases microgametes (rather like sperm) which fertilise the macrogamont which then forms a zygote (i). The zygote then forms oocysts (j, k, l) which either infect other villi in the host (thin-walled oocysts) (l) or are excreted in faeces (1) to begin the cycle all over again in another host (thick-walled oocysts) (k).

Cryptosporidiosis

In most people cryptosporidiosis is short lived and self-limiting (i.e. it clears up on its own), but it can be very serious indeed, and sometimes fatal, in immunocompromised (e.g. AIDS) patients. The body's immune system is very important in fighting *Cryptosporidium* infection and the production of antibodies against the parasite is the mechanism by which the body is able to quickly get the infection under control. If the immune system is not working properly (e.g. in AIDS) antibodies aren't produced quickly enough and in sufficient quantity and therefore the *Cryptosporidium* grows uncontrollably in its human host, causing significant physiological problems that can persist for a long time because the body is unable to eliminate the parasite.

The symptoms of cryptosporidiosis are diarrhoea, cramps, nausea and fever (caused by the immune response) and appear 2–10 days after infection (i.e. as long as it takes for enough *Cryptospridia* to grow to cause a physiological effect). Highly contaminated food will produce symptoms quickly because the *Cryptosporidium* dose is large. In severe cases (e.g. in AIDS patients) dehydration due to water loss with diarrhoea can be a serious problem.

If the *Cryptosporidium* is not cleared from the body quickly it can migrate to other organs (this happens in AIDS patients) including lungs, middle ear, stomach, biliary tract and pancreas. Infection of the biliary tract leads to cholesystitis (inflammation of the gall bladder) and cholangitis (inflammation

Table 5.1 Incidence of cryptosporidiosis in the USA. (Data from HIvasa *et al.* (2005), *MMWR Surveillance Summaries*, **54** (SS01), 1–8; http://www.cdc.gov/mmwr/preview/mmwrhtml/ss5401a1.htm.)

Year	Reported cases
1999	2,769
2001	3,787
2002	3,016

of the bile duct) and infection of the pancreas leads to pancreatitis (inflammation of the pancreas) – these infections can be serious.

Cryptosporidiosis has an estimated incidence of the order of 4 in 100,000 population across the USA (Table 5.1), but the reported cases show an incidence closer to 1/100,000 because many people do not report short-lived gastrointestinal infections and therefore all of the cases are not included in the incidence data.

Giardia

Giardia lamblia (also known as *G. intestinalis* and sometimes incorrectly called *Lamblia intestinalis* and *G. duodenalis*) is a flagellate protozoan that infects the small intestine causing giardiasis. It is primarily water borne, but food washed in contaminated water and eaten raw can transmit *Giardia*. Similarly food contaminated with infected faeces by poor hygiene practices can cause giardiasis. Giardia can infect many mammals, including dogs and farm animals. Such species are an important reservoir of infection because they live in close proximity to humans.

In addition, *Giardia* can infect many wild animals (e.g. beavers); people do not usually come into contact with infected beavers; however, if you are hiking in the countryside in North America, infected beavers might have defecated in the water you might decide to drink – if so it won't be long before the tell-tale smelly symptoms of giardiais develop (see *Giardiasis*) – this is why giardiasis is known as Beaver Fever in the USA and Canada.

Giardia is a very primitive organism; aspects of its biochemistry closely resemble prokaryotic biochemistry and there is speculation about its evolutionary origins involving a very early branch from the procaryotes. One aspect of this primitive biochemistry is anaerobic fermentation – *Giardia* ferments sugars (e.g. glucose) to ethanol and carbon dioxide (just like yeasts do). Other aspects of this primitive anaerobic biochemistry are probably responsible for the production of bad-egg-smelling hydrogen sulphide (H_2S), a characteristic of giardiasis sufferer's flatulence (see *Giardiasis*).

Giardia is thought to be the most common non-viral intestinal infection worldwide.

Giardia lifecycle

Dormant *G. lamblia* cysts are ingested with food or water. Passage of the cysts through the digestive tract stimulates them to release active *Giardia* trophozoites in the intestine. The trophozoites asexually reproduce in the

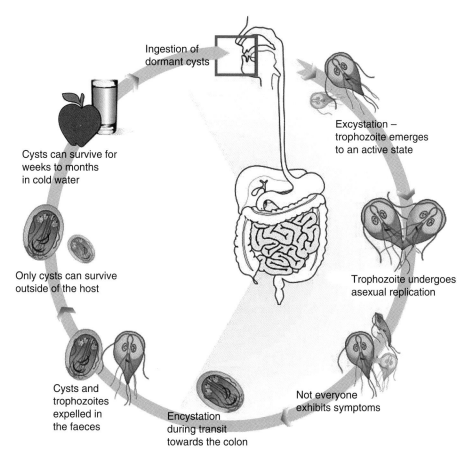

Ingestion of dormant cysts

Cysts can survive for weeks to months in cold water

Only cysts can survive outside of the host

Cysts and trophozoites expelled in the faeces

Encystation during transit towards the colon

Not everyone exhibits symptoms

Trophozoite undergoes asexual replication

Excystation – trophozoite emerges to an active state

Figure 5.13 The lifecycle of *Giardia lamblia*. (Diagram drawn by Lady of Hats, downloaded from http://en.wikipedia.org/wiki/File:Giardia_life_cycle_en.svg.)

intestine and as they travel towards the colon they form cysts which are excreted with the faeces and contaminate the aquatic environment, where they can survive for many months before contaminating food or drinking water and beginning the cycle all over again in some unsuspecting consumer (Figure 5.13).

Giardiasis

Giardiasis (Beaver Fever) is a self-limiting (clears up on its own) disease that is not serious in most people. However, its symptoms can be somewhat of a problem! There are a plethora of possible symptoms, but the main ones are diarrhoea, malaise and excessive, often foul smelling, gas. Indeed the foul smelling gas might emanate from both ends of the digestive system and in severe cases giardiasis sufferers have been induced to vomit by the foul smell and taste of their own flatulence!

The immune system is important in the self-limiting nature of giardiasis and therefore immunocompromised people (e.g. AIDS sufferers) are affected far more seriously by *Giardia* infections than people with a normal immune response.

Table 5.2 Giardiasis incidence in the USA. (Data from CDC, *MMWR Surveillance Summary*, **49** (7), 1–13; *incidence calculated using USA $pop^n = 3 \times 10^8$.)

Year	Giardiasis incidence/100,000 pop^n
1992	4.3
1996	9.3*
1997	9.5*

The incidence of giardiasis is very much greater in areas (e.g. the developing world) where sewage systems are poor and involve direct disposal of faecal matter into water courses that might be used for drinking or washing food in. It is estimated that the prevalence (i.e. percentage of a population suffering from a disease at a particular time) of giardiasis in developing countries is 20–30% compared with 2–7% in the developed world, which equates to about 2.5 million cases annually in the USA (Table 5.2). Despite this high predicted incidence the number of cases reported is much smaller because in many people giardiasis is short lived and therefore they do not go to their doctors. Studies in the USA in which faeces samples have been randomly examined for *Giardia* spores showed a prevalence of 7.2%, which agrees with the predicted prevalence for the developed world (data from http://www.giardiasis.org/Prevalence.aspx). Many of the faeces samples examined must have come from people who did not develop giardiasis or recovered quickly from the disease and so did not seek medical help and therefore did not become a national giardiasis statistic.

Sarcocystis

The very small number of reports of human *Sarcocystis* (from the Greek *sarx* meaning flesh and *kystis* meaning bladder; the organism forms a cyst (which looks like a bladder) in the flesh of its host) infection suggests that the diseases caused by this genus of protozoa are rare. However, the symptoms of infection are usually mild and therefore it is likely that many sufferers do not report their illness and so they are not included in the disease statistics. This is supported by random faeces investigation which shows a much greater incidence of human infection than the disease statistics suggest.

There are two species involved in human infections – *Sarcocystis hominis* and *S. suihominis*. Part of the lifecycle of *Sarcocystis* involves the formation of cysts in the muscles of its host. Pigs and cows are hosts and if humans consume undercooked pork or beef (*Sarcocystis* is killed by high temperatures) they can become infected.

Sarcocystis lifecycle
This is a complex lifecycle; to make it easier to follow look at the lifecycle diagram (Figure 5.14) – I have used the same numbers to show the lifecycle stages in the description below.

Sarcocystis bradyzoite (i.e. slowly reproducing encysted protozoan)-containing cysts are present in the muscles of infected pigs (pork) and cows

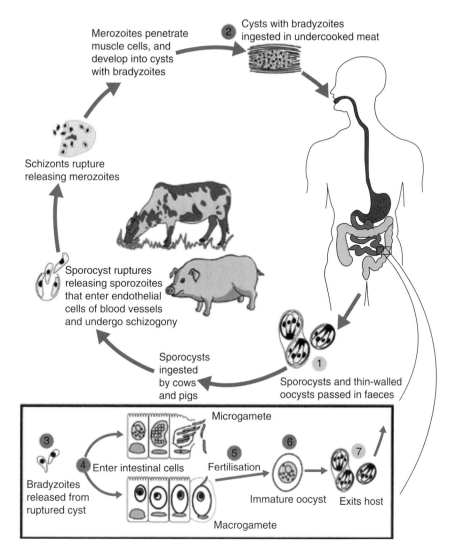

Figure 5.14 *Sarcocystis* lifecycle. (Adapted from a diagram produced by the Centers for Disease Control and Prevention (CDC), Atlanta, USA.)

(beef) (2). When cyst-containing pork or beef are eaten the chemical environment of the gastrointestinal tract stimulates the cysts to rupture and release the bradyzoites (3) which enter cells of the small intestine of their new host and either form macrogametes (similar to ova) or microgametes (similar to sperm) (4). The microgametes fertilise the macrogametes (5) (see *Cryptosporidium parvum/hominis* lifecycle above for more discussion about micro- and macrogametes) which form oocysts (6) which exit the human host in their faeces either as oocysts *per se* or as sporocysts (i.e. thicker-walled, more resilient oocysts) (7).

If human faeces is used to fertilise pasture or humans defecate in areas that cattle and pigs feed, oocytes or sporocysts might be eaten by cattle and/or pigs. On passage through the animal's gut oocysts/sporocysts rupture to release sporozoites (i.e. elongated cells formed by oocyte division; approx. size $20 \times 10 \, \mu m$) and the sporozoites enter the endothelial cells (i.e. lining cells) of the blood vessel walls where they reproduce asexually (schizogony) to produce many merozoites (i.e. cells formed by asexual reproduction). The merozoites invade and penetrate muscle cells and form cysts with bradyzoites inside. The muscle is then eaten by humans and the lifecycle begins again.

Human *Sarcocystis* infection
The symptoms of *Sarcocystis* infection include anorexia (i.e. poor appetite), nausea, abdominal pain and distension, diarrhoea and vomiting and begin 3 hours to a week after eating infected food; the symptoms usually clear up within 36 hours, but some reports show symptoms persisting for 3 weeks or even 5 years – there are very few studies of human cases and so information about symptoms and time to onset is scant and perhaps unreliable.

Toxoplasma

Toxoplasma gondii is the only member of the genus that causes food-borne human disease (toxoplasmosis). It can infect almost any warm-blooded animal and contacts with the meat or faeces from infected animals are the routes of transmission to humans. The closer we live to a particular animal species the more likely we are to be infected via the faecal route. This is why domestic cats are often associated with human cases of toxoplasmosis. Consider a typical pet cat scenario: the cat goes out into the garden to defecate, as part of its normal behaviour it buries its faeces in the ground and in the process its paws get contaminated with faeces. The cat runs back home, bounds in through its cat door and leaps onto the kitchen work surface to greet its adoring owner. The owner prepares a sandwich on the work surface that the cat has just contaminated with the faeces from its paws. If the cat was infected with *T. gondii* the owner's sandwich might well be contaminated ... and soon its consumer will be infected too!

Cats are not the only *Toxoplasma* transmission culprits; eating undercooked meat from infected animals (especially pork, lamb and venison) also accounts for many infections.

Human infection with *T. gondii* is very common; it has been estimated that one third of the world's population is infected at any point in time and studies carried out by the Centers for Disease Control in the USA show that approximately 11% of women of childbearing age are infected with *T. gondii*. *T. gondii* can be passed from mother to her child during pregnancy (i.e. vertical transmission).

Toxoplasma gondii lifecycle
The lifecycle of *T. gondii* is complex (Figure 5.15) because the parasite has a diverse array of hosts, but all paths lead to humans which, along with the environmental prevalence of *Toxoplasma*, explains why human infections are so common.

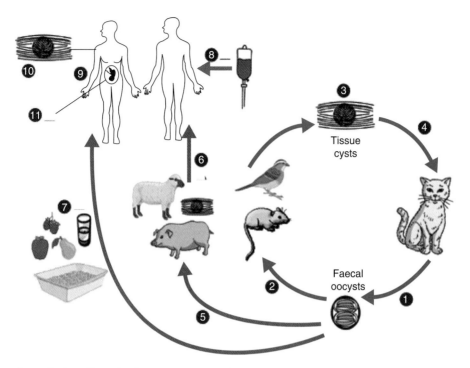

Figure 5.15 Lifecycle of *Toxoplasma gondii*. (Adapted from a diagram produced by the Centers for Disease Control and Prevention (CDC), Atlanta, USA.)

To make it easier to follow this complex lifecycle, look at the lifecycle diagram in Figure 5.15 – I have used the same numbers to show the lifecycle stages in my description below.

As outlined above, cats are an important vector of *T. gondii*; an important facet of *Toxoplasma*'s lifecycle is the cat/rodent/bird sub-cycle (see green arrows on the lifecycle diagram) in which birds and rodents develop *Toxaplasma* tissue cysts (3) and when eaten by cats the cysts develop in the cat and oocysts are excreted in faeces (1). These can be picked up by other birds and rodents (2) to re-begin the sub-cycle, but might also be ingested by pigs or sheep (5) in which *Toxoplasma* muscle cysts are formed and then the pig and sheep meat might be eaten by humans (6) and thus the human consumers might get infected. Also, as discussed above, because some people share their homes with their pet cat(s), if their kitchen hygiene is poor food might get contaminated with cat *Toxoplasma* faecal oocytes and result in human toxoplasmosis (7). *Toxoplasma* can also be transmitted by blood (e.g. used in transfusion) from infected humans (8).

In humans (9) the *T. gondi* does exactly what it does in cats, rodents and birds; its muscle cysts (e.g. in undercooked or raw pork) are activated by the pH changes and enzymes in the digestive system to form tachyzoites (rapidly multiplying tissue phase) which seek muscle and nerve tissue where they form cysts (bradyzoites) (10). There is an extra complication in pregnancy (11) because the tachyzoites can cross the placenta in the blood and infect the developing fetus. Therefore a child from a *T. gondii* infected mum might also be infected and develop toxoplasmosis.

Toxoplasmosis

There are two phases of toxoplasmosis – acute and latent. Acute toxoplasmosis is when the infected person has symptoms of muscle aches and pains (partly because *T. gondii* is forming cysts in the muscles) and influenza-like symptoms (because an immune response has been stimulated by the infection). During this phase the oocysts in cat faeces-contaminated food or tissue cysts in pork, lamb (meat from young sheep) or mutton (meat from old sheep) are activated and begin to infect their human host and migrate to muscle or neurological tissue. In people with an active immune system the migration to tissues (especially neurological tissue) is controlled well, but in people who are immunocompromised (e.g. AIDS sufferers) severe tissue infestation can occur, leading to significant neurological effects including brain (encephalitis) and/or eye (necrotising retinochoroiditis) damage. Congenitally infected children can also develop severe neurological symptoms like those seen in immunocompromised people.

Take home messages

- Food-borne parasitic diseases are common, particularly in developing countries where water might be contaminated and food hygiene might be poor.
- There is a broad array of different parasitic species that cause food-borne illness.
- Toxoplasmosis is a common food-borne parasitic illness of the developed world.
- All parasites in food are killed by thorough cooking.

Further reading

Dawson D (2005) Foodborne protozoal parasites. *International Journal of Food Microbiology*, **103**, 207–227.

Hlavsa MC, Watson JC & Beach MJ (2005) Cryptosporidiosis surveillance – United States 1999 – 2002. *MMWR Surveillance Summaries*, **54**, 1–8, http://www.cdc.gov/mmwr/preview/mmwrhtml/ss5401a1.htm.

Nawa Y, Hatz C & Blum J (2005) Sushi delights and parasites: the risk of fishborne and foodborne parasitic zoonoses in Asia. *Clinical Infectious Diseases*, **41**, 1297–1303.

Ortega YR (2006) *Food-borne Parasites*. Springer, Heidelberg.

Chapter 6

Bovine Spongiform Encephalopathy (BSE)

Introduction

Bovine spongiform encephalopathy (BSE or Mad Cow disease) was first seen in cattle in England in 1986; within a few years it had developed to epidemic proportions and scientists and consumers were worried about the possible effects of human consumption of BSE-infected beef. Everyone's worst fears were realised when a case of a human disease very similar to BSE was identified in 1992. The disease was an unusual early onset form of a rare brain disease called Creutzfeld-Jacob disease (CJD) which was later shown to be linked to eating BSE beef – it was termed new variant Creutzfeld-Jacob disease (nvCJD).

The political furore that developed following the identification of nvCJD led to the collapse of the British beef industry and worldwide bans on the importation of British beef. The ramifications of BSE were (and are) unprecedented in the food safety world; never before (or since) had a new food-borne disease led to such a huge and extreme worldwide response.

BSE is caused by a very strange causative agent called a prion. The BSE prion is a protein; it is not alive, but can reproduce in animal cells by changing specific host proteins into BSE prions. This is ingenious and was unheard of until the 1980s.

In terms of food-borne causative agents the BSE prion is unique, which is why it warrants a chapter of its own. In this chapter I will take you through the discovery of BSE and nvCJD and the effects they had on the beef industry, and the implications for food safety as a whole.

The history of BSE

In 1986 I was working at the UK Ministry of Agriculture, Fisheries and Food's (MAFF; now the Department for Environment, Food and Rural Affairs – DEFRA) Central Veterinary Laboratory (CVL; now the Veterinary Laboratories Agency – VLA) as head of their Toxicology Section. I remember very well being asked

Food Safety: The Science of Keeping Food Safe, First Edition. Ian C. Shaw.
© 2013 John Wiley & Sons, Ltd. Published 2013 by John Wiley & Sons, Ltd.

to look at a cow that had been behaving strangely. It had become unpredictable and walked with a staggering gait. The animal was killed and a post mortem examination carried out. We found no toxicological explanation for the cow's symptoms, but following a great deal of detailed pathological investigation the veterinary pathologists decided that the cow was suffering from some form of spongiform encephalopathy (characterised by a spongy microscopical appearance of the brain) – a disease until then unknown in cattle. A huge national research effort led to the discovery of a new disease – BSE, or Mad Cow disease as the newspapers liked to call it.

Having identified a new disease the next question was what causes it? There were many contenders: a bacterium, a virus, perhaps exposure to a chemical (but our toxicological investigation suggested this was not the case), or an inherited biochemical disorder. Again a huge research effort resulted in the elimination of viruses and bacteria because the causative agent was very heat stable. Studies in mice showed that they developed BSE-like symptoms when their brains were injected with a homogenate of BSE cow brains even if the cow brain homogenate was heated to above 100°C. At the time bacteria and viruses were not thought to be able to survive such a high temperature (we now know that some deep-sea bacteria that inhabit the vents of submarine volcanoes live at temperatures above 100°C). So this left an inherited biochemical disorder as the cause of the disease – this could not explain the epidemiology of the disease because the affected cattle were not necessarily genetically related (we now know that there *is* a rare inherited form of BSE). In order to answer the crucial question of what causes BSE, a major epidemiological research project was undertaken. This aimed to determine the source of the 'infection' by looking closely at the pattern of the development of BSE from the first few cases identified in 1986.

The epidemiology of BSE in England

BSE was officially identified in November 1986, but it is very likely that the first case of the disease occurred in April 1985 or even earlier. The first cases were from southern England and were associated with cattle feed that incorporated meat and bone meal (MBM) from a particular rendering plant. MBM is the ground-up remains of farm animal (including cattle) carcases after the meat has been removed and the fat (tallow) extracted. The bones and a small amount of attached tissue are dried and ground and incorporated into animal feed as a source of nutrients (e.g. calcium phosphate from bones and proteins from the tissues) – the use of MBM in cattle (and other ruminant animals') feed is now banned in most countries.

This link to MBM was a turning point in the identification of the source of the BSE causative agent, but it was difficult to understand why BSE only appeared in 1985 as MBM had been used in cattle feed for very many years. The answer to this conundrum lay in the procedures that renderers used to make MBM; the renderers that supplied the MBM for the feed manufacturers that produced the cattle feed that the epidemiologists had associated with the early BSE cases had changed their MBM manufacturing process. Traditionally, MBM was made from hydrocarbon solvent-treated (to extract the tallow – fats used for food manufacture and frying) animal carcases followed by heat drying of the remains then grinding to produce MBM powder that was added to animal feed products. In order to speed up tallow extraction/MBM production the solvent extraction was replaced by a continuous throughput heat process which melted out the tallow.

It is now known that the BSE causative agent is destroyed by the solvents formerly used for tallow extraction, but not by the temperatures used to melt out the tallow. So a simple change in a manufacturing process allowed the BSE causative agent to survive and be incorporated into cattle feed and thus 'infect' other cattle. Please note that I am being very careful to use the term *causative agent* to describe what causes BSE and to only use 'infect' or 'infection' in inverted commas because BSE is not caused by a living organism, as we will see later.

Spongiform encephalopathies

The spongiform encephalopathies (SEs) are a well-known group of, often rare, diseases characterised by the spongy (hence *spongiform*) microscopical appearance of the brains (ἐγκέφαλος (enkephalos) meaning brain; ἔπαθον (epathon) meaning suffer) of their sufferers.

The following are examples of SEs:

- *Creutzfeldt-Jakob disease (CJD)* – a human neurodegenerative disease first reported by Hans Creudfeldt in 1920 and then independently in 1921 by Alfons Jakob.
- *Kuru* – a human disease that occurs only in Papua New Guinea caused by ritual cannibalism. There is a clue here to the causal mechanism of BSE which was used by scientists when they were unravelling how BSE spread from animal to animal and then to humans (as nvCJD). Cattle were given MBM from other cattle in their food and humans ate BSE beef – this is akin to cannibalism in the Papua New Guinean Fore tribe (*Kuru* means 'shake' in Fore tribe language; shaking is a symptom of the disease kuru).
- *Scrapie* – a disease of sheep first described (but not understood) in 1732. Stanley Prusiner (University of California, San Francisco) won the Nobel Prize in Physiology/Medicine in 1997 for describing the causative agent of scrapie and linking this to BSE and CJD/nvCJD. He called the agent a prion – derived from a contraction of the words *proteinaceous* and *infectious* because prions behave like infectious proteins.

We now know that the SEs are caused by prions. The above diseases were all important in helping scientists to unravel the cause and transmission of BSE.

Prions

Prions are medium-sized proteins with molecular weights in the region of 35–36,000 Da (i.e. about 200 amino acid residues). They are found in most cells in the body, but their function is not fully understood, although they are thought to play a role in cell-cell communication particularly in the brain. The prions that normally occur in cells are termed PrP^c (PrP = prion

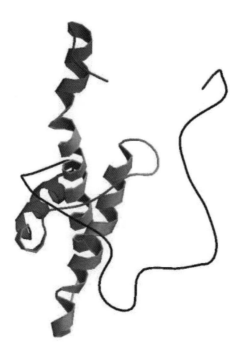

Figure 6.1 The molecular structure of human PrP^C showing its predominant α-helix protein conformation. (From Ilc *et al.* (2010) *Plos One*, **5**, e11715–e11715, downloaded from the Protein Data Bank at http://pdbbeta.rcsb.org/pdb/explore/explore.do?structureId=2KUN.) (To see a colour version of this figure, see Plate 6.1.)

protein; C = cellular). The prions that cause BSE and scrapie are mis-shaped; they have refolded to form a very water-insoluble protein that tends to form conglomerates (i.e. they bunch together) in cells; these prions are termed PrP^{SC} (SC = scrapie) – even though this term relates historically to scrapie it is used to denote the BSE and CJD prions too. Clearly PrP^C has a very important function in cells because when its shape (conformation) changes it has a devastating effect on cell function. I will discuss what causes the conformational change PrP^C→PrP^{SC} and the effects this has later.

PrP^C is quite water soluble and has a predominance of α-helix in its conformational structure (Figure 6.1; Plate 6.1), whereas PrP^{SC} is very water insoluble and has a predominance of β-pleated sheet protein conformation (Figure 6.2; Plate 6.2). Therefore the conformational change from PrP^C→PrP^{SC} involves conversion of α-helix to β-pleated sheet; in essence the amino acid sequence of the two prion forms is the same, it is only the way the amino acid chains are folded in the protein structure that differs (i.e. they are isoforms). This conformational change leads to a huge change in the physicochemical properties of the protein (e.g. change in water solubility) that significantly affects the biological properties of the two prion forms.

PrP^C can convert to PrP^{SC} under specific circumstances (Figure 6.3). This is the basis of the SEs (including BSE and nvCJD) and will be discussed in detail later in this chapter.

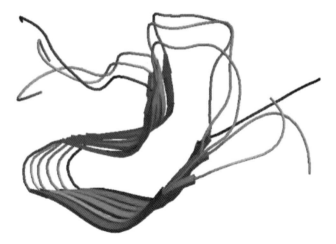

Figure 6.2 The molecular structure of a PrPᵗ conglomerate forming a fibril, showing its predominant β-pleated sheet conformation – each of the different coloured structures is a PrPˢᶜ molecule; they have aligned to form a water-insoluble fibril. (From Van Melckebeke *et al.* (2010) *Journal of the American Chemistry Society*, **132**, 13765–13775, downloaded from the Protein Data Bank at http://pdbbeta.rcsb.org/pdb/explore/explore.do?structureId=2KJ3.) (To see a colour version of this figure, see Plate 6.2.)

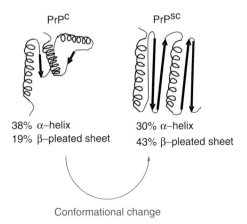

Figure 6.3 Schematic representation of the conformational structures of PrPᶜ and PrPˢᶜ showing their different α-helix (curly lines) and β-pleated sheet (arrows) conformational makeup and the conversion of PrPᶜ→PrPˢᶜ. (Modified from a drawing made by Margaret Tanner. Reproduced with permission.)

The symptoms of BSE

Alteration of the conformational structure of prions in the brain is the important determinant of symptoms, because it is in the central nervous system that the biochemical function of prions seems to be most important - although still not understood. For the normal function of the brain prions must function to help brain cells to communicate with each other properly. It is important that brain cells communicate to transfer information from one cell to another and therefore pass on ideas and actions to allow decision making, thoughts to be generated and actions to be communicated to different parts of the body, so that the animal (whether it be a cow or a person) can function properly. So, it is easy to imagine the central nervous system chaos that will occur if this communication goes wrong ... even only slightly wrong.

We will discuss later exactly what goes wrong, but to understand the origins of the symptoms of BSE it is necessary to have a rough idea of how important the brain is in the behaviour and function of an animal. If the normal, functional prions in a cow's brain change their conformation (and therefore their function) this will affect the brain's activity. What process is affected depends on the physical location of the damaged prions in the animal's brain; if the prions in the cow's brain's mood centre are altered the cow's mood will change; if the prions in the cow's brain region that controls movement of its limbs are altered the cow will not be able to walk properly. The sequence of prion changes in the brain determines the sequence of brain-related symptoms – I'll discuss this in greater detail later.

The symptoms of BSE are almost always the same and occur in exactly the same sequence, which supports the underlying progression of prion changes across the cow's brain. The first signs that farmers usually notice are strange behavioural changes in one of their cows. Farmers know their animals well – particularly dairy cattle that are milked daily and each has a very definite and recognisable character – and therefore they notice changes in their animals' behaviour. For example, a normally placid cow might get aggressive when milked. Later the BSE cow develops a characteristic stance – it appears to stand staring into space with its legs apart as if it finds it difficult to stand upright easily – a bit like the feeling when you might have had a few too many alcoholic drinks! Then the cow finds it increasingly difficult to walk; it staggers and falls as if the ground beneath its feet was slippery. Eventually the animal is unable to stand. No one knows what happens next because it would be inhumane to keep a BSE cow alive longer to see how the symptoms progress and so all animals – even experimental animals – have been slaughtered before symptoms progress further.

BSE cases in the UK

There was a rapid increase in BSE cases following the first confirmed case in November 1986 (Figure 6.4). The peak was reached in 1992 – there were more than 700 new cases in that year, followed by a rapid decline; by 2000 there were very few new cases recorded. During the period 1986–2010 there have been 184,155 cases, and in the early days of the epidemic much of the meat from affected cattle went into the human food chain.

BSE transmission and the origins of PrP^{SC}

I have discussed how normal prions (PrP^C) can be converted to BSE prions (PrP^{SC}) by an apparently simple conformational change, and have outlined the significance of this change to the brain's function and the

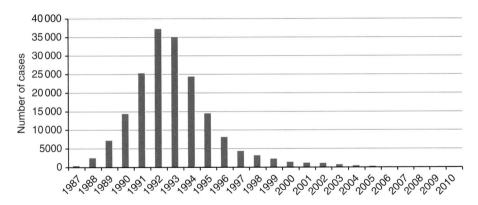

Figure 6.4 BSE incidence in the UK. (Data from World Organisation for Animal Health, Paris, http://www.oie.int/eng/info/en_esbru.htm.)

animal's behaviour and function. But what causes the prions in a cow's brain to change from $PrP^c \rightarrow PrP^{sc}$?

A great deal of wonderful research carried out over a very short time stimulated by the BSE epidemic led to some truly fantastic revelations and answered the above question very elegantly. Who would have thought that a protein could cause a disease like BSE? In fact when Stanley Prusiner first suggested this it was received with disbelief by conventional-thinking scientists; some people thought that Prusiner's ideas were ridiculous. The 'infectious protein' theory of BSE is now fully accepted as is evidenced by Prusiner being awarded the Nobel Prize in 1997. This is a great example of why scientists should explore outside conventional thoughts if they are to change the way we think - never forget that!

Back to the question what causes prions in a cow's brain to change from $PrP^c \rightarrow PrP^{sc}$? In summary: a cow eats feed containing PrP^{sc} (from MBM from a BSE cow), and the PrP^{sc} is absorbed intact in the cow's intestine and slowly migrates via the lymphatic system and the spinal cord to the cow's brain. When the PrP^{sc} reaches the brain it induces the brain's PrP^c to change its conformation to PrP^{sc}. This is an induced conformational change and is a well-understood biochemical phenomenon. But how does PrP^{sc} cause PrP^c to change its conformation to form another PrP^{sc}? This question took a great deal of research to answer, but we now understand exactly what happens.

When a PrP^{sc} molecule comes into contact with a PrP^c molecule it literally pulls the PrP^c molecule into the PrP^{sc} conformation. Very recent research has shown that the interaction that changes $PrP^c \rightarrow PrP^{sc}$ does not involve individual prion molecules, but rather complex molecular conglomerates; I will describe it in terms of simple molecular interactions here because it is easier to understand.

The PrP^{sc} molecules originating from the cow's feed eventually arrive in the cow's brain via the spinal cord - don't forget they are not living and therefore their transfer from the digestive tract where they are absorbed to the brain is pure diffusion and whether they get to the brain or not is pure chance. The time taken to get from the digestive tract to the brain is as long as it takes to diffuse the distance they have to travel. In a cow the

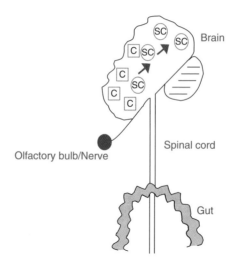

Figure 6.5 BSE prion (PrP^{sc} – represented as SC in this diagram) originating from the cow's feed is absorbed in the gut, diffuses via the lymphatic system up the cow's spinal cord to its brain where it induces a conformational change in native prion (PrP^c – represented as C in this diagram) resulting in changes in brain function where PrP^{sc} is formed. (From Shaw IC (2004) In: Werner KJ, Devine C & Dikeman M (eds) *Encyclopaedia of Meat Sciences*. Elsevier, Amsterdam, pp 846–854; © I.C. Shaw.)

distance is long (cows are big animals) and it takes about 12 years for a PrP^{sc} molecule to go from the gut to the brain. In a smaller animal the time would be correspondingly shorter – if mice (i.e. very small animals) are fed PrP^{sc} it takes about a year for the PrP^{sc} to get to their brains and for them to begin to show the symptoms of 'BSE'. The time taken for symptoms to develop following ingestion of PrP^{sc}-contaminated food is loosely termed the 'incubation period'.

When the PrP^{sc} arrives in the brain it interacts with native PrP^c and induces the conformational change to form PrP^{sc} (Figure 6.5). As the PrP^{sc} is formed the brain begins to misfunction. The PrP^{sc} first arrives in the region of the brain that controls mood; therefore mood is the first function to change – the first change that farmers notice. The wave of PrP^c→PrP^{sc} change progresses across the brain. When the wave of PrP^c→PrP^{sc} change passes through the cow's brain leg control region the animal displays changed gait and begins to stagger when it tries to walk.

In summary, PrP^{sc} from the cow's food gets to its brain and changes native PrP^c→PrP^{sc} which has a significant effect on brain function. But where did PrP^{sc} come from in the first place? The simple answer is from MBM derived from a BSE cow. But if you extend the question *ad infinitum* there must have been a cow sometime before the first case of BSE that somehow made PrP^{sc} to initiate the process. Indeed that is exactly what scientists think happened.

Prions are proteins that are synthesised, like all other proteins, from a DNA template comprising a DNA base code (i.e. gene) that is translated at the ribosome via messenger RNA (mRNA) to a sequence of amino acids in a protein. A change in the sequence of bases on the DNA that codes for a prion would lead to a prion with a different amino acid sequence being formed. This is the process of mutation. It is thought that a cow somewhere in England about 12 years before the first case of BSE (i.e. about 1974) underwent a mutation in its prion gene that led to a change in amino acid sequence that resulted in the prion incorrectly folding to form β-pleated sheets rather than α-helices, i.e. PrP^{SC}. This cow ended up in the rendering plant and its MBM was used to make cattle feed. The cows that ate the feed containing the newly created PrP^{SC} 12 years later developed BSE, but many of them were slaughtered before the symptoms appeared and added more PrP^{SC} to the cattle food chain, so amplifying the incorporation of PrP^{SC} into cattle feed ... eventually a cow lived long enough to show the symptoms of BSE.

An important factor in this story is the change in MBM production process (see *The epidemiology of BSE in England*). The removal of hydrocarbon solvents from the process and their replacement with a continuous heat process to render the tallow from the cows' carcases meant that the PrP^{SC} could survive the process. PrP^{SC} is stable up to 186°C and the rendering temperature was far below this. Indeed meat processing and cooking for human consumption never reaches temperatures anywhere near 186°C which means that human consumers of BSE cows eat PrP^{SC} too - but that's another story (see below).

The risk to human consumers of BSE beef – nvCJD

The BSE epidemic in the UK led to a furore about what the implications to the consumer of BSE beef might be. No one knew the answer. Scientists were cautious because as we began to understand the aetiology of BSE we could see that it might be possible for humans to develop a similar disease since we too have brain PrP^{C}; the fact that hundreds of years of consumption of scrapie sheep (scrapie was first described in 1723) (see *Spongiform encephalopathies*) – which is also a prion disease – had not led to human scrapie made some of us think twice and modify our risk assessment accordingly. On the other hand, the politicians flatly denied that humans would be affected by consuming BSE beef – their opinion was based on hope rather than any science whatsoever! (see *The politics of BSE and implications for food safety worldwide*).

Conjecture was rife about whether beef was safe to eat, and, as a result, the consumption of beef in the UK declined markedly as people decided that the risk of eating beef was not worth the benefit – this is an excellent example of the power of perceived versus actual risk (see Chapter 2, *Risk perception*).

Then in 1995 what everyone had been dreading happened; a man in his 40s developed a very unusual case of CJD. His disease was unusual because the symptoms were slightly different to those of 'normal' CJD, but, very much more importantly, the age of onset was incredibly young. CJD is a very rare disease (incidence = 1/1,000,000 population/year), but almost always shows its first symptoms in people older than 50 years. A case of CJD in a person in his 40s was unprecedented. Between 1995 and 1996 there were eight deaths from this new form of CJD and their age range was 18–31 years (median 29 years) (Figure 6.6). Speculation that this new form of CJD was a human manifestation of BSE was rife. The disease was given the name

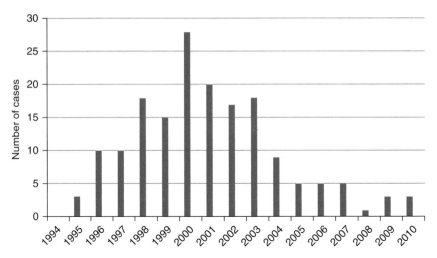

Figure 6.6 nvCJD incidence in the UK. (Data from the CJD Surveillance Unit, Edinburgh, http://www.cjd.ed.ac.uk/figures.htm.)

variant (v)- or new variant-CJD (nvCJD). The symptoms of nvCJD were very similar to those of BSE cows – loss of memory, confusion, mood changes, difficulty walking, loss of coordination which progressed in humans to dementia and death (of course these symptoms/effects were not seen in cattle because they were humanely slaughtered before their disease progressed this far). Later cases were in a teenage girl and a 28-year-old man which further supported the characteristic early onset of nvCJD. We now know that nvCJD is the human form of BSE and is contracted by eating PrP^{SC}-containing beef.

Indeed epidemiological investigations linked the nvCJD cases to their consumption of high risk meat, i.e. cuts of meat that were likely to have higher levels of PrP^{SC} (e.g. cow's brains) (see *A case of nvCJD* and *BSE risk to human consumers and risk management*).

The 'incubation period' for nvCJD (i.e. the time from consumption of BSE beef to onset of symptoms) was estimated to be about 8 years – this is the time it would take consumed beef PrP^{SC} to reach the brain. This 'incubation period' is clear if you compare the peaks of the BSE epidemic and cases of nvCJD (Figure 6.7).

A case of nvCJD

A 28-year-old woman was examined by doctors at St Thomas' Hospital, London, in March 1995. She had recently started a new job and was worried because she was becoming increasingly forgetful. By April 1995 her loss of memory was much worse and she was getting very confused and disorientated. The woman had a previous medical history which pointed to thyroid disease and so she was admitted to the hospital and treated accordingly.

Electroencephalography (EEG) showed an abnormal brain wave pattern, but other means of investigating the brain (e.g. magnetic resonance imaging;

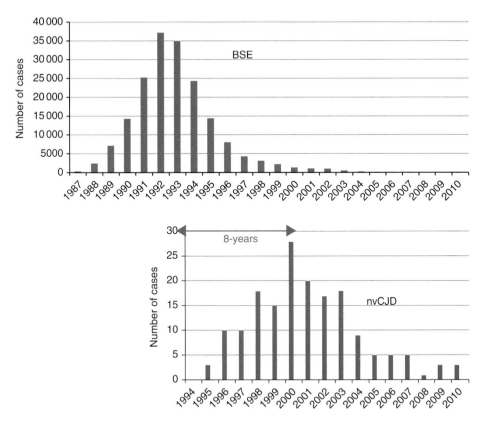

Figure 6.7 The difference between the case number peaks of the BSE epidemic and nvCJD is 8 years which is the approximate 'incubation period' for nvCJD.
(Data from the CJD Surveillance Unit, Edinburgh, http://www.cjd.ed.ac.uk/figures.htm, and World Organisation for Animal Health, Paris, http://www.oie.int/eng/info/en_esbru.htm.)

MRI) showed no abnormalities. She returned home, but was admitted to her local psychiatric hospital in June 1995 because she had continued to deteriorate mentally. By this time her behaviour was childlike, her short-term memory had deteriorated further and she experienced hallucinations. She was drowsy and confused. She was treated with the antipsychotic drug chlorpromazine, but did not respond.

She was transferred to St Thomas' Hospital in July – she was thin and pale, did not respond to simple commands and was incontinent. Exhaustive studies, including a consideration of her medical history for drug use and other activities, were unable to explain her rapid mental deterioration.

In September 1995 she underwent a brain (frontal lobe) biopsy which led to a diagnosis of CJD. This was an unusual case of CJD because of the early age of onset.

It was not possible to attribute a cause to this case, but we know that it was nvCJD which is associated with the consumption of BSE beef.

Reference to case report: Britton *et al.* (1995) *Lancet*, **346**, 945-948.

BSE risk to human consumers and risk management

As the UK's BSE saga unfolded and an understanding of the disease developed (but before the first case of nvCJD) a risk management strategy was developed to minimise the possible effects on consumers of BSE beef. This strategy had two effects:

(1) It reduced the risk of consumers suffering any untoward effects of consuming BSE beef.
(2) It made consumers feel better when they knew that their government had introduced well thought out methods to protect them.

It was not possible to measure PrP^{SC} quickly in beef and therefore it was not possible to withdraw PrP^{SC}-contaminated beef from the market. It was, however, possible to determine in long complex experiments which tissues in BSE cows contained PrP^{SC}. These experiments involved injecting homogenised tissues from BSE cows into mouse brains and waiting a year or so to see if the mice developed BSE-like symptoms. Such experiments showed that the high risk tissues were central nervous system (brain and spinal cord) and lymphatic tissue (e.g. lymph nodes from the intestine). The UK government banned the sale of high risk tissues (e.g. cattle brains) (Offals Ban 1989) as a means of reducing the BSE risk to human consumers. In addition, as a means of preventing further cases of BSE, and eventually wiping out BSE altogether, the government banned the feeding of MBM to ruminant animals (e.g. cows) in 1988.

However, the risk management strategies took several years to take effect because the long 'incubation periods' of BSE and nvCJD meant that many human consumers had eaten BSE beef between the onset of BSE (the first PrP^{SC}-contaminated meat could have been consumed as long as 12 years before the first BSE case was identified; i.e. 1974) and the Offals Ban (1989). The risk to the consumer steadily rose during this period as BSE became more common and therefore the chances of consuming PrP^{SC}-contaminated beef rose concomitantly.

Interestingly, the 'actual' risk of contracting nvCJD from eating beef in the UK was low and, to some extent, the furore that developed around the issue was unwarranted. But BSE was a new disease that could cause a terrible, inevitably fatal disease in beef consumers which struck at the heart of Britain's psyche – its national dish is roast beef! The public perception of the risk was far greater than the 'actual' risk.

It is very difficult to assess the risk of contracting nvCJD from beef numerically because of the long incubation period of the disease and the uncertainties about the proportion of the UK's beef supply that was PrP^{SC} contaminated. However, an indication of risk can be gleaned by assuming everyone in the UK eats beef (UK population = 61,800,000) and using the number of nvCJD deaths (i.e. 28) during the year of its peak in 2000. This gives us a risk of contracting nvCJD from eating beef in the UK of 1 in 2,222,222. I must emphasise that this is a woefully inadequate calculation that most likely underestimates the risk (because not all British people eat beef) and there is an additional huge error in the calculation because we don't know which year the nvCJD cases consumed the PrP^{SC}-containing beef that caused their death. Nevertheless this simplistic calculation shows that the risk is very low indeed. In fact most people would not be worried about a risk this low – the risk of being killed in a car accident in the UK in 2007 was 1 in 20,833 (i.e. 107 times the risk of contracting nvCJD) and most people don't think twice about driving!

The politics of BSE and implications for food safety worldwide

The UK's BSE epidemic and the terrible reality of nvCJD being caused by consumption of beef from BSE cattle led to a worldwide re-think of food safety strategies. To some extent this was an over-response to a very low risk scenario – just think how many people in the UK consume beef and the fact that only 170 consumers contracted nvCJD between 1995 and 2010. The actual risk of the BSE epidemic to the consumer was overshadowed by the perceived risk fuelled by the terrors of nvCJD. All of this overreaction to a terrible new disease led to governments around the world acting quickly, decisively and extremely to ban UK beef imports. This had a huge effect on an already ailing UK beef industry – it almost collapsed completely. Even 20 years later some countries maintain their UK beef bans or implement other policies to protect their population from a vanishingly small risk. For example, New Zealand does not permit anyone who was in the UK during the BSE epidemic to donate blood for transfusion for fear they will transmit PrP^{SC} to blood recipients; this is ridiculous when set in a risk context. Consider a situation in which you were seriously injured in a car accident and you needed blood urgently; would you worry about receiving a unit of blood from someone who had lived in the UK during 1986? This is a classic perceived risk versus actual benefit anomaly (see Chapter 2) – the risk of contracting nvCJD from blood derived from someone who lived in the UK during the BSE epidemic is negligible, but the benefit of receiving blood as a life-saving measure is enormous. It is only the New Zealand government's perception of the risk that overrides a sensible risk assessment being made. This approach is a good example of the irrational, non-risk-based assessment of the risks associated with importing UK beef to the health of the importing nation.

On the other hand, the BSE epidemic made many food safety legislators in nations around the world re-think their approaches to assuring the safety of their populations and led to the introduction of measures to reduce foodborne risk and prevent unexpected events (like BSE) from being missed until it was too late. So some good came out of the UK's disaster.

BSE incidence around the world

The UK was not the only country that had cases of BSE – there have been cases in the many other countries, but always far fewer than in the UK (Table 6.1).

Many of the cases could be traced back to the UK via imported cattle feed. Several (e.g. a case in Switzerland) had no links whatsoever with the UK and are thought to be examples of mutations in the PrP^{C} gene that led to the formation of PrP^{SC} and therefore provide a little evidence for the

Table 6.1 Cases of BSE around the world. (Data from World Organisation for Animal Health, Paris, http://www.oie.int/eng/info/en_esbmonde.htm.)

Country	Number of cases 1989-2010
Austria	7
Belgium	133
Canada	19
Czech Republic	30
Denmark	16
Finland	1
France	870
Germany	419
Greece	1
Ireland	1,647
Israel	1
Italy	144
Japan	36
Liechtenstein	2
Luxembourg	3
Netherlands	86
Poland	69
Portugal	1,069
Slovakia	24
Slovenia	8
Spain	758
Sweden	1
Switzerland	464
UK	184,155
USA	2

mutation theory of the origins of BSE (see *BSE transmission and the origins of PrP^{SC}*).

Take home messages

- BSE is caused by a protein prion.
- The BSE prion (PrP^{SC}) has a different conformation (shape) to the normal prion (PrP^{C}) found in cells.
- nvCJD in humans is linked to consumption of BSE beef.
- BSE was caused by cattle being fed BSE prion-contaminated meat and bone meal (MBM).
- BSE probably originated as a result of a mutation in the prion gene.

- The risk of contracting nvCJD from BSE beef is very low.
- The reaction of governments outside the UK to BSE was excessive in light of the risks associated with BSE.
- The BSE epidemic led to changes in food legislation around the world.

Further reading

Prusiner SB (ed.) (2004) *Prion Biology and Diseases*, 2nd edn. Cold Spring Harbor Laboratory Press, New York. This is the most comprehensive and authoritative text on prions and the diseases they cause.

Shaw IC (2004) Prions and viruses. In: Werner KJ, Devine C & Dikeman M (eds) *Encyclopaedia of Meat Sciences*. Elsevier, Amsterdam, pp. 846-854.

Zeidler M, Stewart GE, Barraclough CR, *et al.* (1997) New variant Creutzfeldt-Jakob disease: neurological features and diagnostic tests. *Lancet*, **350**, 903-907.

Chapter 7
Chemical Contaminants

Introduction

When food is farmed, processed or manufactured many chemicals are used in the process; they often remain in the food and therefore are consumed. These chemicals fall into four categories: pesticides, veterinary medicines, additives (e.g. colouring agents and preservatives) and environmental contaminants. In addition, there are many natural chemicals that are clearly not additives but might still have an adverse effect on the consumer (e.g. solanine produced by green potatoes). In this chapter I will deal with only food contaminants that result from farming practices. Natural toxins (Chapter 8), additives (Chapter 11) and environmental contaminants (Chapter 9) will be dealt with later.

Farmers have a tough job trying to grow their crops amongst a myriad other plants (weeds) that compete for the same space, and with thousands of creatures (e.g. insects) that see the farmer's crop as an excellent food source. With this competition, most farmers resort to using chemicals (i.e. pesticides) to kill their natural opponents in order to give their crops a chance of thriving, thus providing them with a good crop yield and so making their farming business viable. We must not forget that farmers have to make a living out of their produce.

Farmers who produce meat, milk and eggs also have problems; their animals get sick and need to be treated to make them well again so that they can efficiently produce meat, milk or eggs. To maintain healthy livestock, the farmer calls upon the expert services of a veterinarian who prescribes the medicines necessary to keep the farmer's animals healthy.

Some farmers use more extreme means to increase their meat/egg production; they use hormones or other chemicals (e.g. antibiotics) that increase meat production by stimulating tissue growth. The benefit of this practice is purely to the farmer's financial bottom line.

All of these practices – using pesticides, veterinary medicines and growth promoting chemicals and fertilisers – leave residues of the

Food Safety: The Science of Keeping Food Safe, First Edition. Ian C. Shaw.
© 2013 John Wiley & Sons, Ltd. Published 2013 by John Wiley & Sons, Ltd.

chemicals and their metabolites in the animals' tissues and therefore end up in the meat, eggs, milk, grains and vegetables that we eat.

Most countries have strict legislation governing the use of pesticides, veterinary medicines and growth promoting chemicals in animals and crops intended for human consumption to minimise the risk of residues of these chemicals in food to the consumer. An important facet of such legislation is the withdrawal or withholding time; this is the time that must elapse between the use of the chemical (e.g. a pesticide) and the crop, meat, milk or eggs being used for human consumption, and is a means by which human exposure to the chemicals is minimised.

In order to increase their yields, or to grow crops in difficult locations, farmers use fertilisers; fertilisers can be natural (e.g. animal manure) or synthetic (e.g. superphosphate) and there is no doubt that they significantly enhance the growth of crops and pasture, with the ultimate benefit of producing more fruit, vegetables and grain crops or fattening animals faster. On the negative side, some fertilisers increase levels of 'natural' chemical ions (e.g. nitrate; NO_3^-) in food. There is increasing concern about the effects of such chemicals on human health. For example, nitrate forms nitrosamines when it reacts with food components in the intestine and nitrosamines cause cancer.

Finally, there are natural chemicals in soils and aquatic systems that can end up in our food. Some of these (e.g. cadmium; Cd) are of considerable concern to the consumers of specific foods (e.g. oysters) from countries where environmental Cd levels are high, e.g. volcanic countries like Italy and New Zealand.

The use of pesticides, veterinary medicines and fertilisers not only has an impact on the consumer via their residues in food, but also has a significant impact on the environment. Pesticides do not distinguish between insect pests (e.g. aphids) and beneficial insects (e.g. bees). Even though arguably the environmental impact is far greater than the impact on the consumer, I will only discuss food-borne impact in this chapter because this book is about food safety not the environmental impact of farming.

Both the environmental impact and the toxicological impact of residues on the consumer have led to the emergence of the Green Movement which advocates 'organic' farming (i.e. very restricted use of pesticides, veterinary medicines and fertilisers). There is no doubt that organic farming significantly reduces the environmental impact, but there is still significant debate about the benefits, if any, of organically farmed food to the consumer (see Chapter 14).

Pesticides

Pesticides have been used for many thousands of years. Thousands of years ago the Chinese used an extract of chrysanthemum flowers to kill insects on their crops; the chrysanthemum species they extracted their insect killer from was likely to be *Chrysanthemum cochinium* and the toxin they extracted

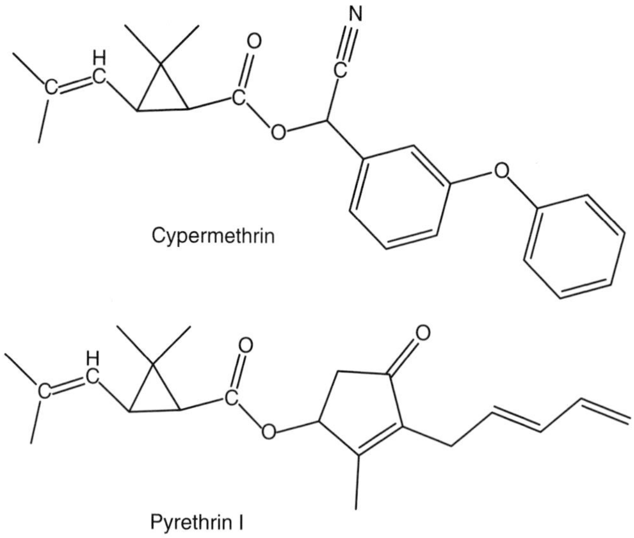

Cypermethrin

Pyrethrin I

Figure 7.1 Pyrethrin I from pyrethrum, the chrysanthemum extract used as an insecticide by the Chinese thousands of years ago, compared with cypermethrin a modern pyrethroid insecticide ... times haven't changed much!

was pyrethrin (named from pyrethrum, the old name for chrysanthemum), the forerunner of modern-day pyrethroid insecticides (Figure 7.1).

There was little development of pesticides until the use of toxic metals in the late 18th, 19th and early 20th centuries. For example, Bordeaux mixture used by the French wine growers consisted of copper sulphate ($CuSO_4$) plus lime (calcium hydroxide, $Ca(OH)_2$) and became a widely used and very effective insecticide and fungicide – indeed it is still used in vineyards and by other fruit growers today.

In the 1950s scientists in the UK discovered the pesticidal activity of a chemical that had been synthesised a century earlier; the chemical was dichlorodiphenyltrichloroethane (DDT; Figure 7.2). DDT began the new generation of pesticides – organic pesticides (do not confuse the use of the word 'organic' here with 'organic' in the Green Movement context) – and soon became the most heavily used and effective pesticide of all time. DDT spawned a whole class of pesticides – the organochlorine (OC) pesticides. DDT's use grew enormously until Rachel Carson, a scientist from the USA, suggested that it might be having a significant negative impact on the environment. In her book *Silent Spring* (published in 1962) she painted a doom and gloom scenario of the use of pesticides such as DDT. Her predictions were along the right lines and led to an enormous research effort to produce a new generation of more specific, less environmentally persistent, pesticides – these were the organophosphates (OPs) (Figure 7.2).

OPs are very effective insecticides that interfere with the generation of nervous impulses in insects (and in fact in any species that has a cholinergic nervous system, including humans) and, as well as being effective at lower concentrations than DDT, do not persist in the environment for as long because they are degraded by soil and other environmental bacteria. In this respect

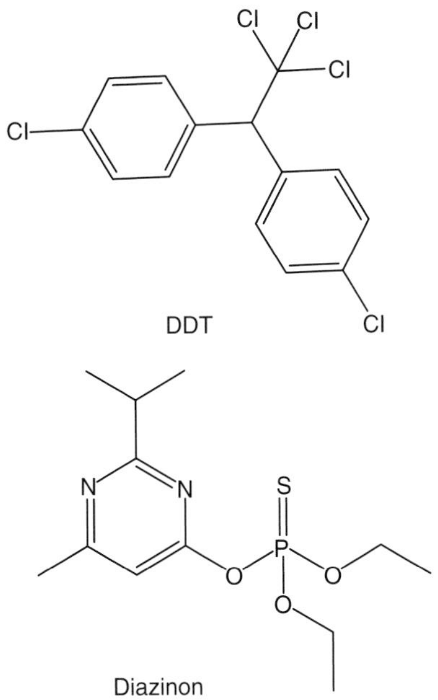

Figure 7.2 The molecular structures of the organochlorine pesticide dichlorodiphenyltrichloroethane (DDT) and the organophosphorus pesticide Diazinon.

alone OPs were a great improvement over DDT which persists in the environment for hundreds of years – its environmental half life is of the order of 50–100 years. As a result of this, DDT was banned worldwide in the mid-1970s to early 1980s (except for use in eradicating mosquitos that transmit malaria). DDT was so widely used, it has been suggested that every cell in every creature and plant on the earth harbours at least one molecule of DDT!

OPs are still extensively used today and the human health implications of their residues in food are the subject of much scientific debate – I will discuss this in greater detail later in this chapter.

The search for less persistent (Figure 7.3) and less toxic pesticides resulted in a return to a previous millennium. In the 1980s a new generation of pesticides, the pyrethroid insecticides, were introduced. They have a lower environmental impact because they are rapidly degraded by UV light and environmental bacteria. On the downside they have a severe aquatic environmental impact because they are selectively very toxic to fish.

The search for the Holy Grail of pesticides continues and the agrochemicals industry has turned its attention to novel approaches to trapping and killing insect pests using pheromones – insect-specific volatile chemicals secreted to attract a mate or mark a path. Such chemicals have little if any impact on the environment and might well be that Holy Grail we have been pursuing for the past 40 years.

This little trip (Figure 7.3) through the development of insecticides illustrates the general philosophy of pesticide development. The development of herbicides and fungicides has followed a similar path, but as we will see later, modern herbicides are less of a concern from the human impact of residues in food point of view because they are designed to interfere with specific aspects of plant biochemistry (e.g. some are plant hormone analogues, like

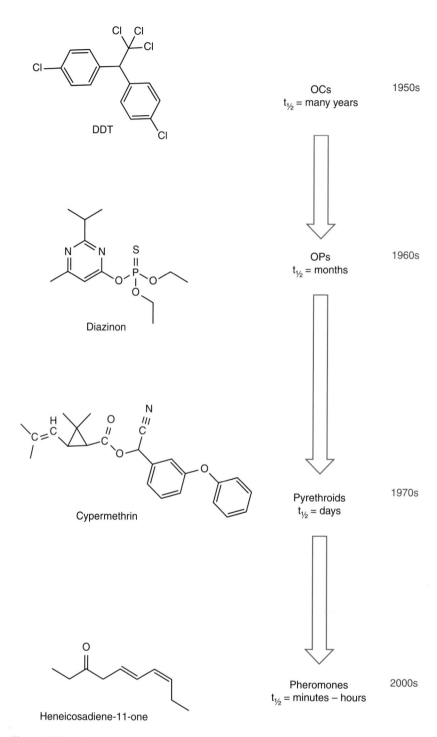

Figure 7.3 The evolution of pesticides to produce more specific, less persistent compounds.

2,4-dichlorophenoxyacetic acid; 2,4-D) and therefore have little or no effect on human biochemical systems. Fungicides are more of a concern because they often target cell division and so can also affect human cells.

From the residues point of view the more modern pesticides are far less of a problem because they do not persist in animal and plant systems for as long as their earlier counterparts. So as we move towards more specific, pest-targeted pesticides their residues risk diminishes.

This rather optimistic conclusion relates only to countries that regulate the use of pesticides in farming and that use new generation pesticides on their farms. Developing countries might still use old generation pesticides (e.g. DDT) and not have legislation to control their use and withholding periods. As more of our food is grown in developing countries (e.g. China) we must be vigilant to ensure that we are not being exposed to unacceptable pesticide residues. Fortunately most countries operate import monitoring programmes to ensure that imported food complies with homeland standards ... but not everything is monitored.

I will deal with the individual classes of pesticides separately and use examples to illustrate the residues' issues, concerns and risks.

Pesticide residues in food – assessing risk to the consumer and making sure farmers use pesticides properly

If a particular pesticide is used during the growth of a crop it will contaminate the surface of the crop and form surface residues. On the other hand, if it is a systemic pesticide (i.e. absorbed into the cells of the crop plant) it will be taken up by the plant and form 'internal' residues. The net effect is the same: the crop is contaminated with residues of the pesticide(s) that has (have) been used in its production. The risk of such residues to the consumer of food made from the crop depends on the residence time of the pesticide residue and the residue concentration in the food.

Less stable pesticides (e.g. pyrethroids; Figure 7.3) degrade quickly and therefore present less of a residue problem because the residues quickly disappear, whereas stable pesticides (e.g. OCs) can form very persistent residues and therefore are more problematic from a consumer health risk perspective.

When pesticides are approved for use (licensed), an important part of the approvals process involves determining the residence time of pesticide residues in/on crops and setting a time between the use of the pesticide and harvesting when it is safe to eat the crop – this is the withdrawal or withholding time. When the pesticide is used, it must be used in compliance with the withdrawal time. Most countries operate policing schemes to ensure that pesticides are used properly and that residues in food are minimal so that consumers are not put at unacceptable risk.

The policing schemes usually involve residues monitoring programmes that allow both population level exposures to pesticides to be monitored and, if residues exceed statutory limits (e.g. because withdrawal times have not been observed by farmers), that the farm on which the crop was grown can be traced to ascertain why. Such investigations can, and do, lead to prosecution which sends a strong message about the importance of using pesticides properly.

In order to police residues, statutory levels are necessary to measure the residues in a particular crop or food against. If the residues level exceeds the statutory residues level the regulatory authority will investigate further. These statutory levels fall into two categories: those set without toxicological parameters purely for trading purposes, i.e. Maximum Residue Level (MRL), and those based on toxicology that can be used to ensure that consumers are not harmed by eating foods containing the particular pesticide residue, e.g. Acceptable Daily Intake (ADI).

The MRL (see Chapter 2, *Maximum Residue Level (MRL) and Maximum Level (ML)*) is determined following the use of the pesticide on a crop according to the label specifications, then measuring the residues of the pesticide in the crop. The residue's concentration found in the crop is the MRL. There is no toxicological consideration in setting the MRL, it is purely a trading standard and is used to ensure that crops traded around the world are being produced using pesticides properly. If the residue of a pesticide exceeds the MRL the crop cannot be traded.

The ADI (see Chapter 2, *Acceptable Daily Intake (ADI)*) is an important measure of toxicity and is the amount of (in this case) a pesticide that can be consumed every day for the consumer's entire lifetime with no adverse effects. This is an extreme safety parameter because it is almost inconceivable that anyone would consume a particular pesticide every day for their entire life, but it does give a huge safety margin for pesticide regulators and consumers alike. If a pesticide residue exceeds the ADI in a crop or a food, most regulatory authorities would insist on the food being withdrawn from the market and a full investigation carried out.

Residues of pesticides are sometimes found in foods that were produced without the use of pesticides (e.g. organic crops) because with the widespread use of pesticides (particularly persistent pesticides like OCs – see *Organochlorine pesticides*) environmental contamination occurs which leads to residues in crops even when the pesticide has not been applied directly to the crop concerned. In addition, when pesticides are used spray drift (e.g. pesticide spray is blown by the wind) from one field to another can occur. Most countries have control measures in their pesticide use regulations which prohibit the spraying of pesticides when it is windy; this safeguard minimises spray drift, but sometimes farmers do not obey the rules, which can lead to crops in neighbouring fields being contaminated and accumulating residues of pesticides that have not been used in their production (Table 7.1).

Pesticides are not only used during the growing of crops, they are also used post harvest to prevent damage to stored crops; wheat stored in grain silos before being ground to make flour is very susceptible to attack by grain weevils and is often treated with pesticides to control the weevils. Residues of such pesticides can persist and end up in the bread and cakes made from the flour made from the wheat stored in the silo. This explains why pesticide residues are often at higher concentrations in brown or wholemeal breads because the flour used to make these breads includes the outer coat of the grain (i.e. testa or bran) and since the grain was treated post harvest most of the pesticide is on the outside of the grain (Tables 7.1 and 7.2).

An example of a post-harvest insecticide is pyrimiphos-methyl (Figure 7.4); it is used to treat stored grain to prevent insect damage and for this reason pyrimiphos-methyl residues are commonly found in bread (Table 7.2), but in studies I have been involved in they rarely if ever exceed the MRL and never

Table 7.1 Residues of pesticides in bread in the UK for the period 1989–1996, showing that organic bread can contain pesticide residues and that residues are greatest in breads made from wholemeal and brown flours which include the outer grain coat (testa). (Data from Working Party on Pesticide Residues Annual Report 1996, Ministry of Agriculture, Fisheries & Food, London, p. 23.)

Bread type	Percentage with pesticide residues
White	12
Brown	23
Wholemeal	27
Multigrain	24
Organic	15

Table 7.2 Residues of pyrimiphos-methyl in UK bread in 1996 showing that wholemeal bread is more likely to have residues. (Data from Working Party on Pesticide Residues Annual Report 1996, Ministry of Agriculture, Fisheries & Food, London, p. 40.)

Bread type (number of samples analysed)	Pyrimiphos-methyl residue concentration range (mg/kg)	Percentage of samples positive
White (48)	Not found	98
	0.06	2
Brown (49)	Not found	98
	0.06–0.1	2
Wholemeal (48)	Not found	77
	0.05–0.2	23

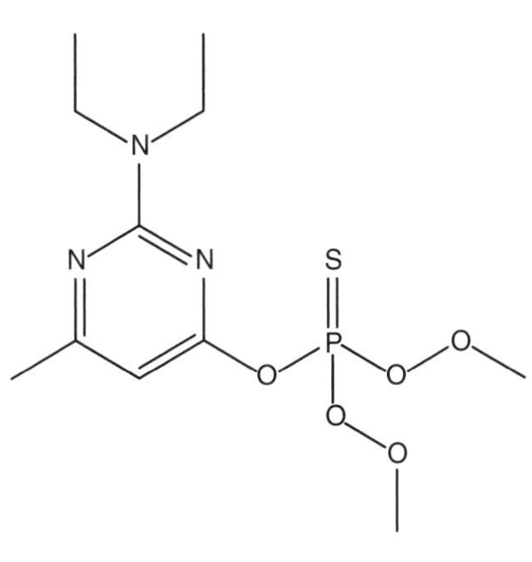

Figure 7.4 The OP pyrimiphos-methyl which is often used as a post-harvest treatment to stop insects damaging stored grain.

exceed the ADI. This means that even though there are residues of pyrimiphos-methyl in some breads it is legal to sell the bread and safe to eat it. Pyrimiphos-methyl is high hazard, but because the dose in bread is small the risk is low.

Risk/benefit assessment for pesticides

The risk to the consumer associated with pesticide residues in food is different to other food-related risks we have discussed before because in this case the consumer bears the risk and the food producer (e.g. farmer) receives the benefit (i.e. financial and ease of production). For this reason it is difficult to carry out a conventional risk/benefit assessment – I will discuss the risk of pesticide residues to the consumer in more detail at the end of this chapter (see *Dietary intake and risk to human consumers*).

Insecticides

Organochlorine pesticides (OCs)

Most OCs are not, or only very slowly, broken down by environmental conditions (e.g. UV light) or biological systems (i.e. metabolism); in addition they are very lipophilic (i.e. they dissolve in lipids or fats) and therefore persist for a long time in the environment because they are not degraded and dissolve in the lipid-based membranes of cells (Figure 7.5) where they can stay for many, many years. For this reason our environment is contaminated with OCs and they inevitably end up as residues in our food because either the food (e.g. plant crops) is grown in contaminated soil or farm animals graze contaminated pastures. Also, since some OCs are still permitted for use in special circumstances (e.g. the World Health Organisation (WHO) permits DDT use to control the Anopheles mosquito that spreads malaria) this means that DDT is still entering environmental systems. Unfortunately the clandestine use of DDT intended only for malaria control undoubtedly occurs, which explains the appearance of DDT residues in food in certain crops grown in geographical regions of illegal DDT use.

Since most OCs were banned under the Stockholm Convention on Persistent Organic Pollutants (2001) the problem with OC residues in food should be diminishing, but because the OCs have such long environmental half-lives the rate of diminution will be very slow. However, by careful management of the use of farm land it is possible to reduce OC residues in food faster than the rate of decrease (i.e. $t_{1/2}$) of environmental levels of a particular OC. For example, New Zealand has a particular problem with DDT residues because DDT was used extensively in the 1960s for control of grass grub (*Costelytra zealandia*). Grass grub is a beetle larva that eats grass roots and therefore interferes with the growth of pasture; since New Zealand's economy depends heavily on primary agricultural produce, control of grass grub is essential. DDT worked very well, but left significant environmental resides which remain today. However, measuring DDT contamination of farm land and not using land with higher concentrations of DDT for meat and milk production have resulted in a significant reduction in the exposure of New Zealanders to DDT residues in food (Figure 7.6). This is an excellent example of dietary intake management (i.e. risk management).

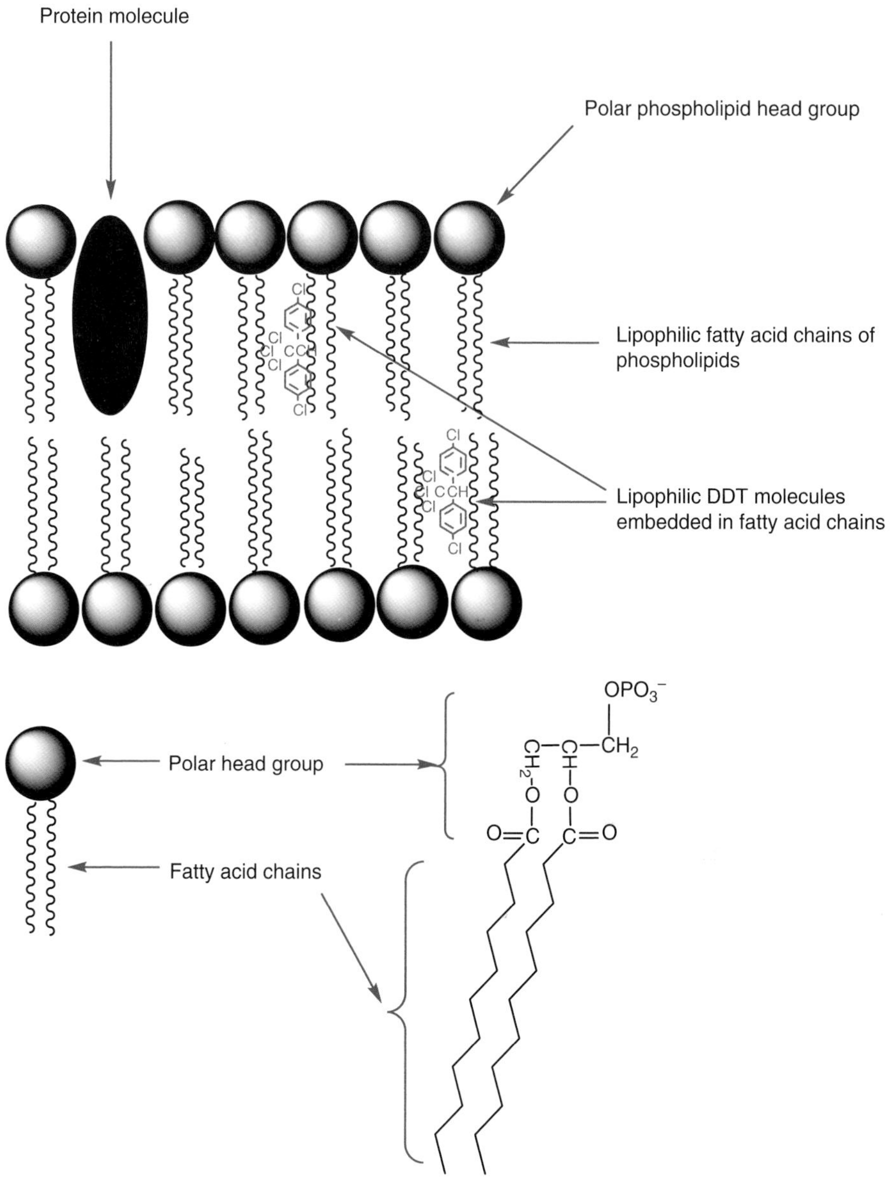

Figure 7.5 Schematic representation of a cross section of a cell membrane showing DDT embedded in the highly lipophilic internal structure of the membrane bilayer (top) and the molecular structure of a phospholipid molecule (bottom) showing how it is stylised in my membrane drawing.

Mechanism of action of the OCs

Surprisingly, the detailed mechanisms of action of all of the OCs are not fully understood; this is probably because they were discovered and introduced as pesticides a long time ago and have largely been withdrawn from use which means that there is not much incentive to study them any longer.

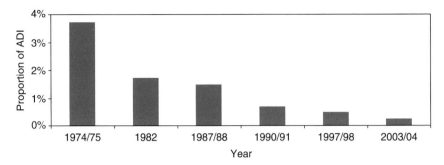

Figure 7.6 Dietary intake of DDT in New Zealand males aged 19–24 shown as a percentage of DDT's ADI. (Data from 2003/04 New Zealand Total Diet Survey, New Zealand Food Safety Authority, Wellington (2005), p. 29. With permission from MAF.)

In general terms they interfere with the generation of nerve impulses in their insect targets by perturbing the neuronal cell membrane proteins that pump ions (e.g. sodium; Na^+) in and out of nerve cells as part of the process of generating a nervous impulse.

There is not enough space to go into detail about nervous impulses here (if you want to know more about the cell membrane read Chapter 9, Membranes and Membrane Transport, in *Biochemistry* by R.H. Garret & C.M. Grisham, Brooks/Cole, Boston; see also Chapter 3, *How botulinum toxins inhibit neurotransmission*), but in very simple terms nerve cells (neurones) carry messages from the brain to other parts of the body (e.g. motor neurones carry messages to 'tell' muscles to contract and therefore to cause limb movement) or from the periphery to the brain to 'report back' on what is happening (sensory neurones). Sensory neurones might, for example, detect temperature to prevent you getting burnt. The messages that all neurones carry are electrical in nature; they are simply exchanges of ions across the neurone membrane - positive ions in/negative ions out - which leads to a wave of depolarisation along the nerve membrane; this is a nervous impulse.

Since a single neurone does not stretch all the way from the brain to the periphery (e.g. a muscle in your big toe) the bundles of neurones that link end to end to form nerves have to communicate with each other to transfer the nervous impulse - the electrical wave of depolarisation - from one to the other (Figure 7.7). They do this across a gap between the two neurones' ends called a synapse. The nervous impulse is transferred across the synapse by a chemical messenger called a neurotransmitter which is stored in membrane-bound vesicles in the presynaptic neurone. The neurotransmitter is released from the neurone terminus on the side that the impulse originated (e.g. on the brain side for motor neurones), and it diffuses across the synapse and binds to a receptor protein in the membrane on the other side of the synapse (the postsynaptic receptor). When the neurotransmitter binds to the postsynaptic receptor to cause ions to be pumped across the postsynaptic membrane, the wave of depolarisation (and therefore the nervous impulse) is re-initiated. All of this happens very fast.

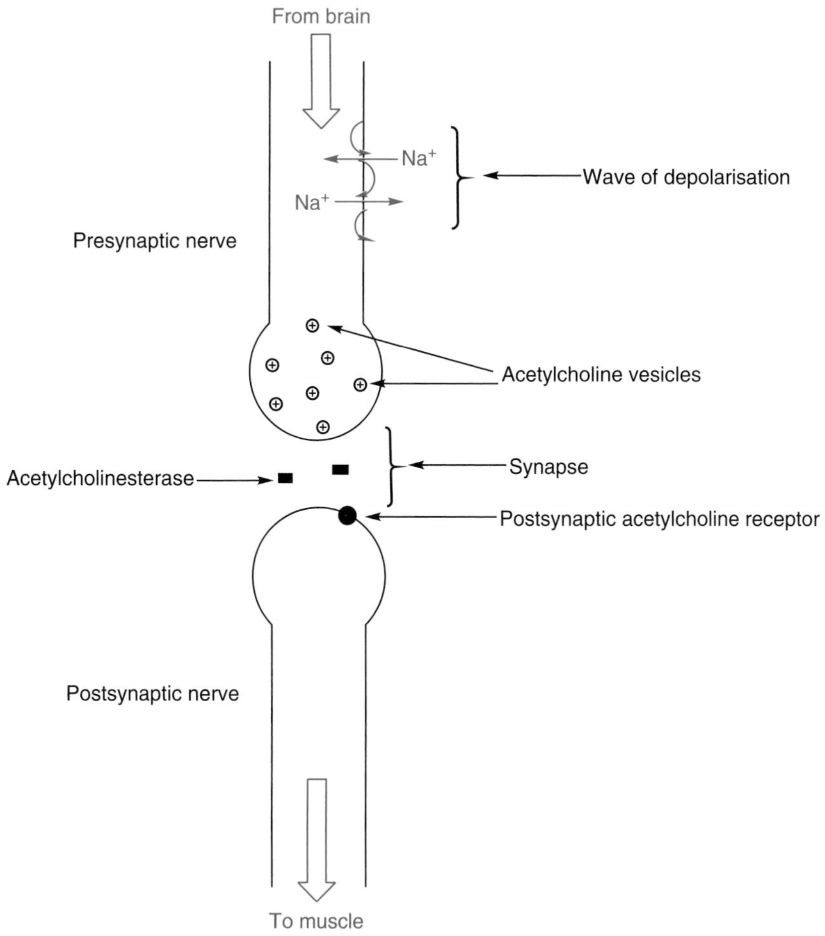

From brain

Na⁺

Na⁺

Wave of depolarisation

Presynaptic nerve

⊕
⊕
⊕   ⊕   ⊕
⊕
⊕

Acetylcholine vesicles

Acetylcholinesterase

Synapse

Postsynaptic acetylcholine receptor

Postsynaptic nerve

To muscle

Figure 7.7 Schematic representation of a cholinergic nerve system. The impulse from the brain causes a wave of depolarisation along the presynaptic neurone's cell membrane; when it reaches the nerve terminus it causes the release of acetylcholine from the presynaptic vesicles. The acetylcholine carries the impulse message across the synapse, binds to the postsynaptic acetylcholine receptor and re-initiates a wave of depolarisation which eventually arrives at a muscle and makes it move. Acetylcholinesterase in the synapse destroys the acetylcholine just as soon as it drops off the postsynaptic receptor.

Once the neurotransmitter has re-initiated a single wave of depolarisation it must be destroyed, otherwise nervous impulses would be amplified as they travelled towards their destination. There are enzymes in the synapse that break down the neurotransmitters as soon as they leave the postsynaptic receptor, or there are carrier proteins that pick up the used neurotransmitter and take it away from the synapse quickly so that it cannot interact with the postsynaptic receptor for a second time.

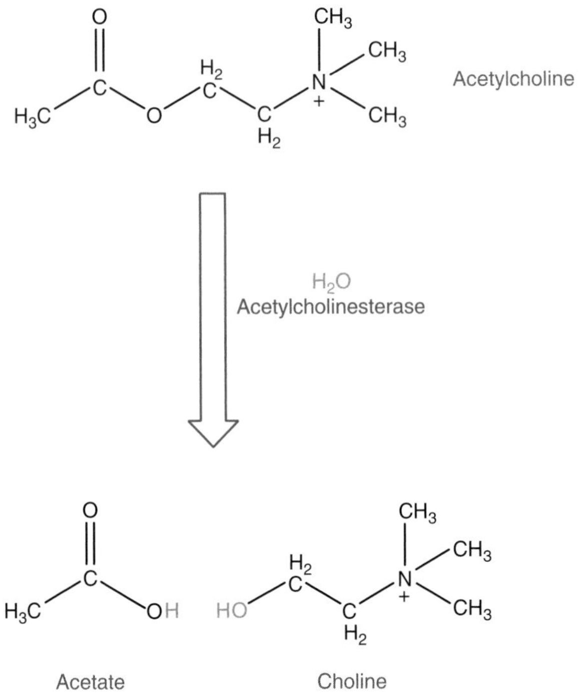

Figure 7.8 Acetylcholinesterase in the synapse breaks down acetylcholine to choline and acetate, so preventing multiple impulses being initiated by a single acetylcholine molecule.

An example of a neurotransmitter is acetylcholine; its postsynaptic receptor is called the acetylcholine receptor and the synaptic enzyme which quickly breaks down acetylcholine is acetylcholinesterase (AChE; Figure 7.8). This nerve type is referred to as cholinergic.

OCs interfere with the membrane ion transport proteins that initiate the wave of depolarisation. Other pesticides are designed to interfere with different facets of the neuronal system ... but more of that later.

If you were an insect and you were sprayed with a pesticide that interfered with the nerve impulses to your wings you would fall from the sky and eventually die!

Metabolism/degradation of OCs

OCs are very stable and are only very slowly broken down by animals' metabolic systems and in the environment (e.g. by UV light). Their environmental half-lives ($t_{1/2}$) are very long; indeed there have been widely varying estimates of DDT's environmental $t_{1/2}$ ranging from 50–150 years.

DDT is degraded in the environment, albeit very, very slowly to p,p'dichlorodiphenyldichloroethene (DDE). The presence of DDT residues *per se* indicates relatively recent use of DDT. Usually the breakdown products p,p'-DDE and DDD are found due to its historical use (Figure 7.9).

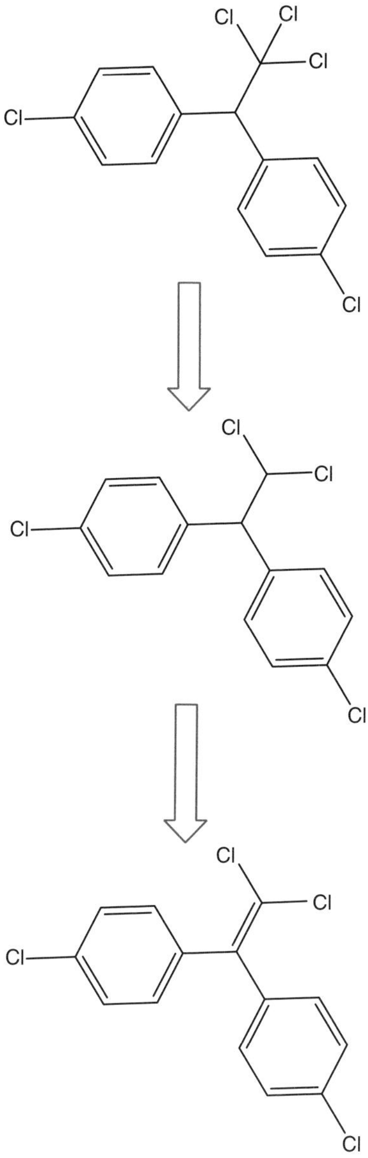

p,p′-Dichlorodiphenyltrichloroethane
DDT

p,p′-Dichlorodiphenyldichloroethane
DDD

p,p′-Dichlorodiphenyldichloroethene
DDE

Figure 7.9 The environmental degradation of DDT. Total residues in food are usually measured, i.e. DDT + DDD + DDE. The presence of DDT residues indicates recent use; the presence of only DDE indicates historical use.

OC residues in food

DDT

As discussed above, OCs are very persistent in the environment and therefore very low concentrations occur in many fatty foods (e.g. butter) and since there are concerns about the toxic effects of DDT (it is a cancer suspect agent) on consumers, DDT residues are often measured as part of food surveillance schemes – particularly in fatty foods. The UK has an excellent pesticide monitoring programme; DDT residues in butter sold in the UK illustrate well that residues are transferred internationally (Table 7.3) when products are imported. In this case New Zealand's DDT residues (see *Organochlorine pesticides* and Figure 7.6) appear in butter exported to the UK.

Residues surveillance in New Zealand shows a downward trend in DDT levels in fatty foods due to good pasture management (Figure 7.6 and Table 7.4).

Table 7.3 DDT residues in butter sold in the UK – the MRL for DDT = 1 mg/kg and therefore none of the samples exceeds the MRL. (Data from Working Party on Pesticide Residues Annual Report 1996, Ministry of Agriculture, Fisheries & Food, London, p. 86.)

Origin of butter sample purchased in the UK (number of samples analysed)	DDT residue concentration range (mg/kg)	Percentage of samples in range
UK (36)	Not found	58
	0.01–0.06	42
New Zealand (7)	Not found	43
	0.04–0.07*	57
Denmark (24)	Not found	100
Finland (2)	Not found	100
France (1)	Not found	100
Netherlands (2)	Not found	100

* Present as *p,p'*-DDE (a metabolite of DDT which shows that the residues represent historical use of DDT, not current use).

Table 7.4 Total DDT (i.e. DDT + metabolites) residues in food in New Zealand showing the downward trend with time. (Data from 2003/04 New Zealand Total Diet Survey, New Zealand Food Safety Authority, Wellington, p. 28. With permission from MAF.)

Food	Total DDT residue concentration (mg/kg)			
	1987/88	1990/91	1997/98	2003/04
Bacon	0.03	0.02	0.006	0.005
Beef	Trace	0.01	0.002	0.005
Lamb	0.03	0.03	0.01	0.006
Butter	0.07	0.04	0.02	0.02
Cheese	0.07	0.02	0.008	0.008

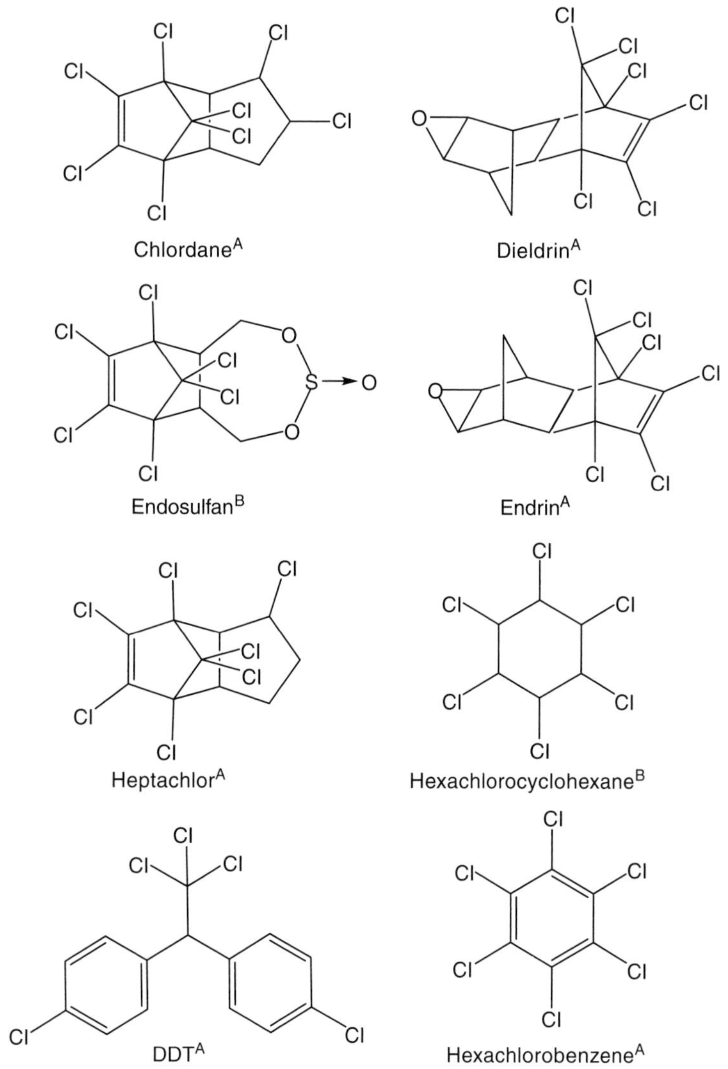

Figure 7.10 OCs that still occur as residues in food even though some have been banned (A) and others are still permitted (B).

Other OCs

DDT is not the only OC that has been, or is still, used in food production; Figure 7.10 shows other OCs and their diverse, but highly chlorinated and very hydrophobic, molecular structures.

The issue relating to the use of banned pesticides by developing countries is illustrated well by residues found in rabbit meat imported into the UK from China in 1997; 75% of the samples analysed ($n = 36$) contained residues of DDT ($n = 7$) or hexachlorocyclohexane (HCH) ($n = 20$) or both, albeit below the MRLs (data from Working Party on Pesticide Residues Annual Report 1997, Ministry of Agriculture, Fisheries & Food, London, p. 133).

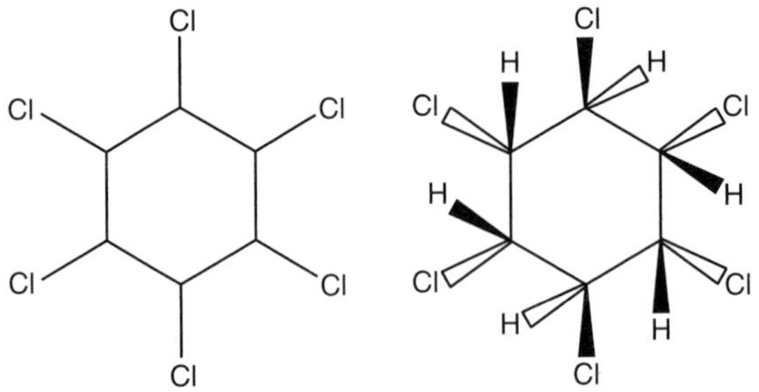

Figure 7.11 The general molecular structure for the hexachlorocyclohexanes (left) – since the carbon ring is unsaturated there are many possible orientational combinations of the H and Cl at the carbon atoms, hence the numerous isomers of HCH. γ-HCH, the most active insecticidal isomer, is illustrated (right) – the solid bond indicates that it comes towards you (i.e. out of the page) and the open bond goes away from you (i.e. into the page).

γ-HCH

HCH is a good example of an OC insecticide that is permitted in agriculture in some countries for very specific uses (e.g. for growing sugar beet in the UK), but is banned in others. There has been considerable controversy about its toxicity to consumers. At very high doses it is carcinogenic, but is very unlikely indeed to ever reach sufficiently high doses as residues via food to cause cancer. Nevertheless people are concerned because, despite the risk being very low if not negligible, they do not want to be exposed to a carcinogen, however low the dose (this is another perception of risk issue – see Chapter 1, *Risk perception*). HCH occurs as five isomers – α-HCH, β-HCH, γ-HCH, δ-HCH and ε-HCH – the γ-isomer (Figure 7.11) is the most active insecticide; the other isomers occur as impurities in the commercial products (trade names: Lindane, Gammexane).

A case of γ-HCH residues in milk in the UK

There was a very interesting and equally worrying case of γ-HCH residues in milk in the UK in 1995. γ-HCH is often found in milk at very low concentrations because milk has a high fat content and γ-HCH is fat soluble. In addition, γ-HCH is quite persistent (but very much less so than DDT) in the environment and so cows can consume residues in their feed. This is an example of food chain contamination – cattle feed might contain residues of γ-HCH; cows eat the feed and accumulate the residues in their fatty tissues, then they excrete γ-HCH in their milk which humans consume. So, the γ-HCH from the environment ends up in people via cows. However, the γ-HCH residues situation in milk in the UK in September 1995 was extreme (Figure 7.12); the γ-HCH concentration exceeded (by a long way) the MRL (0.008 mg/kg) and approached the ADI (8 μg/kg body weight/day). A risk assessment for this situation follows:

- Highest milk residue concentration = 0.05 mg/kg.
- Assuming someone consumed 1 L of milk per day this would contain about 0.05 mg γ-HCH.

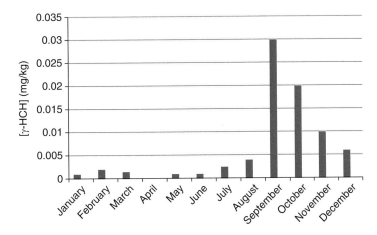

Figure 7.12 γ-HCH residues in milk in the UK in 1995 – in September the MRL was exceeded and the ADI almost reached. (Data from Working Party on Pesticide Residues Annual Report 1996, Ministry of Agriculture, Fisheries & Food, London, p. 21.)

- Mean human weight = 70 kg.
- Therefore the dose would be 0.7 µg/kg body weight/day, i.e. 10% of the ADI (8 µg/kg body weight/day), which means to exceed the ADI on a SINGLE day a consumer would have to drink 10 L milk – this is not very likely.

Despite the above risk calculation it was decided to err on the side of safety and therefore milk was temporarily withheld from the market to protect the consumer.

The summer of 1995 was very warm and dry in the UK and therefore pasture growth was low because of lack of rain. This meant that feed manufacturers had to import some of the components of their proprietary cattle feed to meet the demand because farmers needed more cattle feed to keep their cows well fed in the absence of grass. It is likely that feed component imports from countries that permitted the use of γ-HCH contributed γ-HCH residues to the cattle feed. This led to a food chain contamination that resulted in excretion of γ-HCH in milk.

OCs in humans
Since we consume residues of OCs in our food and OCs are very lipophilic we too build up residues in our bodies. Measuring OCs in human fat or human milk gives a good indication of our exposure to these fat soluble molecules and shows that nearly everyone studied had been exposed to DDT, HCH and hexachlorobenzene during their lives (Table 7.5).

It is possible to compare exposures to OCs around the world by comparing human OC residues data from different countries. DDT residues in human milk illustrate this well (Table 7.6). It is interesting to note that levels in developing countries are significantly higher than in the developed world (illustrated by the USA and West Berlin in the examples in Table 7.6), which suggests that either environmental DDT levels are higher because of past

Table 7.5 OCs in human post-mortem fat showing that most people are exposed to OCs during their lives. (Data from Working Party on Pesticide Residues Annual Report 1996, Ministry of Agriculture, Fisheries & Food, London, p. 129.)

OC	Percentage of samples with measurable OC residues (>0.01 mg/kg)
Chlordane	53
DDT	99
Dieldrin	59
β-HCH	98.5
γ-HCH	3
Heptachlor	30
Hexachlorobenzene (HCB)	94

Table 7.6 DDT residues (measured as DDT + metabolites) in human milk from women during their first lactation from different countries. (Data from Shaw et al. (2000) *Environmental Science and Pollution Research* 7, 75–77.)

Country	Human milk (DDT) (mean or range, mg/kg)
USA	0.02–0.04
Germany	
West Berlin	0.8
East Berlin	2.3
Thailand	0.7
Vietnam	0.02–4.2
Papua New Guinea	0.4
India	7.2–13.8
Indonesia	0.4–17.7

high use, or DDT is still in use and direct exposure has led to the human milk residues. Many of these countries have endemic malaria and therefore it is possible that DDT is being used as part of a WHO-approved Anopheles mosquito control programme.

The case of Germany is very interesting because the data were collected before German unification and reflect two very different jurisdictions – West Berlin had (as Germany now has) comprehensive and efficiently policed pesticide use legislation which prohibited the use of DDT; on the other hand, East Germany probably had less stringent regulations. This illustrates well that regulations are very important in protecting populations. You might wonder why DDT concentrations in milk from West German women are higher than the USA (an example of a well regulated country just as we would expect West

Germany to have been); it is possible that food and animal feed trading across the former East/West German border led to import of DDT residues from the East into West Germany.

What will happen to OC residues in food (and people) in the future?

Since the use of OCs is declining worldwide because of bans and severe restrictions on their use, the OC residues problem will slowly diminish. The word *slow* is important here because of the extremely long environmental and animal body half-lives of OCs and their metabolites. There is little that we can do to speed this up, but as I discussed above it is possible to manage human exposure and therefore minimise the risks to health.

Organophosphorus (OP) and carbamate pesticides

I will deal with the the organophosphorus (OPs) and carbamate pesticides together because they have similar molecular structures, chemical and biological properties.

The OPs and carbamate pesticides are very different to the OCs. They have a different toxicity profile because their mechanisms of action are different and they are very much less persistent in both the environment and animals' bodies because they degrade in the environment and are metabolised by animals.

Mechanism of action and toxicity of OPs and carbamate insecticides

All OPs and carbamates have the same mechanism of action and common structural features (Figure 7.13). They inhibit synaptic acetylcholinesterase (see *Mechanism of action of the OCs*; Figure 7.14) and result in multiple waves of depolarisation being initiated by a single acetylcholine molecule. This causes tetany (i.e. rapid succession, uncontrolled muscle contractions) – if a flying insect is sprayed with an OP it can no longer control its wing beats and it drops out of the air and dies.

OPs and carbamates are very toxic. The chemical precursors of the OPs are incredibly toxic nerve gases that have been used in chemical warfare and terrorist attacks (e.g. sarin; Figure 7.15). At high doses they can be fatal; at low doses their neurological effects might be cumulative. Indeed very recent research suggests that repeated low dose exposure to OPs results in memory loss and behaviour changes in exposed people – this also applies to children exposed *in utero* via their mother's exposure to OPs. Mothers' exposure to OPs during pregnancy has been implicated in the development of attention deficit hyperactivity disorder (ADHD) in their children in later life.

Degradation/metabolism of OPs and carbamates

OPs and carbamates have relatively (compared to OCs) short environmental and animal body $t_{1/2}$s because they both degrade under the physical conditions presented by the environment (e.g. UV light) and are extensively metabolised by animal and microbiological (soil is rich in microorganisms) systems. The environmental $t_{1/2}$ of Diazinon, for example, is approximately 8 days.

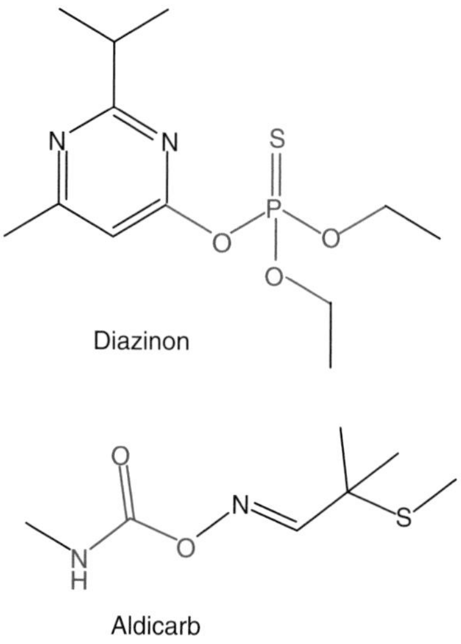

Diazinon

Aldicarb

Figure 7.13 The molecular structures of an OP (Diazinon) and a carbamate (Aldicarb) pesticide showing (in green) the chemical group that gives them their names – OPs are phosphoric or thiophosphoric acid esters and carbamates are carbamic acid derivatives.

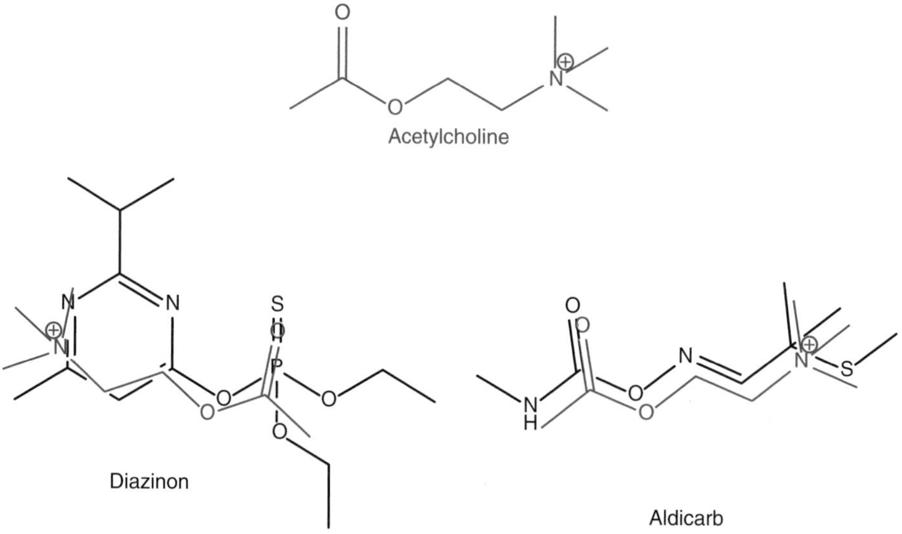

Acetylcholine

Diazinon

Aldicarb

Figure 7.14 The OPs and carbamate insecticides inhibit acetylcholinesterase because of their structural analogy to its substrate, the neurotransmitter acetylcholine – I have superimposed the OP Diazinon and the carbamate, Aldicarb on acetylcholine to show their molecular similarities.

The metabolic pathways of the OPs are complex with six or more metabolites (Figure 7.16). This plethora of metabolites is an indicator of the ease by which the compounds are degraded. All except the oxidation of the OPs to their respective biologically active (i.e. inhibit acetylcholinesterase) oxo-form (e.g. diazoxon, see *Mechanism of action and toxicity of OPs* above) results in

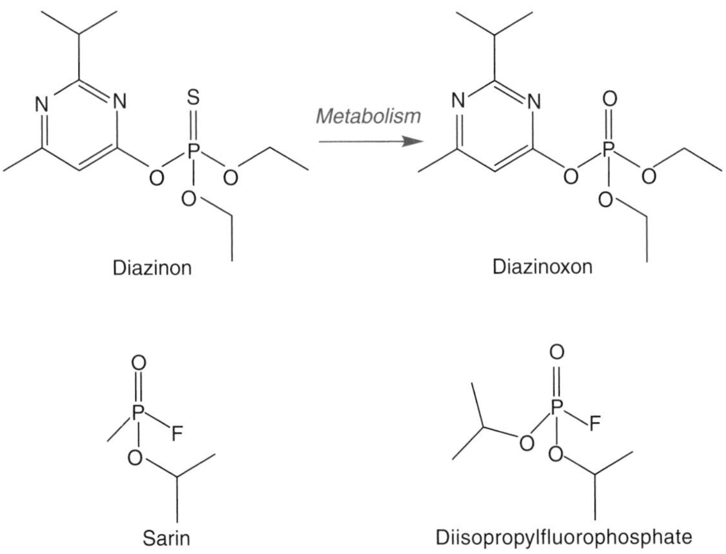

Figure 7.15 The molecular structures of the OP Diazinon and its biologically active metabolite diazinoxon, the incredibly toxic nerve gas sarin and the potent acetylcholinesterase inhibitor diisopropylfluorophosphate. They all inhibit aceylcholinesterase by a process of enzyme phosphorylation mediated by the organophosphate group; the oxo-OPs are more potent acetylcholinesterase inhibitors than the thio-OPs.

reduction in toxicity. Food residues monitoring programmes usually only measure the parent compounds.

OP and carbamate residues in food

OPs and carbamates are still used in agriculture around the world even though there are many, perhaps better, alternatives (e.g. pyrethroids) available now. The reason for this is cost, since many of the newer pesticides are more expensive than the older OPs and carbamates. In addition, the OPs and carbamates are very effective and relatively long acting. It is arguable that poorer countries' farmers have little alternative but to use the cheaper OPs and carbamates.

Because they are so widely used, OP and carbamate residues are very commonly found in crops and the food made from them. And because of the toxicity of the OPs/carbamates there is concern amongst consumers about the effect of such residues on their health – I will return to this later.

Phorate residues in carrots

Carrots are susceptible to many pests (e.g. the carrot root fly (*Chamaepsila rosae*)) and therefore farmers often use a battery of different pesticides, including OPs and carbamates, to protect their carrot crops. The UK's pesticide residues surveillance programme (see *Residues monitoring programmes* for details of how surveillance programmes are run) included a survey on residues in carrots in 1998 and found residues of eight different OPs and carbamate pesticides in carrots (Table 7.7). Some individual carrots had residues of more than one pesticide. All of the residues were below their

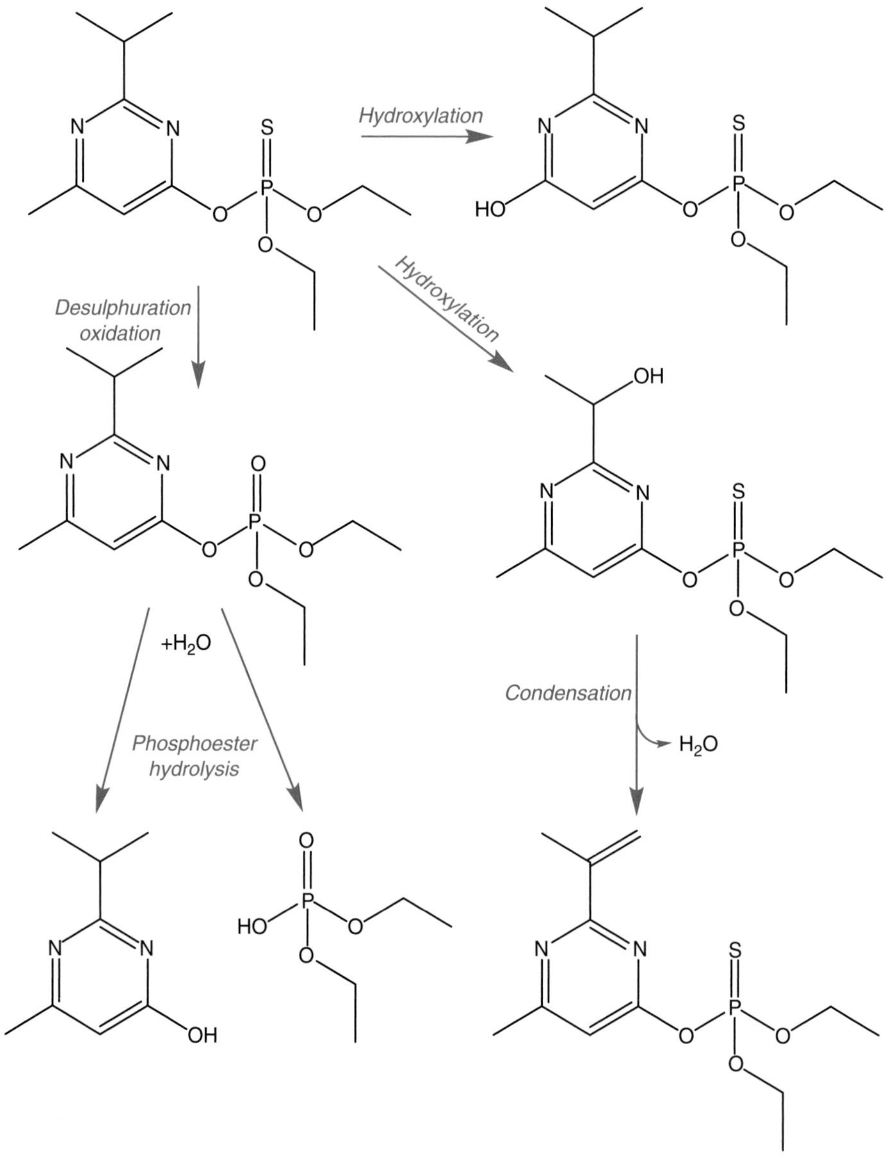

Figure 7.16 Metabolic/environmental degradation pathways for the OP Diazinon.

respective MRLs and ADIs and so the risk to the consumer was very low; this illustrates the breadth of use of pesticides in carrot growing.

The pesticide used to grow a particular crop varies according to pest pressure at the time, or perhaps the marketing prowess of the agrochemicals company salesman. This is illustrated well by phorate's use in carrot production. Table 7.7 shows that in 1998 the incidence of phorate residues in carrots was low (1.5%, $n = 66$). Just a few years before the situation was very different (Table 7.8). The problem with phorate residues in carrots was, at least in part, attributed to the morphology of a growing carrot. Phorate was applied as

Table 7.7 Pesticide residues in carrots (n = 66) in the UK in 1998; not all of the pesticides in this study were OPs or carbamates, but I have included them all for completeness. MRLs used are those in operation in 1998. This shows that a broad array of pesticides are used in carrot growing and that consumers are exposed to a wide range of residues, but that none exceeded their MRLs. (Data from Working Party on Pesticide Residues Annual Report 1998, Ministry of Agriculture, Fisheries & Food, London, pp. 124–125.)

Pesticide	Percentage with residues	Residue concentration/ range (mg/kg)	MRL (mg/kg)
Chlofenvinphos[OP]	9	0.01–0.1	0.5
Dithiocarbamate[C]	3	0.05	0.5
Iprodione[F]	42	0.01–0.04	10
Pendimethalin[H]	18	0.01–0.04	No MRL set
Phorate[OP]	1.5	0.07	0.2
Quinalphos[OP]	6	0.02–0.03	No MRL set
Triazophos[OP]	3	0.08–0.1	1
Trifluralin[H]	3	0.01–0.02	No MRL set

OP = organophosphate.
C = carbamate.
H = herbicide.
F = fungicide.

Table 7.8 Pesticide residues in carrots (n = 63) produced in the UK in 1995; not all of the pesticides in this study were OPs or carbamates, but I have included them all for completeness. MRLs used are those in operation in 1998. This shows that the OPs were extensively used in carrot production in 1995 and resulted in unacceptably high residues frequency and concentrations – Phorate and Triazophos were a particular problem. (Data from Working Party on Pesticide Residues Annual Report 1995, Ministry of Agriculture, Fisheries & Food, London, p. 32.)

Pesticide	Percentage with residues	Residue concentration/ range (mg/kg)	MRL (mg/kg)
Chlofenvinphos[OP]	9.5	0.01–0.1	0.5
Iprodione[F]	24	0.01–0.07	No MRL set
Pendimethalin[H]	6	0.01–0.03	1.0
Phorate[OP]	22	0.01–0.1	0.2
Pirimiphos-methyl[OP]	2	0.01	1.0
Quinalphos[OP]	16	0.10–0.1	No MRL set
Triadimefon[F]	2	0.01	No MRL set
Triazophos[OP]	51	0.01–0.2	1.0
Trifluralin[H]	8	0.01–0.03	No MRL set

OP = organophosphate insecticide.
H = herbicide.
F = fungicide.

Figure 7.17 A carrot from my garden; the arrow shows the indentation where the leaf stalks meet the root – it is here that phorate (and other pesticides) can accumulate. This was a particular problem with phorate because it is used as a granular preparation – one granule falling into the root/stalk indentation would mean that the particular carrot would have a very high residue of phorate. (Photograph by the author.)

granules and sometimes a granule lodged in the dip around the leaf stalks of the carrot plant (Figure 7.17), thus resulting in a high residue level in the carrot.

Residue cocktails in citrus fruit

Citrus fruit is another example of a crop that has a lot of insect pests which means that citrus farmers use complex mixtures of pesticides, including OPs, to ensure a saleable crop. A UK survey of imported mandarins and clementines in 1997 showed residues of 14 different pesticides; some individual fruits had several different residues. No MRLs were exceeded, but several pesticide residues came quite close (e.g. thiabendazole; residue = 5.5 mg/kg; MRL = 6 mg/kg) (Table 7.9).

Pyrethroid insecticides

The pyrethroid insecticides can be either synthetic or natural. The natural ones are extracts (often crude) of the pyrethrum plant (a type of chrysanthemum) and are used as treatments for insect pests on home produce or perhaps by some organic farmers. Nevertheless they contain a mixture of pyrethroids very similar in molecular structure and toxicity to the synthetic pyrethroids (Figure 7.18).

The synthetic pyrethroids were developed by the agrochemical industry based on the chemical structure of the natural pyrethrins. They have been designed to be more active insecticides than their natural counterparts and to degrade rapidly in the environment. This means that their environmental impact should be low and they should degrade quickly enough in the crop growing period so as not to leave residues. Human exposure to pyrethroid insecticides is very low because their residues are indeed rarely found. This is

Table 7.9 Pesticide residues in citrus fruit (mandarins and clementines) ($n = 34$) in the UK in 1997; not all of the pesticides in this study were OPs or carbamates, but I have included them for completeness. MRLs used are those in operation in 1998. This shows that a broad array of pesticides are used in citrus growing and that consumers are exposed to a wide range of residues, but that none exceeded their MRLs. (Data from Working Party on Pesticide Residues Annual Report 1997, Ministry of Agriculture, Fisheries & Food, London, pp. 146–147.)

Pesticide	Percentage with residues	Residue concentration/ range (mg/kg)	MRL (mg/kg)
2-Phenylphenol[F]	73.5	0.04–5.7	12*
2,4-D[H]	53	0.02–0.6	No MRL set
Azinphos-methyl[OP]	6	0.1–0.2	2
Carbendazim[F]	12	0.1–0.6	5
Chlorpyriphos-methyl[OP]	9	0.1–0.2	0.3
Dimethoate[OP]	3	0.02	2
Fenthion[OP]	3	0.03	No MRL set
Imazalil[F]	100	0.3–2.6	5
Malathion[OP]	3	0.02	2
Pirimiphos-methyl[OP]	15	0.02–0.08	2
Tetradifon[OC]	9	0.02–0.09	No MRL set
Thiabendazole[F]	82	0.2–5.5	6

OP = organophosphate insecticide.
OC = organochlorine insecticide.
H = herbicide.
F = fungicide.
*Permitted limit (PL) not MRL. PLs are set (in the UK) for chemicals that are covered by the Miscellaneous Food Additives Regulations 1995 rather than the Food & Environmental Protection Act 1985 (which regulates pesticides) even though they might also be used as pesticides.

a success story. Unfortunately the success of the synthetic pyrethroids was marred when after they had been in use for a short time large numbers of fish died in rivers into which pyrethroid run-off had drained – pyrethroids were blamed for this environmental impact. We now know that pyrethroids are peculiarly toxic to fish and so their use in the vicinity of waterways is now controlled in most countries.

The pyrethroids have low acute toxicity and no long-term effects have been proved (although there are concerns about possible long-term effects of some pyrethroid metabolites; see Chapter 9).

Mechanism of action of pyrethroid insecticides
Like other insecticides the pyrethroids work by interfering with neurotransmission. The pyrethroids have a mechanism of action similar to that of the OCs; they interfere with the Na^+ pump in the neurone membrane which means that nervous impulses cannot be generated.

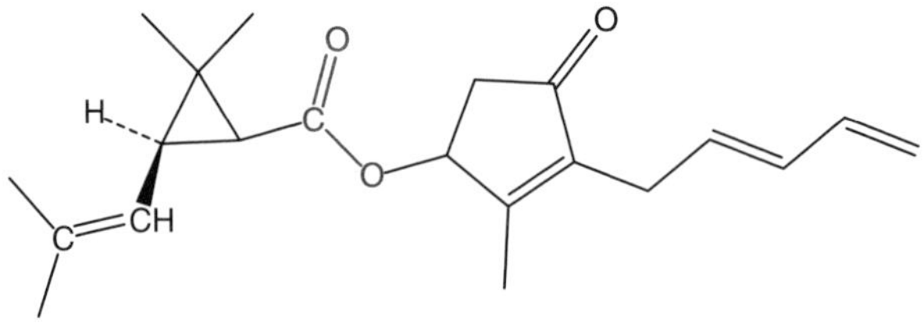

A Pyrethrin

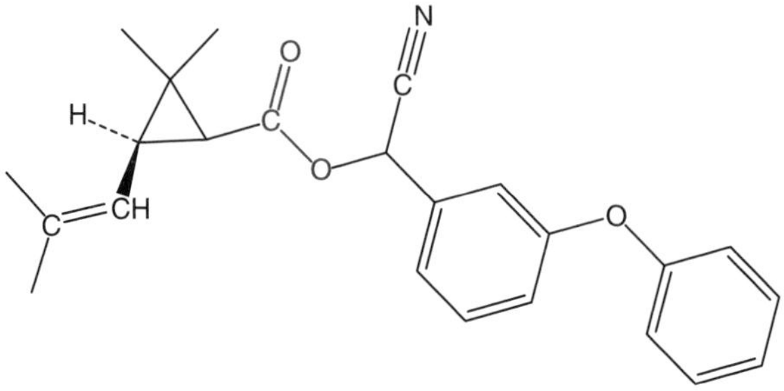

Cypermethrin

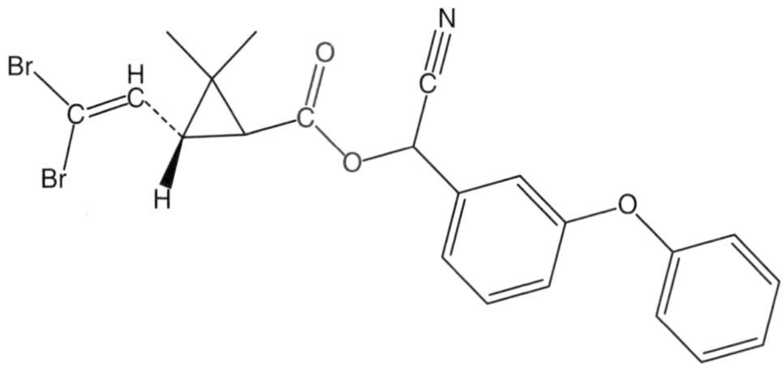

Decamethrin

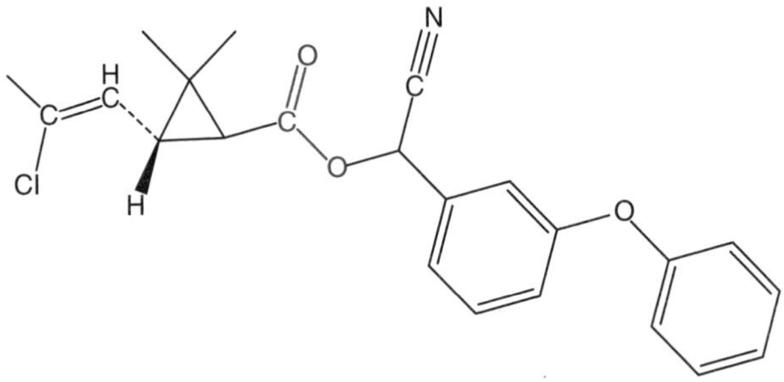

Permethrin

Figure 7.18 Three synthetic pyrethroid insecticides and the natural pyrethroid pyrethrin showing the similarity of structure, specificity of the special arrangements of groups around the cyclopropane ring and the unstable ester bond (in green) which leads to their short environmental $t_{1/2}$s and rapid metabolism in animals (including humans).

Degradation of pyrethroid insecticides

As discussed above, the short environmental $t_{1/2}$s of the pyrethroids (cypermethrin environmental $t_{1/2}$ = 5–30 days) is due to their chemical instability; particularly of the central ester bond (Figure 7.18) which breaks under the influence of pH changes and UV light. Similarly they are very susceptible to metabolism by mammalian enzymes. Esterases (enzymes that cleave ester bonds to form an alcohol and a carboxylic acid) are ubiquitous in mammalian systems and therefore almost immediately after absorption by a mammal the central ester bond is broken to release the diphenylether and cyclopropyl halves of the pyrethroid molecule (Figure 7.19). These breakdown products do not interfere with the generation of nerve impulses and have remarkably low toxicity (although there are concerns about the long-term toxicity of the diphenylether and its metabolites (see Chapter 9).

The pyrethroids are degraded quickly in the environment and metabolised quickly in animals (including humans) to form many metabolites.

Pyrethroid insecticide residues in food

Since the pyrethroids are so rapidly degraded by environmental systems (including the creatures that inhabit them) it is not surprising that pyrethroid residues are rarely detected in crops or food produced using them. However, it is likely that residues of one or more of the pyrethroid metabolites/degradation products would be present on crops and perhaps in food, but these are not measured as part of national pesticide surveillance schemes. Some pyrethroid metabolites are now known to be toxic (see Chapter 9) and therefore, perhaps, their residues in food should be measured.

Pyrethroids are often used to control aphids on grain crops and therefore if we were to find pyrethroid residues we might expect them in foods made from grains (e.g. bread). In a UK study carried out in 1998 the pyrethroids cypermethrin, deltamethrin and permethrin were analysed in bread ($n = 243$). No pyrethroid residues were found above their analytical detection limit of 0.05 mg/kg (data from Working Party on Pesticide Residues Annual Report 1998, Ministry of Agriculture, Fisheries & Food, London, pp. 42–43). In 1997 a study was carried out in the UK in which the same three pyrethroids were analysed in fruit breads ($n = 25$); on this occasion 16% of the bread samples had residues of cypermethrin between 0.02 and 0.1 mg/kg (data from Working Party on Pesticide Residues Annual Report 1997, Ministry of Agriculture, Fisheries & Food, London, p. 53) – this suggests that the residues were in the fruit (e.g. raisins – dried grapes) in the bread rather than the flour. Indeed pyrethroids are used to control aphids on grape vines. The drying process used to make raisins from grapes would have concentrated any pyrethroid residues present which might explain this result. A UK study on sultanas (dried grapes) found residues of the pyrethroid λ-cyhalothrin in 17% (1 of 6) of the samples analysed.

Looking through 10 years of UK pesticide residues monitoring reports shows that pyrethroid metabolites are uncommon, which makes the point that the pyrethroids are not a residue problem – I did find examples though after a lot of searching; in 1998 2% (one sample, $n = 45$) of edible podded peas (sugar snap peas or mange tout) contained 0.2 mg/kg cypermethrin and the same year 31% ($n = 21$) of lettuces analysed also contained cypermethrin – one was very close (1.1 mg/kg) to the MRL (2 mg/kg) (data from Working Party on Pesticide Residues Annual Report 1998, Ministry of Agriculture, Fisheries & Food, London, pp. 57 and 65).

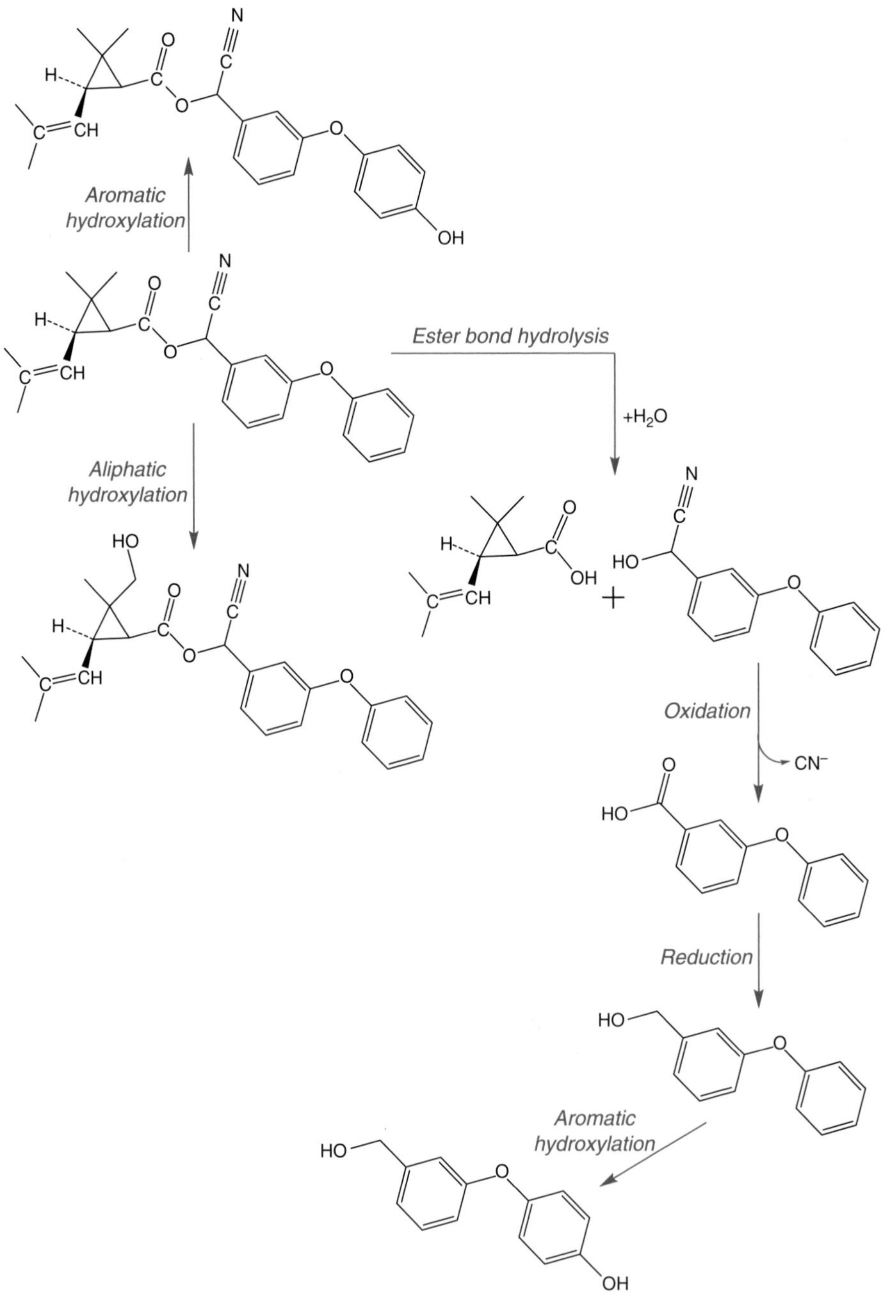

Figure 7.19 A metabolic/degradation scheme for cypermethrin showing the chemical reactions involved. Some of the reactions occur readily in the environment (e.g. ester bond hydrolysis), others are enzyme catalysed in animals (including humans) and microorganisms (e.g. cytochrome P_{450}-catalysed hydroxylations). (Based on a metabolic pathway proposed by Cremlyn (1990), *Agrochemicals*. John Wiley & Sons, Ltd., Chichester.)

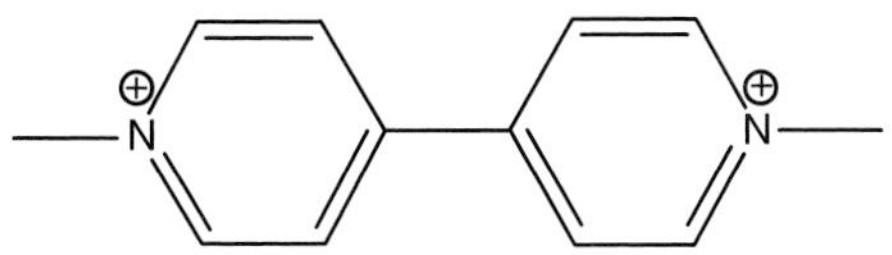

Figure 7.20 The herbicide paraquat works by inhibiting photosynthesis – an excellent way to kill green plants; it does this by accepting electrons (via the positive ring nitrogen) and preventing electron transfer which is an important part of the mechanism of photosynthesis ... the problem is that electron transport is important in animal respiration too which is why paraquat is lethal to humans and no longer permitted for use as a pesticide.

Herbicides

Herbicides interfere with the biochemistry of plants and thus kill them. Modern herbicides are designed to target plant-specific biochemistry and therefore they are not toxic to humans except at very high doses (i.e. far greater than food residues levels could achieve). Some of the old (many are now banned, e.g. paraquat, introduced in 1961; Figure 7.20) herbicides, how-ever, interfere with key mammalian biochemical pathways (electron transport in the case of paraquat) which makes them very toxic to animals (including humans). Many modern herbicides (e.g. 2,4-dichlorophenoxyacetic acid – 2,4-D) are plant hormone mimics. 2,4-D mimics the plant growth hormone auxin (Figure 7.21) – which makes them very much less toxic to animals (including humans). They kill plants by perturbing their hormone-mediated growth processes.

Herbicides are extensively used in agriculture to kill weeds and allow crops to grow more efficiently without competition and achieve good yields. We would expect to find herbicide residues in crops on which they have been used. You might wonder how herbicides can be used to more efficiently pro-duce plant crops – surely they would kill the crop as well as the weeds. Many herbicides are specific to either broad-leaved (dicotyledonous) plants (e.g. dandelion) or narrow-leaved (monocotyledonous) plants (e.g. grasses), or they target a specific stage in a plant's growth (e.g. pre-emergent herbicides stop seeds germinating). So it *is* possible for farmers to selectively kill some of the weeds growing amongst their crops.

Herbicides are also used for non-weed-killing purposes. For example, for wheat to achieve a good price at market it must have exactly the right moisture content; farmers use herbicides (e.g. glyphosate) to accelerate the wheat's development process when the weather is warm so that the crop dries out in the field and can be harvested at optimum moisture content.

Low concentrations of the auxin-mimicking herbicides (2,4-D and trichloro-phenoxyacetic acid – 2,4,5-T) are also used to change a crop's (e.g. citrus fruit) maturation rate and hopefully synchronise production – it prevents fruit drop in citrus. Residues will result from these uses too.

From a human health perspective most herbicide residues in food are likely to be of very little concern because of their non-mammalian mechanisms of action and consequent low toxicity.

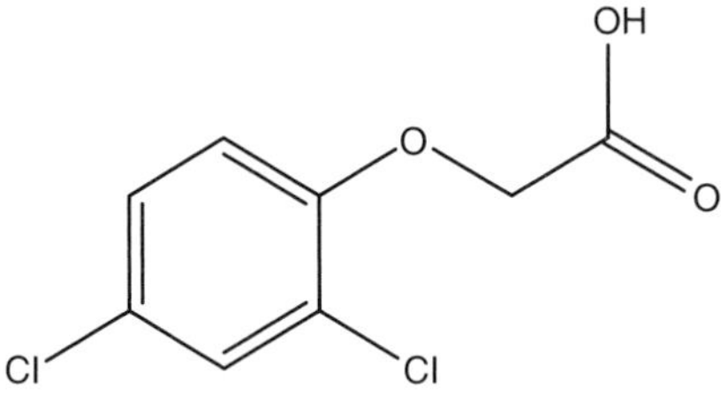

2,4-Dichlorophenoxyacetic acid
2,4-D

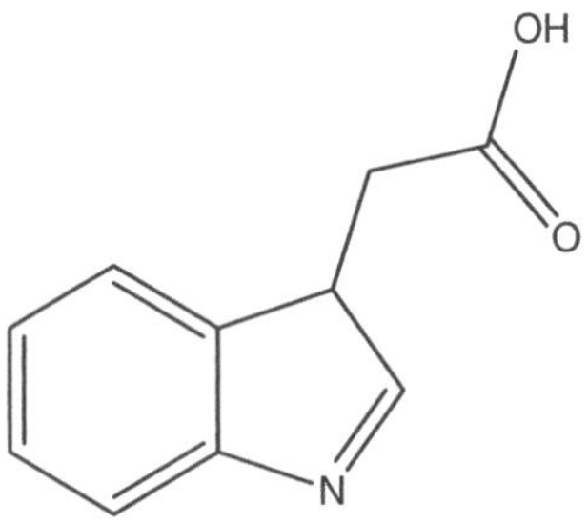

Indoleacetic acid
Auxin

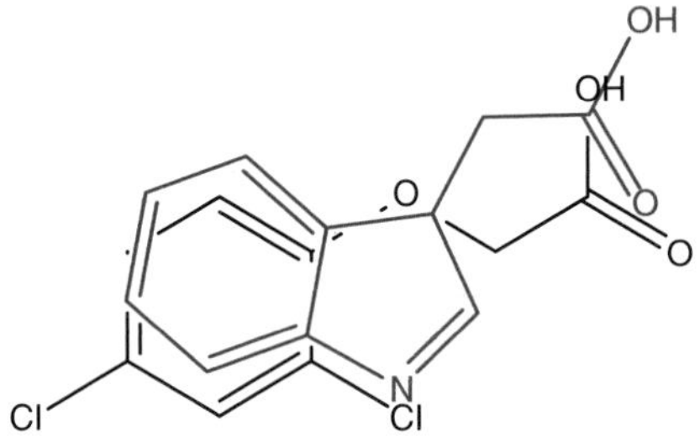

Figure 7.21 Molecular structures of the herbicide 2,4-D and the plant growth hormone auxin showing their molecular similarities. 2,4-D binds to and interferes with the auxin receptor, killing the plant. 2,4-D is not very toxic to humans because it does not affect animal biochemistry.

Herbicide residues in crops and food

Herbicides are often not analysed for in pesticide residues monitoring programmes because of their low impact on the consumer and the high cost of analysis - toxicologists and regulators have to decide how best to spend their money when designing residues monitoring programmes so they tend to focus on chemicals that have potential human health implications (e.g. insecticides). However, in crops on which they are used extensively (e.g. 2,4-D in citrus production) they are often included. For example, a study in the UK found 2,4-D residues (range 0.03-0.5 mg/kg) in 57.5% of oranges analysed ($n = 66$) (data from Working Party on Pesticide Residues Annual Report 1998, Ministry of Agriculture, Fisheries & Food, London, pp. 129-130). This suggests it is quite likely that if you eat an orange you will also consume some 2,4-D. But don't worry, this minute dose of 2,4-D is unlikely to have any effect on you, but if you were a plant you would need to worry!

Look at the carrot residues in Tables 7.8; you will find residues of the herbicides Trifluralin and Pendimethalin (Figure 7.22) in some (3-18%) of the carrots analysed. Trifluralin (banned in the European Union in 2008 because it is carcinogenic) and Pendimethalin are pre-emergent herbicides - they

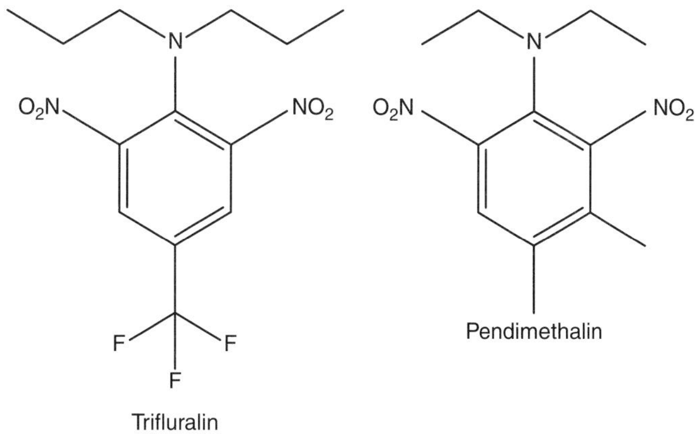

Figure 7.22 The herbicides Trifluralin and Pendimethalin. Trifluralin is carcinogenic at high doses in experimental animals and was banned by the EU in 2008.

prevent seeds germinating – and are used by carrot growers to stop annual weeds growing after the carrots are established.

The increase in production of genetically modified (GM) crops (see Chapter 10) is likely to increase herbicide residues in food because some GM crops are designed to be resistant to herbicides (e.g. glyphosate-resistant corn) which allows herbicides to be used to kill weeds but not the crop. Clearly this is likely to result in greater herbicide use which will, in turn, increase herbicide residues in crops and food.

Fungicides

There are many fungi that infect plants, including crop plants, and many fungi that grow on stored crops (e.g. peanuts). If you have ever tried to grow courgettes (zucchini) you will know that they often develop white powdery mildew (a fungus) on their leaves and when they are infected they grow slowly and do not produce many courgettes. Farmers are presented with the same problems, but on a much larger scale. It is for this reason that fungicides are extensively used in agriculture – in fact fungicides account for about 20% of all pesticides used.

Fungicides include a broad array of different chemicals (including inorganic compounds, e.g. $CuSO_4$) with diverse molecular structures. They are all designed to interfere with aspects of fungal biochemistry, so inhibiting fungal growth. Some do this by interfering with nucleic acid replication or protein synthesis, so preventing fungal cell division and important aspects of cell function. This means that some fungicides are potentially toxic to humans; for example, fungicides that interfere with nucleic structure and function are mutagenic, carcinogenic and/or teratogenic (cause birth defects) in animals. However, most fungicides have low acute toxicity (Table 7.10) because their mechanisms of action interfere with processes in cells that take a long time to manifest (e.g. nucleic acid damage). In addition, some are not well absorbed from the gastrointestinal tract and therefore their potential human toxicities might not be realised; nevertheless there is

Table 7.10 Acute toxicities (LD_{50}) for some commonly used fungicides compared with the OP insecticides Diazinon and Propetamphos. The fungicides have very low acute toxicities – it would take about 240 g of iprodione to kill a human.

Pesticide	LD_{50} [oral; rat] (mg/kg body weight)
Imazilil	343
Vinclozolin	10,000
o-Phenylphenol	2,480
Zyneb	1,800–8,900
Iprodione	4,000
Thiabendazole	3,100
Diazinon	250
Propetamphos	82

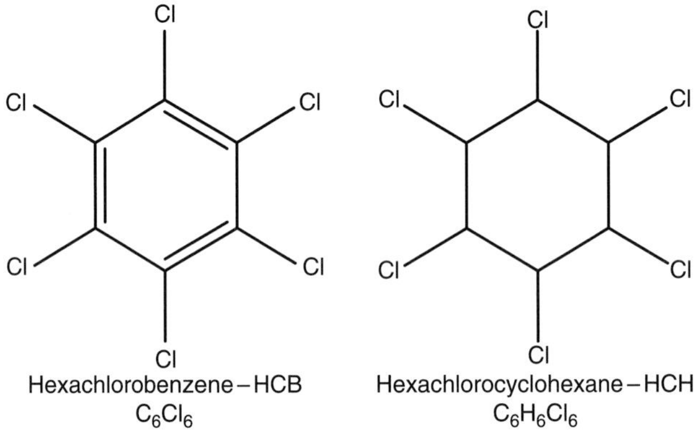

Hexachlorobenzene – HCB
C_6Cl_6

Hexachlorocyclohexane – HCH
$C_6H_6Cl_6$

Figure 7.23 The molecular structures of the fungicide HCB and the insecticide HCH – do not confuse them!

significant debate amongst toxicologists about the long-term effects of human exposure to fungicide residues in food.

Some of the early organic (in the correct sense, i.e. a molecule containing carbon) fungicides (e.g. hexachlorobenzene – HCB; do not confuse this with the insecticide HCH (Lindane); Figure 7.23) have long environmental $t_{1/2}$s and are very toxic to mammals (HCB is immunosuppressive, teratogenic and a cancer suspect agent). HCB was used (its use was banned in most countries in the 1970s) as a seed dressing to prevent fungal attack when the seeds were planted – this would lead to minimal residues in food crops grown from dressed seed, but might result in environmental contamination which might lead to residues in crops grown in the same field. There is a classic case of human toxicity from HCB which occurred in Turkey in the 1950s. People consumed flour made from HCB-dressed wheat during a period of wheat shortage due to

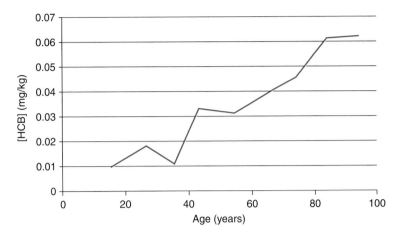

Figure 7.24 HCB concentration in human fat showing that older people have higher HCB concentrations because they were alive when HCB was still used (i.e. before 1970), but even people who were born after 1970, i.e. age 30 years, have HCB residues because it persists in the environment for a long time. (Data from Working Party on Pesticide Residues Annual Report 1996, Ministry of Agriculture, Fisheries & Food, London, p. 35.)

crop failure; they became very ill with symptoms of alopecia, unusual skin pigmentation and photosenstitvity. The disease was called 'new disease' and showed the extreme effects of high doses of HCB in humans. Under normal circumstances residues in food would never lead to this because the dose is far too small.

Despite HCB being withdrawn in most countries in the early 1970s, residues persist in human fat (Figure 7.24); the older the person from whom the fat sample was obtained the greater the HCH concentration because people over 30 years old were alive when HCB was still in use in the UK; the human fat study was carried out in 1996; 1996 – 30 years = 1966. Interestingly, residues were still present in subjects with a mean age of 15.6 years (i.e. born in 1980 – 10 years after HCB use stopped) which suggests a persistent environmental source perhaps delivered via food residues.

HCB has long since been replaced by a new generation of fungicides (Figure 7.25). Since fungal diseases in agriculture are common and rarely go away on their own, fungicides are extensively used by farmers. I will deal with just a few examples of commonly used fungicides here because I do not have enough space to cover them all.

Imidazoles (e.g. Imazilil)

Imidazoles are used both on growing crops and post harvest during storage. They are very commonly used, particularly on citrus fruit; indeed 98% of citrus fruit analysed contained Imazilil in a study carried out in the UK in 1998.

The imidazoles work by inhibiting an important fungal enzyme – demethylase – which removes methyl groups from protein molecules. Methylation of proteins is used by fungi to transfer messages from one cell to another (signal transduction); in the context of fungal growth, methylation is used to transfer

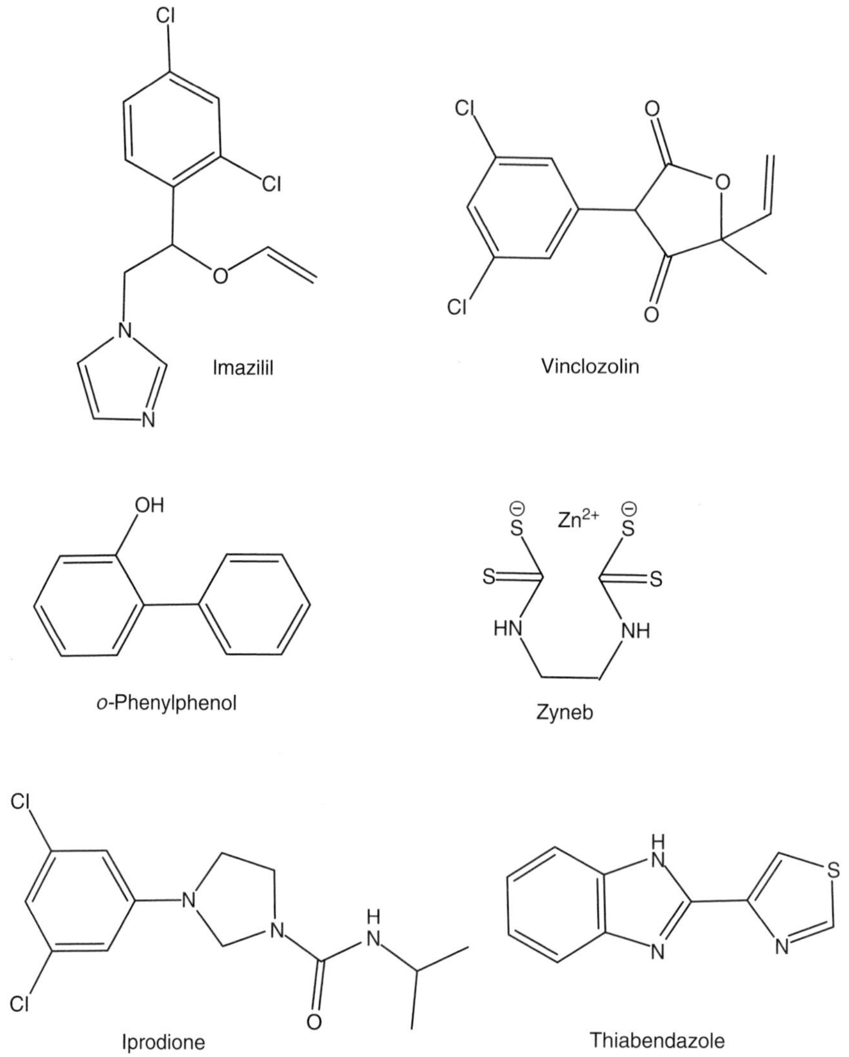

Figure 7.25 Six commonly used fungicides showing their diverse molecular structures.

signals about nutrient sources to allow the fungal cells to grow or move in the direction of a food source (chemotaxis). Clearly. if this aspect of the fungus's physiology is destroyed it will considerably impair its survival.

Benzimidazoles (e.g. Thiabendazole)

The benzimidazoles were introduced in the 1960s and represented a ground-breaking advance in fungicide design, activity and efficacy. They are still used on a very wide variety of crops and are effective against a broad spectrum of fungi.

The benzimidazoles inhibit β-tubulin assembly in mitosis. β-Tubulin is impor-tant in cell division; as the cell prepares to divide, tubulin monomers assemble to form huge protein polymers that form tracks running out from the cell nucleus. These tubulin tracks are the guides that the chromosomes move along when the genetic material is transferred to the daughter cells during cell division. Without tubulin cells cannot divide, so Thiabendazole-treated fungi cannot divide and therefore the infection on a crop plant does not spread.

Phenylphenol

o-Phenylphenol (only the o-isomer is used as a fungicide) was first described in 1939, but was not used as a fungicide until the 1950s. It is used as a surface fungicide to protect fruit during storage – it is a waxy substance that is used to coat the fruit. It is commonly used to protect citrus fruit.

Phenylphenol's mechanism of action is unknown, but it has been suggested that it interferes with mitochondrial metabolism by depleting important thiol levels in mitochondria.

Dithiocarbamates (e.g. Zyneb)

The first dithiocarbamate was patented in 1934 and heralded the organic era in fungicides; before the dithiocarbamates, metal salts (e.g. $CuSO_4$) were the most popular fungicides. The dithiocarbamates were only one step away from the inorganic metal fungicides though, because they are simply metal ions (e.g. Zn^{2+}) ionically associated with charged organic molecules. The metal ions vary from dithiocarbamate to dithiocarbamate:

Maneb	Mn^{2+}
Zyneb	Zn^{2+}
Propineb	Zn^{2+}
Ferbam	Fe^{2+}

The dithiocarbamates are thought to prevent fungal growth on plant surface by the dithiocarbamate metal ion upsetting the redox potential of the fungal cells.

Dithiocarbamates are used to treat a broad array of crops for a broad spectrum of fungi.

Iprodione

Iprodione is a broad-spectrum fungicide used on many different crops. It inhibits histidine kinase, an important osmotic signal transduction enzyme in fungal cells. It allows the fungal cells to move towards their ideal growth conditions (e.g. wet). If their ability to grow towards good growth conditions is inhibited their ability to infect crops is also inhibited.

Vinclozolin

Vinclozolin is a dicarboximide fungicide and was introduced in the late 1970s. It was used on a broad range of crops, but recently concerns about its long-term (chronic) toxicity to humans has led to its withdrawal from many applications. Its mammalian (presumably including humans) chronic toxicity is two-fold: it interferes with the male hormone (androgen, e.g. testosterone) receptor resulting in cellular feminisation (see Chapter 9) and at least one of its metabolites is carcinogenic.

Table 7.11 Fungicide residues on UK lettuces ($n = 70$) in 1995 showing their widespread use and the illegal use of vinclozolin, a fungicide not approved for use on lettuce crops in the UK. (Data from Working Party on Pesticide Residues Annual Report 1995, Ministry of Agriculture, Fisheries & Food, London, pp. 52–53.)

Fungicide	Percentage with residues	Residue concentration/ range (mg/kg)	MRL (mg/kg)
Dithiocarbamates	45	0.1–9.8	5
Iprodione	43	0.08–19.0	10
Chlorothalonil	1	0.02	No MRL set
Propamocarb	37	0.1–27	10
Propyzamide	21	0.01–7.6	No MRL set
Tolclophos-methyl	57	0.01–3.4	No MRL set
Vinclozolin	6	0.03–1.9	5

Concerns about the use of vinclozolin

There was considerable concern about the misuse of vinclozolin in the UK in 1998. Vinclozolin is a very effective fungicide and prevents fungal growth on crops including those for which its use is not approved (i.e. licensed by regulatory authorities; e.g. lettuce). It is tempting for some farmers to use vinclozolin on their greenhouse lettuce crops even though this use is not approved and therefore is illegal. Anyone who has tried to grow lettuces in a greenhouse in winter will know that the crop is susceptible to fungal attack; in fact I have stopped trying to grow lettuce in my greenhouse because the fungi always win! Some farmers soon found that vinclozolin solved their problem and so they used it regardless of its legislative status. This led to residues of vinclozolin on lettuces in the UK and considerable concern about both the breach of regulations and the potential harm that lettuces with vinclozolin residues might cause their consumers (Table 7.11). Eventually the UK authorities tracked down some of the offending farmers and prosecuted them. At the same time there was considerable press interest in the feminising effects of vinclozolin – we will never know whether the farmers began to obey the pesticide regulations because they did not want to be prosecuted or because they were concerned about the effect vinclozolin might have on their masculinity! This story illustrates the importance of pesticide surveillance schemes to protect the consumer.

Veterinary medicines

Our meat is produced from living creatures (i.e. farm animals) and living creatures get ill and need medicines just like we do. If we eat the meat from animals that have been given medicines, we might also consume residues of the medicine and/or its metabolites.

Veterinarians prescribe medicines to treat farm animals in accordance with strict rules (enshrined in veterinary medicines legislation in most countries) that ensure that animals are not slaughtered, their milk drunk or eggs used for human consumption while the medicine or its metabolite residues remain at levels that might cause harm to the consumer – this is the withdrawal time (i.e. the time between giving the animal the medicine and eating it or its products). Since most medicines used in animals would have similar effects on humans it is important that we minimise the consumer's exposure to them for fear of significant pharmacological effects. This might sound farfetched, but I will discuss the β-agonist clenbuterol later (see below) and you will see that some consumers of clenbuterol-treated cattle became very ill and several people died because clenbuterol has pharmacological effects on the heart at very low doses.

You might think that veterinary medicines are only used to treat ill animals, but this is not the case. Some farmers use medicines (e.g. antibiotics) as growth promoting agents; for example, low doses of some antibiotics given over long periods of time stimulate the growth of farm animals, thus giving the farmer a better financial return (see *Growth promoting chemicals*). For this reason residues of veterinary medicines (usually referred to as 'veterinary residues') can be present in animals that have not been treated for an illness.

The concentration of a particular medicine residue in meat depends on which part of the animal is eaten. Medicines are distributed around the animal's body following their administration; some are concentrated in particular tissues because they are metabolised (e.g. liver) and/or excreted (e.g. kidney) there or because this is their site of action (e.g. muscle) and others are simply distributed around the body in the blood and so appear in most tissues. Therefore, the consumer's exposure to veterinary residues depends both on the medicine and which tissues are eaten.

Some pesticides (e.g. insecticides) are also used to treat animal diseases. For example, the OP insecticide Propetamphos used as an insecticide in crop growing is also a veterinary medicine used for the treatment of warble fly (*Hypoderma bovis*) strike and other insect infestations in cattle. Therefore the consumer might be exposed to pesticide residues in both their fruit and vegetables and their meat.

Veterinary residues resulting from the use of medicines to treat farm animal diseases are rarely a problem because they are used sporadically and often only to treat a small proportion of a farmer's animals. Therefore consumer exposure is low. Sometimes, however, if an animal on a farm develops a communicable disease the vet might prophylactically treat all of the susceptible animals on the farm to prevent spread of the disease. This will present a greater potential residues issue, but providing the withdrawal time for the medicine is observed, human exposure will be low and of no toxicological significance.

The route of administration of the medicine is also important. If it is given orally its residues will be distributed around the animal's body quickly if it is absorbed, or will be excreted if it is not (e.g. anthelmintics are often not absorbed – they are used to kill gut parasites) and therefore will not form tissue residues. On the other hand, if the medicine is injected (e.g. intramuscularly) it might leave a very localised, very high residue concentration in the muscle – remember it is the muscle (e.g. steak) that most of us eat when we consume meat. If you happen to eat the piece of muscle with the injection site in it you might get a very high residue dose. The risk of this is very low and the dose is one-off and therefore its effect is likely not to persist; nevertheless injection site residues are taken seriously by meat regulators.

We must not forget that mammals are not the only animals farmed for human consumption. Fish (e.g. trout), shellfish (e.g. mussels) and crustaceans (e.g. prawns) are also farmed and, like their mammalian counterparts, are susceptible to diseases that will affect the farmers' profit and therefore are medicated to either prevent or treat the diseases. In this kind of farming, however, individual animals are not treated, but rather the medicine (e.g. an antibiotic) is added to the water in which the creatures live so that they all get a sufficient dose to prevent or treat the particular disease. And, as with their mammalian counterparts, these animals acquire residues too, and the residues are passed on to their human consumers in just the same way.

In this section I will illustrate veterinary residues with a few examples of commonly used medicines or particularly interesting or worrying examples of human exposure. I will deal with veterinary residues originating from the use of medicines as growth promoting agents (e.g. antibiotics) separately later.

Antibiotics (e.g. the penicillins)

Most of the antibiotics used in human medicine are also used in veterinary medicine and therefore they have been risk assessed for use in humans which means that the tiny doses the consumer might receive from food residues are usually insignificant in a toxicological sense. However, some people are allergic to antibiotics and even if they receive the tiniest dose they will develop a severe allergic response which might lead to anaphylaxis and possibly death. This is one of the reasons antibiotic residues are tightly and carefully controlled by meat regulators; ADIs and MRLs (Table 7.12) for antibiotics are low, which reflects the very low doses of some antibiotics (e.g. benzylpenicillin)

Table 7.12 ADIs and MRLs for some commonly used antibiotics. The ADI (and therefore the MRLs) for benzylpenicillin is very low because some people are allergic to very low doses.

Antibiotic	ADI (μg/kg body weight)	MRLs (μg/kg)
Benzylpenicillin	30	Muscle 50 Liver 50 Kidney 50 Milk 4
Neomycin	60	Muscle 500 Liver 15,000 Kidney 20,000 Milk 500
Sulphadimidine	50	Muscle 100 Liver 100 Kidney 100 Milk 25
Streptomycin	50	Muscle 600 Liver 600 Kidney 1,000 Milk 200

necessary to cause effects in some people. In addition, there are concerns about the development of antibiotic resistant bacterial strains. Such strains might result from continuous low-dose environmental exposure to antibiotics. Indeed, environmental contamination by farm use of antibiotics is thought to be a significant contributor to the development of antibiotic resistant bacteria worldwide. Antibiotic resistant human pathogens are assuming increasing importance in human medicine, particularly in hospitals where patients might have impaired resistance because of their illnesses or treatments (e.g. following surgery); multiple resistance strains (MRS) of human pathogens can cause havoc in hospitals, lead to wards being closed, and deaths of infected patients because MRS-bacterial pathogens do not respond to antibiotic treatment. This is another reason why we must control the use of antibiotics in farming.

Despite these concerns antibiotics are very useful in the treatment of animal diseases and therefore are widely used. For this reason antibiotic residues occur in meat, eggs and milk and are usually measured as part of national residues surveillance schemes.

In addition, don't forget that antibiotics are also used as growth promoting agents and prophylactically to reduce the risk of animals housed in close confines contracting bacterial infections. These are two more reasons to 'watch' antibiotic residues in animal products carefully.

Antibiotic residues in animal products

Trends in antibiotic use vary from year to year and the residues analysed for by regulatory authorities vary similarly in order to keep abreast of residues of the antibiotics that farmers might be using (either legally or illicitly). In general, however, antibiotics are more commonly used in animals that are intensively reared because they are more likely to contract and pass on bacterial infections. This is illustrated well by the UK's veterinary residues surveillance results for 1988 where 28% of pig kidneys had chlortetracycline residues compared to 1% in sheep and 7% in cattle; pigs are usually intensively reared in the UK whereas sheep and cattle are free range (Table 7.13). In addition, tetracyclines (Figure 7.26) have growth promoting properties and might be used for that purpose, particularly in intensive rearing systems.

Table 7.13 Residues of tetracycline antibiotics in kidney samples in the UK in 1988. (Data from Veterinary Residues in Animal Products 1986–1990, Food Surveillance Paper No. 33, p. 24, HMSO, London (1992).)

| Animal | Percentage of samples with antibiotic residues (concentration/range (mg/kg)) | | |
	Chlortetracyline	Oxytetracycline	Tetracycline
Cattle	7 (0.01–0.12)	8 (0.01–6.3)	1 (0.04)
Pigs	28 (0.01–2.0)	8 (0.01–0.11)	6 (0.01–0.09)
Sheep	1 (0.03)	5 (0.01–0.02)	0

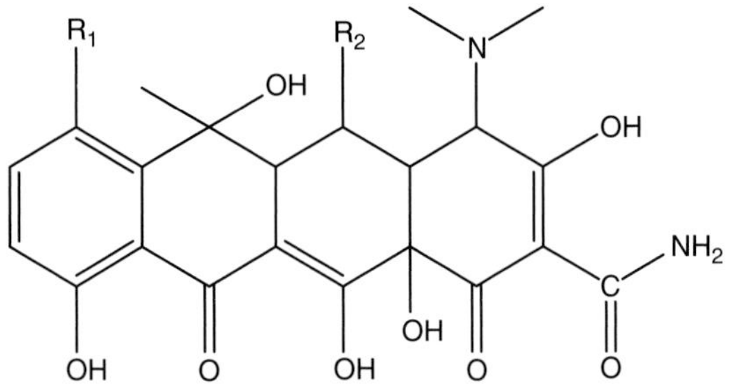

	R1	R2
Tetracycline	H	H
Oxytetracycline	H	OH
Chlortetracycline	Cl	H

Figure 7.26 Molecular structures of the tetracyclines.

Anthelmintics (e.g. Ivermectin)

Anthelmintics (from the Greek ελμινθ (*helminth*) meaning worm) are medicines used to treat worm infestations. I have already discussed food-borne worm parasites in Chapter 5 in the context of human infection; farm animals, just like humans, are susceptible to worm infestations and therefore are often treated, or prophylactically treated, to prevent infections with anthelmintics. Many anthelmintics are poorly absorbed from the gastrointestinal tract; indeed they are designed not to be absorbed because their role is to kill the worms in the intestine; for this reason they are often not of great toxicological significance to human consumers.

As with other medicines, trends in use change as new, and improved, medicines come to market. In the 1970s and early 1980s levamisole was a very commonly used anthelmintic; by the mid-late 1980s and early 1990s the benzimidazoles were the anthelmintics of choice (Figure 7.27). However, by 2000 their use had been largely superseded by the now universally used avermectins (e.g. Ivermectin); this change is reflected by the residues findings in the UK (Table 7.14).

As for other veterinary medicines (unless they are used as growth promoting agents) they are only used in animals that suffer from the disorder they are designed to treat (obviously!) – sheep are susceptible to many worm parasites and therefore are often prophylactically treated with anthelminthics. This is why anthelmintic residues (avermectins and benzimidazoles) were only found in sheep liver in 2000 in the UK.

Antiprotozoal drugs (e.g. Imidocarb)

As discussed above, farmers only use, and vets only prescribe, drugs for animals that need them and therefore residues tend to occur in tissues from animals susceptible to the diseases that a particular medicine is used to treat. The antiprotozoal drugs are no exception. Poultry are very susceptible to a

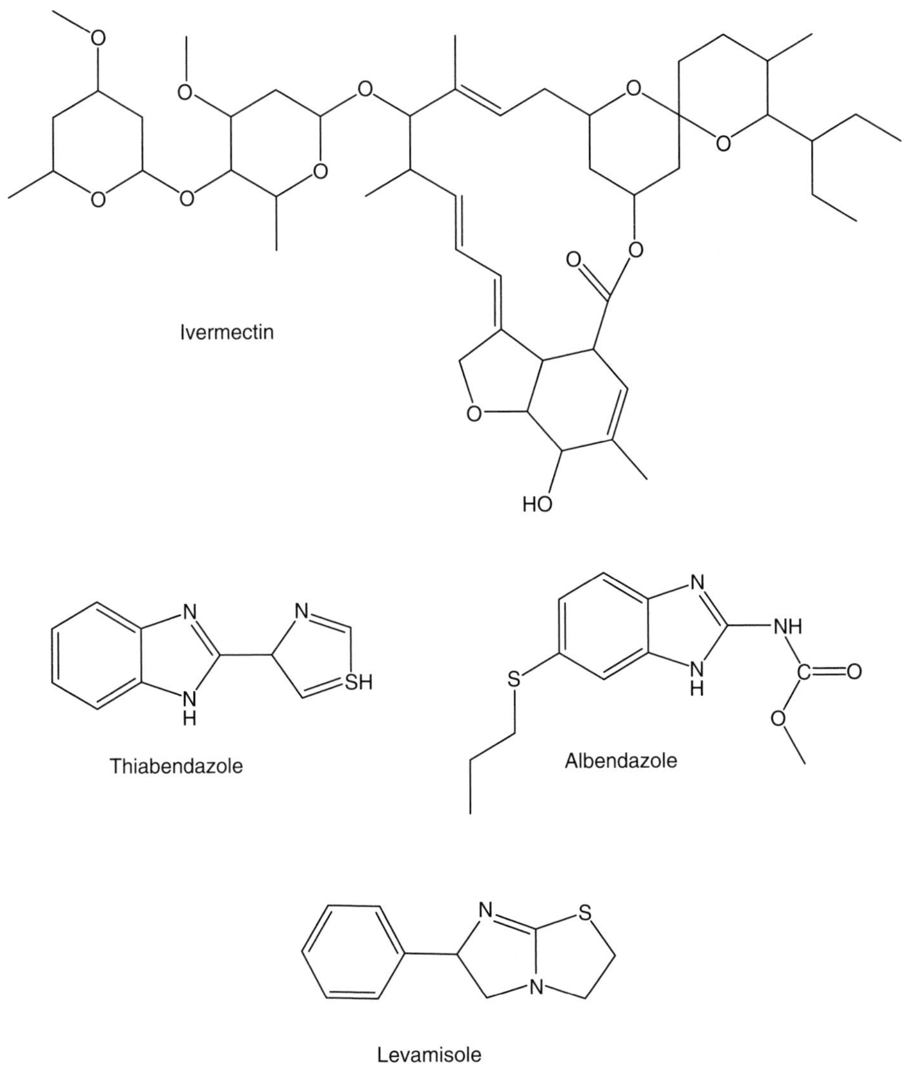

Ivermectin

Thiabendazole

Albendazole

Levamisole

Figure 7.27 Anthelmintics: the avermectin Ivermectin; the two benzimidazoles, thiabendazole and albendazole; and levamisole.

Table 7.14 Anthelmintics residues in liver analysed as part of the UK's surveillance scheme in 2000. (Data from the Veterinary Medicines Directorate (VMD) Annual Report on Surveillance for Veterinary Residues in 2000, VMD, Weybridge, UK, p. 92.)

| Animal | Percentage of samples with anthelmintic residues (concentration/range (mg/kg)) | | |
	Benzimidazoles	Avermectins	Levamisole
Cattle	0	0	0
Pigs	0	0	ND
Sheep	0.7 ($n = 576$) (0.64–2.9)	0.3 ($n = 282$) (0.84)	0

ND, no data.

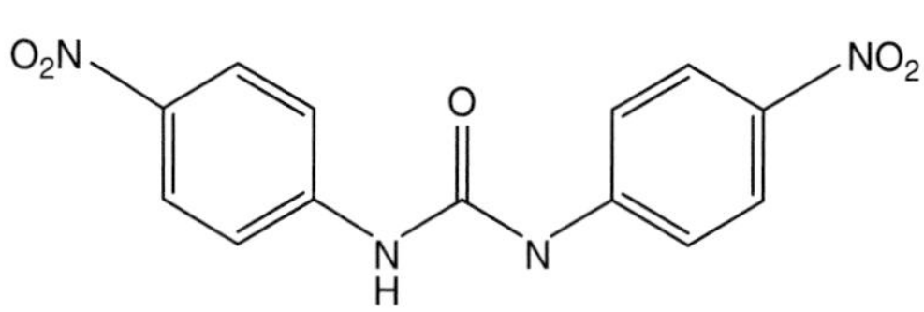

N,N′-di(p-nitrophenyl)urea
DNC

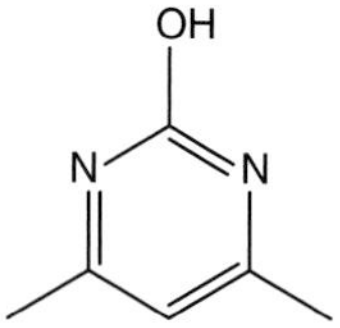

1-Hydroxy-3,5-dimethylpyrimidine
HPD

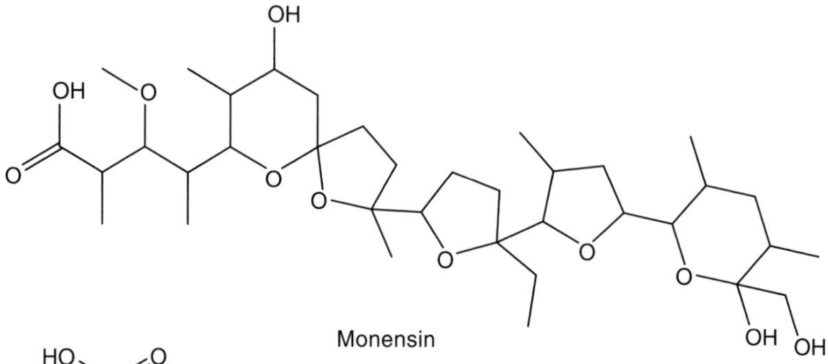

Monensin

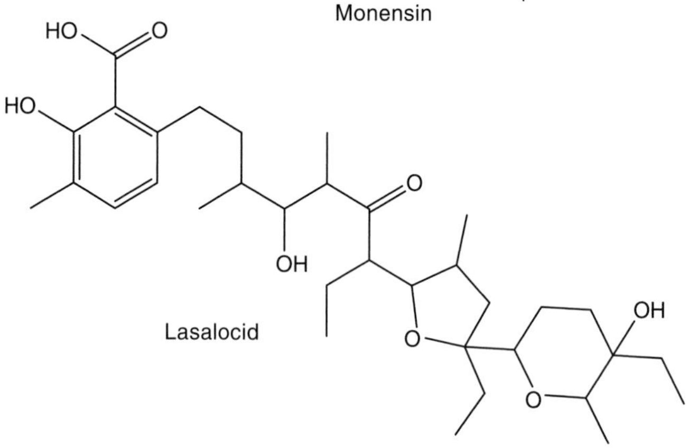

Lasalocid

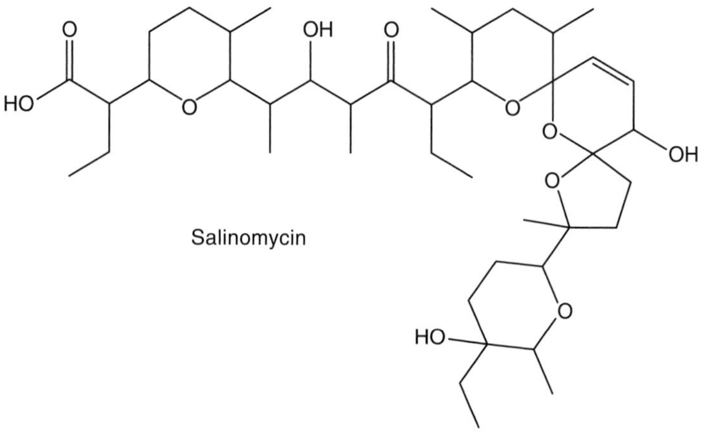

Salinomycin

Figure 7.28 Some coccidiostats showing their diverse and often very complex molecular structures. Nicarbazin is a 50:50 mixture of DNC and HPD.

plethora of protozoal infections because they are usually housed in close confines as part of intensive rearing (e.g. battery) processes and therefore poultry meat is the most likely source of antiprotozoal drug residues.

A particularly important group of protozoal parasites that infect farm animals – particularly poultry – are the Coccidia (*Eimeria* sp.); they cause coccodiosis and are prevented by prophylactic treatment with coccidiostats. I will use the coccidiostats as an example of antiprotozoal drug residues here.

The coccidiostats are a diverse group of drugs that interfere with the development of Coccidia at different stages in their complex lifecycles. Examples of coccidiostats are nicarbazin, salinomycin, monensin and lasalocid (Figure 7.28), but there are many more. They are very commonly used in poultry farming.

In the UK's 2000 surveillance scheme, coccidiostats were analysed in meat samples, particularly poultry (Table 7.15). Considering their widespread use there were very few positive samples, which further illustrates the point that veterinary residues are a minor, if not a negligible, risk to the consumer.

Sedatives and β-blockers (e.g. Azaperone)

Animals get stressed just like humans do. Stressed people often lose weight and look out of condition; animals are the same. For this reason farmers try to reduce their animals' stress levels in order to maximise their weight gain and their condition – this all leads to a better price for the meat at the market. Some animals are more prone to stress than others. Pigs are stressed by being crowded together in intensive units and during their journey to the abattoir. The stress on the way to slaughter is particularly important because it might cause changes to the pigs' muscles which leads to their meat being tough.

Sedatives and β-blockers can be used to reduce stress in animals (just like in people – in fact they are the same drugs that are used to treat humans), but

Table 7.15 Residues of coccidiostats in liver in the UK in 2000. This shows clearly that chickens are the main recipient of coccidiostats, but still the residues frequency and levels are low. (Data from the Veterinary Medicines Directorate (VMD) Annual Report on Surveillance for Veterinary Residues in 2000, VMD, Weybridge, UK, pp. 87, 88 and 92.)

Animal	Coccidiostat (number of samples analysed)	Percentage of samples with residues	Residue concentration/ range (µg/kg)
Cattle	Multiple residues method (43)	0	
Sheep	Salinomycin* (389)	0.3	41
Chicken	Nicarbazin (205)	14	220–2,700
	Lasalocid (218)	0.5	12
Turkey	Nicarbazin (55)	0	
	Lasalocid (50)	0	
Duck	Nicarbazin (11)	0	
	Lasalocid (12)	0	

*These samples were analysed by a multi-residues method but only salinomycin was found.

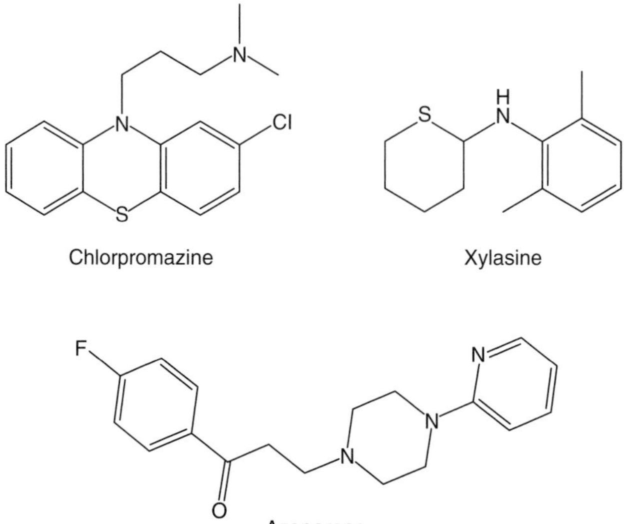

Chlorpromazine

Xylasine

Azaperone

Figure 7.29 Tranquillisers sometimes used illegally by farmers to calm their animals.

it is illegal in most parts of the world to use tranquillisers and β-blockers to reduce stress in farm animals. Nevertheless, some farmers use these methods illegally. Surveillance for tranquillisers and β-blockers in meat is used both to assess consumer exposure and trace farmers who are acting illegally, with a view to prosecuting them.

Examples of tranquillisers used to calm farm animals are chlorpromazine, xylasine and azaperone (Figure 7.29) and the β-blockers most commonly illegally used are clenbuterol, salbutamol and cimaterol (Figure 7.30).

Regulatory authorities not only include approved veterinary medicines in their surveillance schemes; they also look for substances that they suspect might be used illegally. The UK is a good example of this approach; in 2000 they analysed for tranquillisers, β-agonists and clenbuterol specifically (see *Growth promoting chemicals*) in hundreds of tissue samples and samples of processed foods (e.g. pork liver pâté) derived from animals that farmers might illicitly tranquillise. They found no residues in any of the samples. This suggests that illicit tranquilliser use is not widespread and that residues are not an issue for consumers (see also *Clenbuterol*).

Pesticides (e.g. Diazinon)
Pesticides are also used as veterinary medicines to treat and prevent insect and lice infestations in farm animals. In some countries (e.g. the UK) sheep are dipped in a bath of pesticides (e.g. Diazinon) to kill insect and lice parasites. The pesticide might be absorbed through the animal's skin or perhaps the animal takes a gulp of the dip when it is being forced to swim through it. Whichever way the animal receives its pesticide dose (i.e. dermally or orally) tissue residues might result. Since some of the pesticides used in crop protection are the same as those used in animal treatment (e.g. Diazinon) it is not possible to work out by which route the animal acquired its residues (Table 7.16), but from the consumer risk perspective it does not matter; it is the human dose of the pesticide not its source that is important.

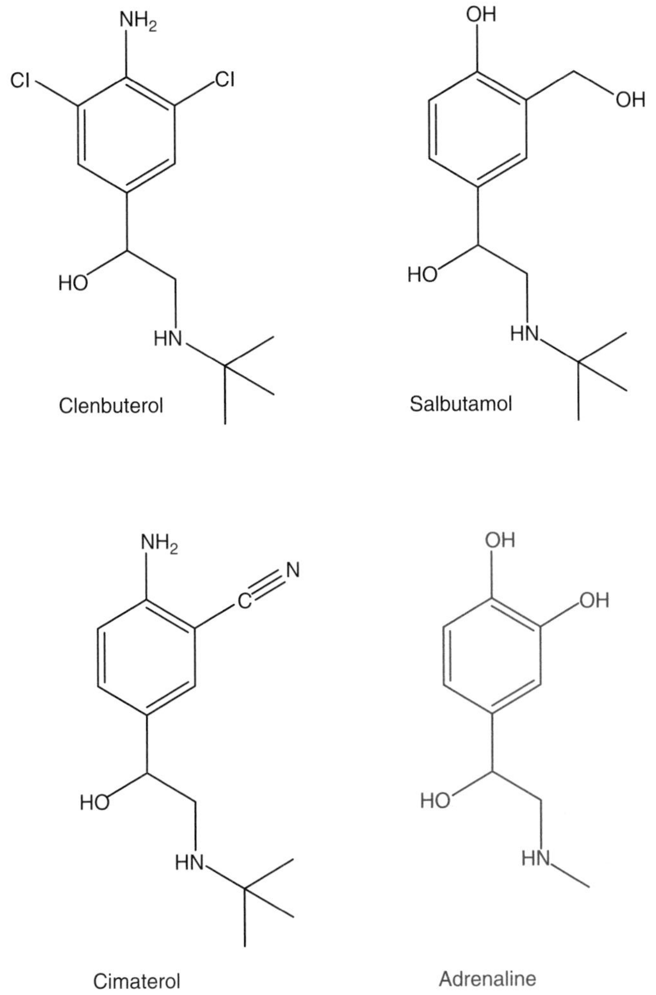

Figure 7.30 The β-agonists clenbuterol, salbutamol and cimaterol showing their structural analogy to the natural β-adrenergic receptor ligand adrenaline. The β-agonists occupy the receptor and stimulate it in the same way that adrenaline does. Occupancy of the receptor increases heart rate and causes bronchodilation, but at the same time relaxes smooth muscle.

Another important source of pesticide residue in animal products is via the animal's food chain. Not only are crops grown to feed people, but also huge amounts are also grown to feed farm animals. These crops are grown using the same pesticides as those grown for human consumption and so they too might have pesticide residues, and when they are eaten by the farm animals they too get a dose of pesticides that might form residues in their tissues.

Pesticide residues cocktails
Risk assessments are usually carried out only for single compounds, partly because it is difficult to assess risks of mixtures without carrying out animal toxicity studies on the same mixture. Clearly, since there are many possible

Table 7.16 Residues of pesticides in lamb imported into the UK from New Zealand* in 1996 (n = 70). It is impossible to know whether the residues came from environmental contamination, insecticide use on crops used to feed the animals or their use as veterinary medicines. Diazinon is used as both a veterinary medicine and a crop insecticide. (Data from Working Party on Pesticide Residues Annual Report 1996, Ministry of Agriculture, Fisheries & Food, London, p. 91.)

Pesticide	Percentage with residues	Residue concentration/ range (mg/kg)	MRL (mg/kg)
DDT	41	0.01–0.9	1
Chlordane	1	0.02	0.05
Dieldrin	1	0.01	0.2
Diazinon	1	0.05	0.7

*69 samples were from New Zealand; one sample was of unknown origin.

Table 7.17 OP residues in oranges imported into the UK in 1995 showing the large range of OPs used in orange production. (Data from Working Party on Pesticide Residues Annual Report 1995, Ministry of Agriculture, Fisheries & Food, London, pp. 60–61.)

OP	Residue concentration/ range (mg/kg)	Percentage with residues	MRL (mg/kg)
Chlopyriphos	0.05–0.3	7	0.3
Dimethoate	0.06	1	2
Ethion	0.08–0.2	3	2
Fenitrothion	0.05–0.2	5.5	2
Fenthion	0.08–0.1	3	2
Malathion	0.07–0.2	4	2
Mecarbam	0.1	1	2
Methidathion	0.06–0.9	24	2
Parathion	0.05	1	No MRL set
Parathion-methyl	0.1–0.8	4	0.2
Phosmet	0.08–0.1	3	5
Pirimiphos-methyl	0.07	1	0.5
Triazophos	0.1	3	No MRL set

mixture combinations of pesticides residues in food it would be impossible for studies to be carried out to cover all possible combinations. Despite the difficulties of assessing the risk of mixed residues, or residues cocktails as they are often called, mixtures are the norm and might have different toxicological effects than individual residues – their effects will be at least additive.

Let's explore a typical situation. Orange growers use mixtures of OPs (Table 7.17) to protect their crops from insect attack. The toxicity of each of

Table 7.18 The cocktail of pesticides found on two oranges imported into the UK in 1995. (Data from Working Party on Pesticide Residues Annual Report 1995, Ministry of Agriculture, Fisheries & Food, London, p. 65.)

	Residue concentration (mg/kg)	
Pesticide	Orange 1	Orange 2
Dicloran	0.3	0.3
Endosulphan	0.1	
Fenitrothion	0.08	
Imazalil	0.5	1.6
Tetradifon	0.08	0.06
Thiabendazole	0.05	2.4
Aldicarb		0.1
Chlorpyriphos		0.09
2,4-D		0.4
Malathion		0.1
Phenylphenol		0.2

the residues is used to assess the risk to the consumer and providing an individual residue concentration does not exceed a statutory limit (ADI or MRL) the fruit is deemed safe and tradeable. However, all of the OPs work by the same mechanism (see *Mechanism of action and toxicity of OPs and carbamate insecticides*) and therefore we *should* add up their molecular equivalent (i.e. number of moles of each pesticide in the food we eat) concentrations to assess their risk to the consumer. The cynics amongst us might even suspect that farmers could use this 'loophole' in residues legislation to effectively apply potentially toxic levels of OPs by simply changing the OP they use regularly to ensure that no individual OP residue exceeds its MRL or ADI!

Not only OPs are used in orange production and therefore the residues situation is far more complex. No one knows what the toxic effects of complex (Table 7.18) combinations of OPs and other pesticides might be.

Growth promoting chemicals

Some chemicals (e.g. steroid hormones) are able to influence the physiology of an animal in such a way that the animal converts its food into muscle more efficiently and therefore provides the farmer with a better financial return – human body builders sometimes illicitly use the same approach to enhance their muscles (they take steroids, e.g. testosterone).

Other chemicals are able to influence which tissues the animal forms (these chemicals are called repartitioning agents, e.g. clenbuterol) and can be used to increase the muscle to fat ratio and increase the value of the meat at market. Farmers farm to make money and therefore it seems obvious that they would want to use these chemicals to maximise and enhance their product. The problem is that many of these chemicals are not permitted in some countries. For example, the UK banned the use of hormone growth promoting chemicals in farming in 1981 whereas the USA still permits some of them to be used. The toxicological reason for such a ban is that residues might have effects on human consumers at very low doses – they are hormones and hormones work at nanogram (10^{-9} g) concentrations. But there is also a trade rationale; if a country bans a hormone growth promoter it means that meat from a country that allows the particular growth promoter cannot be imported; this protects the trade of the country that banned the hormone – there's more to residues legislation than consumer safety!

The use of growth promoting chemicals in farming is very controversial, partly because consumers are concerned about the health effects of growth promoting residues in their food and partly because of the animal health and the humanitarian considerations relating to the acceleration of farm animals growth. Some growth promoted chickens put on weight so quickly that their leg bones break because they do not develop at the same rate as the rest of their body and cannot support the bird's weight. Scientists still argue about the risks and benefits of growth promotion – the controversy is by no means over.

Growth promoting drugs

There are two important classes of growth promoters: antibiotics and hormones. The mechanism by which antibiotics cause growth promotion is not fully understood, but they appear to affect gene regulation which results in, for example, some protein synthetic systems being up-regulated. In addition, continuous antibiotic administration might prevent the farm animals contracting low-grade bacterial infections which means that their bodies can divert their efforts to protein production and growth rather than making antibodies to fight infections. The hormone growth promoters are usually steroid hormones, steroid analogues or molecules that fool the animals' bodies into thinking they are steroid hormones because they fit steroid receptors (see Chapter 9 for more discussion about how hormones work). For example, the androgens, their analogues or mimics lead to up-regulation of muscle production.

The hormone growth promoting drugs have evolved from simple hormone analogues (Figure 7.31) – the anabolic (*anabolic* means causing build-up of the body's tissues) drugs – to a more 'inventive' use of chemicals that act via different mechanisms. Many of the compounds are not obviously anabolic if you look at their molecular structure and consider their 'normal' use (e.g. clenbuterol, a β-agonist designed to treat respiratory disorders). Part of the reason for this 'evolution' is that the suppliers of illicit growth promoting drugs want to be one step ahead of the regulators who police the use of such illegal compounds. In parts of the world where some growth promoting drugs are allowed (e.g. the USA) a much more carefully controlled development of compounds by the agrochemicals industry should mean that residues and consumer risk are minimised.

One of the first (introduced in 1947 in the USA) growth promoting drugs to be used was diethylstilbestrol (DES; Figure 7.31); it is an analogue of the

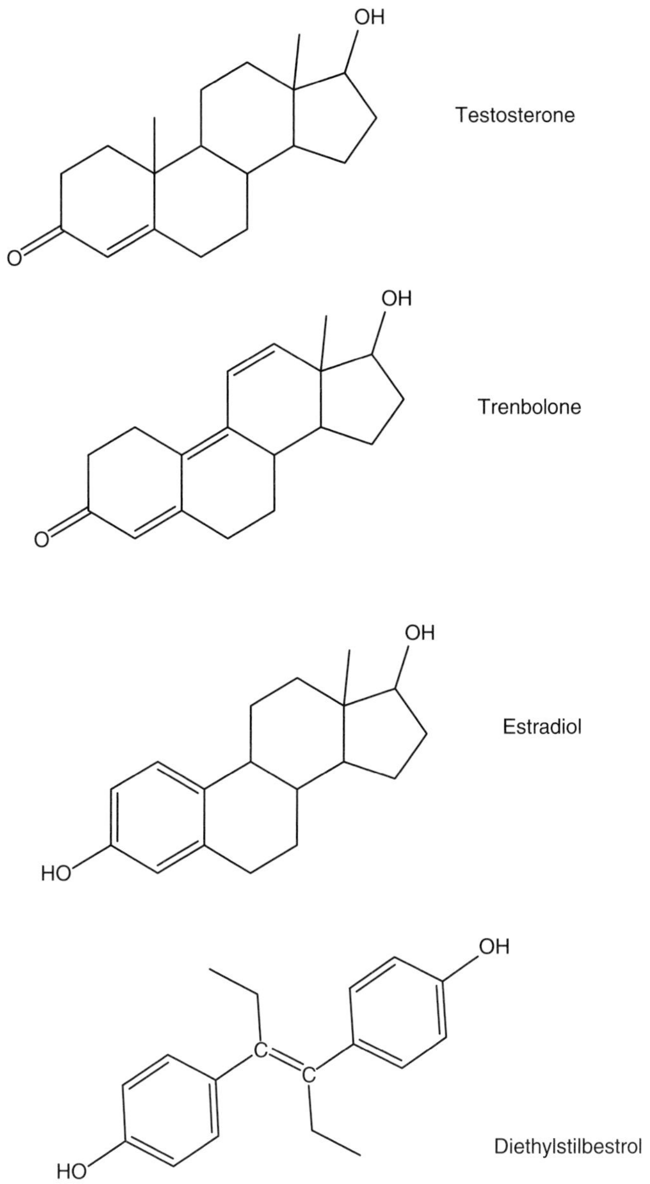

Figure 7.31 The growth promoting agents diethylstilbestrol and trenbolone showing their structural similarities to the naturally occurring female hormone (estradiol) and male hormone (testosterone) respectively.

female hormone 17β-estradiol (Figure 7.32) and fits into the estrogen receptor and activates it, causing cellular feminisation and the development of secondary female characteristics. In fact DES is more active than 17β-estradiol itself (see Chapter 9 for more discussion about the estrogen receptor). You might wonder why a farmer would want to give his animals a chemical that

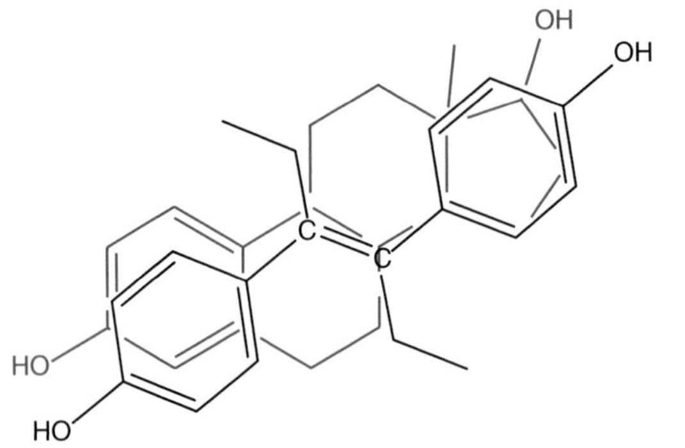

Figure 7.32 Diethylstilbestrol (black) superimposed on estradiol – this clearly shows their molecular similarities and why diethylstilbestrol fits the estrogen receptor and causes feminisation of male animals (e.g. caponisation of roosters).

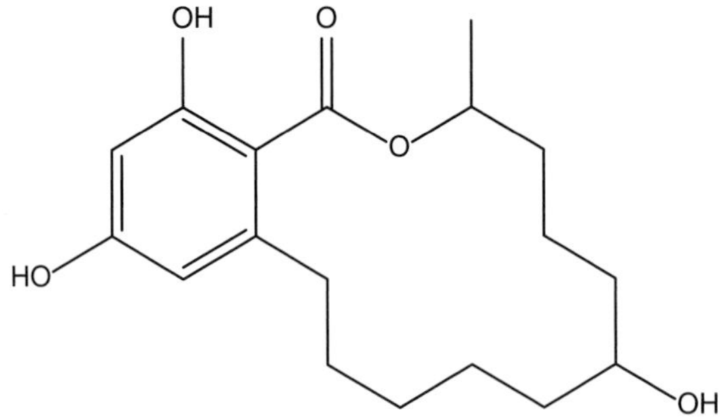

Figure 7.33 The anabolic steroid-mimic zeranol. It does not look like a steroid hormone, but it has the right chemical groups in the right spatial arrangement to fit a steroid receptor.

caused them to develop secondary female characteristics. The female hormone causes fattening – just what the farmer wants. This is particularly important in the poultry industry where it was used to cause feminisation of male hens (roosters; cockerels). The process of physically removing the rooster's testes is termed caponisation; the use of DES was termed chemical or hormonal caponisation. Hormonal caponisation was used for years in the poultry industry until it became clear that male consumers of caponised poultry were showing signs of female hormone effects, and that exposure to female hormones was associated with the development of some cancers (e.g. breast cancer). Studies showed that some consumers might be exposed to a sufficient dose to cause these effects. DES was phased out in the 1970s and is no longer used.

There are many more steroid hormone analogues used to fatten animals; some have molecular structures that are not quite so obviously steroid-like (e.g. zeranol; Figure 7.33), but if you look at them closely they have all of the attributes of a steroid in the correct spatial arrangement.

Clenbuterol

As mentioned above, there are other non-steroid drugs used in growth promotion. I will describe only one example here: the β-agonist clenbuterol (Figure 7.30). β-agonists have the interesting property of diverting metabolism away from lipid (fat) synthesis to muscle development; this is called repartitioning. In the 1990s some unscrupulous farmers, and equally unscrupulous suppliers, found that leaner beef could be produced if clenbuterol was administered to cattle a few weeks prior to their slaughter. Lean beef commands a high price at market and therefore there was a benefit for the farmer – well worth the risk of using an illicit drug… perhaps!

In Spain in 1992, 113 people noticed that their heart rate significantly increased after eating beef. In fact they became quite ill with temporary cardiac arrhythmia. After considerable detective work the regulatory analysts found residues of clenbuterol in samples of beef from Spain. Farmers were using clenbuterol as a repartitioning agent. Following this the European Union (EU) added clenbuterol to its meat residues surveillance programme. The results were quite alarming showing that the use of clenbuterol was widespread. Illicit users and suppliers were traced and prosecutions resulted, and slowly the use of clenbuterol waned. Now we hardly ever find clenbuterol residues in beef produced in developed countries.

In parts of the world where there is less (or no) control over the chemicals used in farming the situation with clenbuterol (and presumably other chemicals too) is very different. In 2006 about 3,000 people in Shanghai, China, were taken ill after eating beef. Their illness was caused by clenbuterol residues. A similar situation occurred in Guangdong, China, in 2009 when 70 people succumbed to clenbuterol meat residues. Clearly the problem has not gone away worldwide. As world food markets open, such incidents are very worrying because residue-containing meat from countries where controls are not in place might enter the markets of countries that have good residues control programmes. This illustrates very well why import monitoring is so important.

Residues of growth promoting agents

The EU carries out extensive surveillance for residues of growth promoting chemicals because their use is banned. In the UK in 2000, 231 broiler hens were analysed for trenbolone; no residues were detected – this is good evidence that the hormone ban has been effective. In the same year, cattle, sheep and pig meats were analysed for estradiol, testosterone, progesterone, nortestosterone, methyltestosterone, zeranol and trenbolone (Table 7.19). Of hundreds of samples analysed positives were found for testosterone in female cattle ($n = 2$), progesterone in male cattle ($n = 7$) and zeranol ($n = 4$). This is very different to the residues situation that would have been found pre-1981 when growth promoters were banned. Clearly the ban is working, but there are still a few itinerant farmers out there!

The situation in other countries is quite different. For example, studies on beef sold in Alexandria, Egypt (Table 7.20), showed that most samples analysed contained trenbolone and DES residues, some contained estradiol while none contained zeranol. These residues findings illustrate clearly different countries' approaches to growth promoter control.

Table 7.19 Growth promoter residues in farm animals in the UK in 2000 – since growth promoting hormones are illegal in the UK surveillance is aimed at finding farmers who are breaking the law; therefore the best sample to demonstrate illicit use is analysed (e.g. bile; many steroids are excreted in bile). (Data from the Veterinary Medicines Directorate (VMD) Annual Report on Surveillance for Veterinary Residues in 2000, VMD, Weybridge, UK, pp. 64 and 68.)

Growth promoter	Percentage of samples[1] with residues above action level			
	Cattle	Sheep	Pigs	Poultry[2]
Estradiol	0	NA	NA	NA
Testosterone	0.5[3]	NA	NA	NA
Progesterone	0.2[4]	NA	NA	NA
Nortestosterone	0	0	NA	NA
Methyltestosterone	0	0	0	NA
Trenbolone	0	0	0	0
Zeranol	2	0	0	0
Stilbenes (e.g. DES)	0	0	0	0

NA, Not analysed.
[1] Samples analysed were liver, bile, urine and/or faeces.
[2] Includes chickens, turkeys and ducks.
[3] Female cattle.
[4] Male cattle.

Table 7.20 Growth promoter residues in chicken from Alexandria, Egypt. (Data from Sadek et al. (1998) Eastern Mediterranean Health Journal, **4**, 239–243.)

Growth promoter	Percentage of samples with residues	Residues concentration range (μg/kg)
Trenbolone	92	0.013–1.21
Zeranol	0	
DES	100	0.006–0.05

Fertilisers

Fertilisers are used universally because of the huge benefit to crop yield and thus the farmers' financial income. Fertilisers can be either natural (e.g. animal manure) or synthetic (e.g. superphosphate – $Ca(H_2PO_4)_2 + CaHPO_4 + CaSO_4$). Their purpose is to enrich nutrient levels in the soil to promote crop growth, but they might also increase the concentrations of these nutrients in the crop too. This can be by either the plant absorbing the nutrient or the plant's surface being contaminated when the fertiliser was applied. Whichever the route the consumer might increase their intake of the particular nutrient. If the fertiliser contamination is on the surface of the crop, washing is likely to remove it, thus minimising the risk to the consumer. Most nutrients are not a problem because a little extra in our diet will do us no harm - some might even do us some good (e.g. calcium). Some nutrients (e.g. nitrate) are toxic at high doses - I will return to nitrate later. In addition to the desirable (from the farmer's perspective) components of fertilisers (e.g. nitrate) there might also be toxic contaminants (e.g. cadmium - Cd) which might also end up in our food (see *Natural environmental chemicals*).

I will use an example - nitrate - to illustrate the implications of fertiliser use on consumer health.

Nitrate – NO_3^-

Nitrate is essential for plant growth and is an important component of many fertilisers, both natural and synthetic. In addition, it is a natural component of our diet. However, high doses have been linked to specific cancers in animal studies and therefore there is concern about high intakes in humans.

Nitrate is linked to cancer because it is reduced to nitrite in the anaerobic environment of the gut, then the nitrite reacts with dietary secondary amines to form highly carcinogenic nitrosamines. Nitrosamines (Figure 7.34) are absorbed from the intestine and can cause cancers in the kidney and/or liver. The question is, what dose of nitrate does a human need to cause cancer? The answer is, we don't know! But what we do know is that in animal studies the higher the nitrate dose the greater the risk of certain cancers.

Knowing that nitrate is associated with cancer we must consider the implications of higher than natural nitrate levels in our crops and food.

The use of nitrate fertiliser has other environmental effects which should also be taken into account when making decisions about its use. It ends up in waterways and lakes and causes excessive algal growth. When the algae have used up the nitrate they die and provide an excellent source of nutrition for bacteria; the bacteria 'eating' the algae consume oxygen and result in low oxygen levels in the lake which kills the fish - this is the process of eutrophication and is a very good reason to control nitrate fertiliser use. Also nitrate from fertilisers percolates through the soil and bedrocks into aquifers from which we take our drinking water. Therefore, nitrate fertilisers can

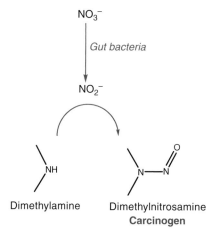

Figure 7.34 The formation of a carcinogenic nitrosamine from dietary nitrate and a secondary amine in the gut.

Table 7.21 Nitrate concentrations in vegetables compared with the nitrate level in bacon – bacon is pork cured in salt and sodium nitrate. It is impossible to know whether the nitrate in vegetables originates from the use of nitrate-containing fertilisers. (Data from Ministry of Agriculture, Fisheries & Food Surveillance Paper No. 3; Nitrite, nitrate and N-nitroso compounds in food: second report. Stationery Office, London, 1992; except * from Kotsonis et al. (1996) In: Klaassen CD (ed.) *Cassaret & Doull's Toxicology, the Basic Science of Poisons*, 5th edn. McGraw-Hill, New York, p. 937.)

Vegetable	Mean [nitrate] (mg/kg)
Spinach	1,631
Beetroot	1,211
Lettuce	1,051
Cabbage	338
Potato	155
Bacon	160*
Swede	118
Carrot	97
Cauliflower	86
Brussels' sprouts	59
Onion	49
Tomato	17

increase our nitrate intake via drinking water as well as food. It is important to remember that the possible effects of nitrate on consumers might not only be food borne. And nitrates in food do not only originate from fertiliser use – nitrates are also used to preserve meats (e.g. bacon; see Chapter 11) (Table 7.21).

The problem, of course, is that we can't distinguish between natural nitrate in food and nitrate from fertilisers; and indeed what does it matter? It is nitrate *per se* that is associated with cancer not only nitrate from fertilisers. Despite this, many countries monitor nitrate in food as part of schemes to assess human exposure and the possible health effects.

Natural environmental chemicals

There are many natural environmental chemicals that end up in our food. Indeed food is made of environmental raw materials. However, some of the natural environmental chemicals that occur as residues in food are toxic and so are analysed as part of surveillance schemes to assess the risk to consumers. In extreme situations governments might issue warnings about consumption of certain foods if regulators have evidence that intakes of a particular natural environmental chemical might cause harm. It is not possible (usually) to control the environmental levels, and therefore food levels, of natural environmental chemicals and therefore the only way to reduce the risk to the consumer is by limiting consumption of foods containing the chemical.

Cadmium in New Zealand shellfish

An excellent example of a government advising its population to limit its intake of a natural food contaminant is New Zealand and Cd. Cd occurs at very low concentration in the earth's crust (0.1 mg/kg) and therefore it usually occurs in food at very low concentrations if at all – average Cd intake in food = 35 µg/day. It is not an essential element and therefore plays no part in metabolism and is not needed for normal growth and development – in short, we don't need Cd in our diet. Cd salts are carcinogens and therefore human exposure is undesirable – so not only do we not *need* Cd in our food, we don't *want* Cd in our food.

As mentioned above, Cd occurs at very low concentration in the earth's crust. However, in some places it occurs at much higher concentration. Cd is a valuable element and was used as a yellow pigment in the paint industry and is still used (although it use is rapidly declining) in Cd batteries. For these reasons there was money to be made mining Cd ores. Cd ore mining releases Cd salts into the environment and often Cd concentrations in the environs of the mines are high, and sometimes the Cd is transported by rivers and streams to distant places. Similarly other metal (e.g. Zinc – Zn) ores are associated with Cd and when these ores are mined Cd is a waste product and is present at high concentrations in the mine tailings. This is another source of environmental Cd. Also volcanoes and geothermal activity release Cd from deep in the earth to the environment in which our crops are grown; therefore volcanic (either past or present) countries tend to have higher environmental Cd levels than non-volcanic countries. Finally, Cd salts are sometimes present as impurities in fertilisers; therefore Cd might be applied inadvertently to the land in which crops are grown as a contaminant of fertilisers.

When Cd occurs in soil (either naturally or as a fertiliser contaminant) it might be taken up by crops. It also gets washed out of the soil by rain and finds its way to rivers, lakes and the sea. Cd in marine systems tends to be associated with silt which might be filtered out and accumulated by mollusc filter feeders (e.g. mussels).

New Zealand is a volcanic country that has poor soils that need fertilisers. For both reasons it has high Cd levels in its river and sea silts. In addition, New Zealand has a well-developed mussel farming industry and sought-after natural oysters (e.g. Bluff oysters) that attract high prices on the international market. Mussels and oysters are both filter feeders. They filter out suspended solids in their marine environments and digest the food component (e.g. algae) of the silt and either dispel or accumulate the other silt components. Since New Zealand coastal silts have high levels of Cd (relative to other countries) the New Zealand oysters and mussels accumulate Cd (Figure 7.35). In fact they accumulate so much Cd that in 2004 the New Zealand Food Safety Authority advised New Zealanders to limit their consumption of mussels and/ or oysters in order to keep Cd intake at an acceptable level. This is an excellent example of a government department controlling intake of a toxic food residue (that cannot be controlled by legislation because it is natural) by limiting consumption of the food(s) rich in the residue.

Cadmium in offals

Cd is excreted in both bile and urine and therefore it is concentrated in the liver and kidney en route to its excretion from the animal's (and human) body.

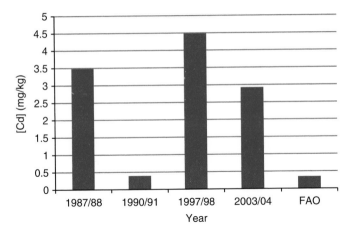

Figure 7.35 Cadmium residues in New Zealand oysters compared with the Food and Agriculture Organisation (FAO) international average for oysters. (Data from 2003/04 New Zealand Total Diet Survey, New Zealand Food Safety Authority, Wellington (2005), p. 42.)

Table 7.22 Cd residues in animal products from two high environmental Cd countries showing that horse which is slaughtered older has higher Cd residues than sheep which is slaughtered younger, and that both countries' liver residues are greater than the FAO average for sheep liver. (Data from [1]Balzan et al. (2002) http://www.lnl.inf.it/~annrep/readAN/2002/contrib_2002/B011_B117T.pdf; [2]2003/04 New Zealand Total Diet Survey, New Zealand Food Safety Authority, Wellington (2005), p. 42; [3]Assessment of Chief Contaminants in Food, Joint FAO/UNEP/WHO Food Contamination and Monitoring Programme, WHO, Geneva (1988).)

Country/animal	Cd residue concentration/range (mg/kg)		
	Liver	Kidney	Muscle
Italy/Horse[1]	38–92	47–1,192	73
New Zealand/Sheep[2]	0.1	NA	0.0017–0.003
FAO[3]	0.03	NA	NA

NA, Data not available.

Therefore, countries with high environmental Cd levels tend to produce offals (e.g. liver and kidney) with high Cd residues because their animals have high dietary Cd intakes. Cd accumulates in the animals' (and human) organs over time; therefore the older the animal is at slaughter the higher the Cd residues. In countries where horse is eaten the horses are not slaughtered until they are several years (at least) old which means that Cd residues in offals are likely to be relatively high (Table 7.22). Italy not only consumes horse offals, but is volcanic – this is the worst combination possible vis à vis Cd residues intake because environmental Cd concentrations are high and Cd is concentrated in offals with time.

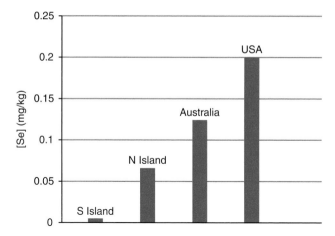

Figure 7.36 Selenium residues in bread from the North and South Islands of New Zealand, Australia and the USA. (Data from 2003/04 New Zealand Total Diet Survey, New Zealand Food Safety Authority, Wellington (2005), p. 56.)

Other natural environmental contaminants in food

There are many other chemicals from the environment that are either absorbed by animals and/or plants or deposited on the surface of plants and end up in our food. Some are toxic at high levels, but essential at low or trace levels (e.g. selenium - Se) and therefore regulators have to make certain that consumers' intakes are at the right level rather than trying to achieve zero intake. For example, Se is thought to be important in protecting cells against some carcinogens and therefore maintaining an appropriate intake could reduce cancer incidence. Se levels are naturally low in the environment of South Island and higher in the North Island of New Zealand (Figure 7.36); this combined with Cd intake being generally high (remember, Cd is a carcinogen) in New Zealand means it is important that Se intake is sufficient to reduce the cancer risk. A major source of Se is wheat flour (e.g. in bread) and therefore wheat grown in North Island is blended with wheat grown in South Island to ensure sufficient Se is present in bread made from the blended grains – this is far less controversial than adding Se to South Island flour. Consumers often do not like regulators adding things to their food because they see this as an erosion of personal choice.

There are other elements like Cd that are toxic and ideally should not be present in our food, for example, mercury (Hg) and arsenic (As); both are found in fish and occur in organic (e.g. methylmercury – $Hg(CH_3)_2$)) and/or inorganic (e.g. Hg^{2+} in $HgCl_2$) forms which have differential toxicity – organic Hg is more toxic than inorganic Hg, whereas organic As is less toxic than inorganic As. This is just the tip of the iceberg regarding natural contaminants, but I hope these few examples have illustrated the issues involved.

Non-agricultural environmental pollutants

There are many organic pollutants that contaminate our food; they are the products of our modern industrial society. They are discarded in effluents from our factories or in our household waste; they go out with our sewerage or via our dustbins, the contents of which end up in landfill where they slowly degrade to release their components into the environment. The myriad chemicals that pollute our environment via these routes might end up in our food. Some of them are very toxic, some are innocuous; most of them we don't even know about!

This subject alone needs a book devoted to it and so I can only scratch the surface of a very complex and controversial issue here. To understand the magnitude of the problem just think about everything you use in a day and what happens to it, then think of the industrial processes that were necessary to produce each of the products you have used and the waste that each of those processes produced ... it all ends up in the environment and we grow our food in the same environment.

I will consider just one product that I used today – my computer (I am typing this book on it as I think about where it came from and where it is going to!). It is made of plastic with electronic bits and pieces inside. The plastics are hydrocarbon polymers, some are chlorinated, some have other elements (e.g. oxygen) in their molecular structures; the electronic wizardry inside is based on silicon with metallic conductors made of copper and other highly conducting metals plus some heavy metals as impurities. The battery that safeguards my data backup is based on lithium (Li). The manufacturing processes for all of these components are complex and undoubtedly polluting. It will take far too long to go through the fates of each component of my computer, so I will concentrate on the main one – plastic. What happens to the millions of tonnes of plastic we dispose of each year? Some goes to landfill, some is recycled, some is incinerated – incinerating chlorinated plastics produces a vast array of complex chlorinated cyclic hydrocarbons depending upon the temperature of incineration. Degradation of chlorinated plastics in landfill sites produces some similar molecules and a great many others (e.g. trichloromethane (chloroform); $CHCl_3$), but much more slowly. A group of low temperature incineration products of chlorinated plastics are the dioxins.

Dioxins are also waste products of some manufacturing processes (e.g. making the pesticide 2,4,5-T). Dioxins are very long lived environmental pollutants that are very fat soluble and therefore they concentrate in animals. So why are dioxins a problem? They occur as residues in food (particularly fatty food) and they are potent carcinogens.

I will use the dioxins to illustrate non-agricultural environmental pollutants, but don't forget there are many more examples.

Dioxins

The dioxins are a family of polychlorinated tricyclic molecules. The individual dioxins differ according to which carbon atoms in their aromatic rings are chlorinated (Figure 7.37). There are many combinations possible which give a large number of different dioxins. They are all toxic – some more than others. They usually occur in the environment as complex mixtures of the different chlorinated molecules (called congeners).

Dioxin residues in food

As mentioned above, dioxins are very fat soluble; for this reason they are found at higher concentration in fatty foods (e.g. butter). We cannot control residues in food because the dioxins are present in the environment whether

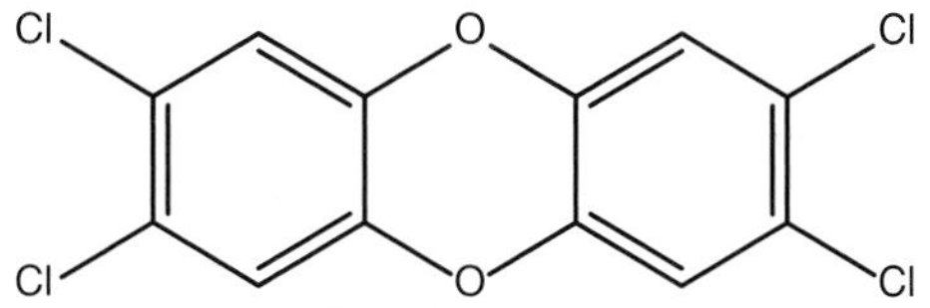

2,3,7,8-Tetrachlorodibenzodioxin

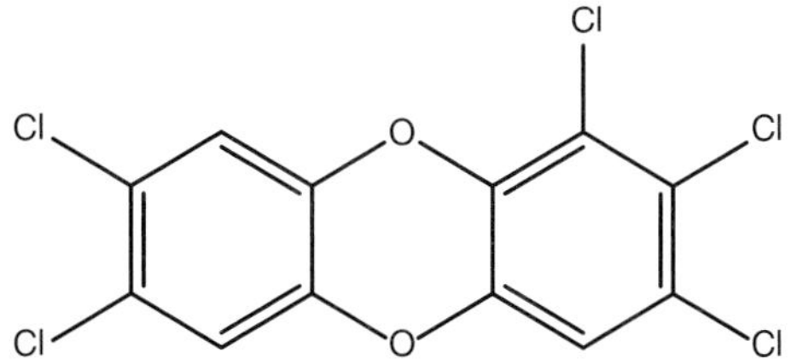

1,2,3,7,8-Pentachlorodibenzodioxin

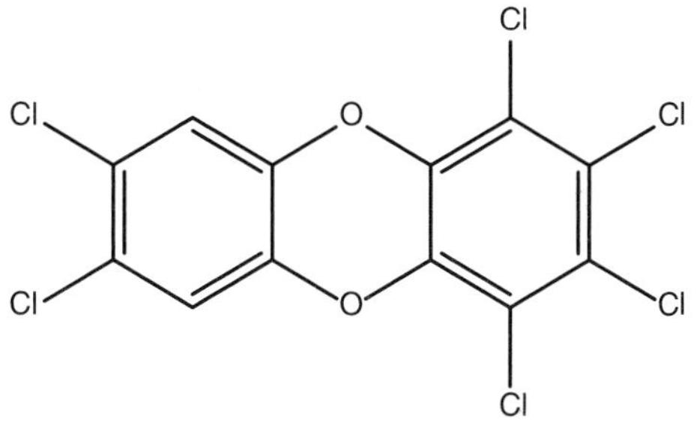

1,2,3,4,7,8-Hexachlorodibenzodioxin

Figure 7.37 Three dibenzodioxin congeners.

we like it or not. We can, of course, reduce residues in the future by minimising dioxin pollution, but this will be a very long, slow process because the dioxins have very, very long environmental $t_{½}$s. We can minimise our exposure to dioxins in food by limiting consumption of foods with significant dioxin residues.

Dioxin residues in food are a significant concern because dioxins are carcinogenic and teratogenic; for this reason the concentrations analysed are very much lower than for most other chemical food residues (i.e. picograms per gram; pg = 10^{-12} g). Since their presence in food has no benefit to the consumer and they are not derived from a process that is beneficial, in theory we cannot accept any risk associated with their residues. This theoretical argument cannot be enforced because we have no control over dioxin residues since they are environmental contaminants with long $t_{½}$s; therefore, regulatory authorities monitor them in order to assess the risk, and develop approaches to minimise consumer exposure (e.g. by removing highly contaminated food from the market – see *Dioxin food contamination scares*) and there are national Maximum Limits for dioxins in most countries or jurisdictions (Table 7.23).

To set this in perspective, 1 pg of 2,3,7,6-tetrachlorodibenzodioxin (molecular weight = 192 Da; 1 pg = 5.2×10^{-15} moles; Avogadro's Number = 6.02×10^{23}) is just 3×10^{7} molecules. Therefore, in Irish cheese (Table 7.24) there are only

Table 7.23 Maximum Limits for dioxins in food in the EU. (Data from 2002 amendment to Commission Regulation EC/466/2001.)

Food	Maximum Limits for dioxins (pg/g fat)
Beef/lamb	3
Pork	1
Poultry	2
Eggs	3
Milk	3

Table 7.24 Dioxin residues in food in Ireland showing the very low concentrations present. (Data from O'Keefe et al. (2001) Food Residues Data Base. National Food Centre, Dublin, Ireland, p. 12.)

[Dioxin] (pg/g fat)	Percentage of food samples with dioxin residues in concentration range*				
	Cheese	Beef	Pork	Poultry	Sheep meat
0.1–0.5	82	74	94	88	23
0.5–1.0	16	19	6	6	50
1.0–3.0	2	7	0	6	27

*Approximate values read from a graph in the Irish Food Residues Data Base 2001 (see above for source).

about 3×10^6 molecules per gram in the majority of samples analysed (based on 0.1 pg/g) - this is a very small number when counting molecules.

Dioxin food contamination scares

From time to time routine dioxin monitoring schemes pick up higher than expected levels. This often leads to interesting investigations into the source of the contamination. One such incident occurred in Germany in December 2010. Levels above the EU MLs were found in eggs and pork. A major investigation found that a German animal feed company purchased fatty acids for two purposes: incorporation into their animal feeds and for other technical purposes not associated with human or animal feed. The two uses have different acceptable levels of dioxins - the technical use allows higher levels than the food use. The 'technical' fatty acids containing high levels of dioxins were used in animal feed manufacture by mistake. This was a huge mistake and led to a very significant embarrassment for the German government because farm animals and their products (e.g. eggs) fed dioxin-contaminated feed accrued dioxin residues far in excess of MLs (Table 7.25).

Initially the German government insisted that the problem was confined to Germany and that it had been contained, but slowly a much more extensive story unfolded. Indeed, over 900 farms in Germany were affected and their

Table 7.25 The highest dioxin levels found in German produce following the 2010 animal feed contamination incident. (Data from Federal Ministry of Food, Agriculture and Consumer Products, Germany, http://www.bmelv.de/cln_173/SharedDocs/Standardartikel/EN/Food/DioxinSummary.)

Product	Highest dioxin residue (pg/g fat)	EU Maximum Limit	Percentage of Maximum Limit exceedance
Eggs	12	3	300
Chicken	5	2	150
Pork	1.5	1	50

produce was not allowed to be sold or processed. High levels of dioxins were found in eggs, chicken and pork (Table 7.25). Fortunately it appeared that no contaminated meat or eggs were exported.

Residues monitoring programmes

There are four types of residues monitoring programmes: National Surveillance Schemes, Total Diet Surveys, Import Monitoring Programmes and Export Monitoring Programmes.

National Surveillance Schemes (NSSs) involve collecting predetermined numbers of samples of specific foods, then analysing them for predetermined pesticide residues. For example, the UK runs a very extensive NSS in which every year the dietary staples (bread, milk and potatoes) are sampled and other fruit, vegetables and animal products are sampled on a rolling programme with the more commonly consumed foods (e.g. carrots) being sampled more often than the less commonly consumed (or consumed in smaller amounts) foods (e.g. figs). This, in my opinion, is the best way to produce a reliable picture of exposure of consumers to residues and allows trace back to the producer of foods exceeding statutory limits. The problem is that NSSs are very expensive.

NSSs are used in conjunction with commodity consumption figures (from National Nutrition Surveys, i.e. how much of different foods an average person consumes) to make dietary intake assessments. It is important to note that NSSs usually sample the entire product (e.g. carrot) and analyse an aliquot of a total homogenate – this often does not represent what is actually consumed (e.g. onions are usually peeled before eating, but the whole onion would be homogenised and analysed for an NSS).

Total Diet Surveys (TDSs) involve collecting exactly what a preselected number of consumers eat. The consumers are asked to prepare two meals instead of one and to submit the duplicate meal for analysis. This gives regulators an idea of residues consumed, but does not usually allow trace back to the producer. New Zealand runs a very good TDS on a 4-year cycle. TDSs are generally cheaper than NSSs (although the cost, of course, depends on the number of samples analysed) while still giving the regulators a measure of consumers' exposure to residues and allows intake risk assessments to be made. TDSs analyse food that has been prepared for eating and therefore they give a much better indication of what the consumer is actually exposed to than NSSs do.

Import Monitoring Programmes, as their name suggests, involve collecting samples of imported foods and subjecting them to residues analysis. They are used for two purposes: to make sure that exporters are complying with international residues regulations (e.g. MRL compliance) and to allow intake risk assessments to be carried out (although most NSSs and TDSs also include imported food in their schemes).

Export Monitoring Programmes again are self explanatory from their name – they are used to ensure that exported food complies with international residues regulations and that the exporting country is not embarrassed and does not risk losing markets for non-compliance. Clearly, these programmes are very important to countries that rely on exports to maintain their balance of payments.

Dietary intake and risk to human consumers

This is a huge issue, and I will only give an overview of the processes used here.

Residues monitoring programmes allow regulators and toxicologists to calculate average dietary intakes of specific chemicals (e.g. a pesticide) by multiplying the residue level in a particular food by the national consumption figure for that food – national consumption figures are produced by nutritional surveys in which people are asked what they eat; they are measured in grams of a particular food consumed per day. The intake figure allows toxicologists to determine the risk to the consumer by comparing intakes to ADIs.

There are many assumptions made in food residues risk assessments; nevertheless they give an indication of the population level impact and allow regulators to make changes if the risk is considered too great – an action might be to stop farmers using a particular pesticide that repeatedly gives cause for concern or to temporarily remove a food with high residues from the market.

In the UK Lindane residues in milk approached the ADI in September 1995 (see *A case of γ-HCH residues in milk in the UK*) and even though to exceed the ADI a consumer would have to drink milk containing above the ADI concentrations of Lindane every day for their entire life the UK regulators decided to withdraw the affected milk from the market while they investigated further. Good regulators err on the side of safety in such situations.

Another case in the UK illustrates changing the approval for a pesticide to reduce unacceptable residue levels. Phorate, an OP used on carrots, led to unacceptably high intakes in the summer of 1995 (see *OP and carbamate residues in food*; Table 7.8). The reason for the high phorate residues in carrots was, in part, due to pest pressure (e.g. carrot root fly), but also phorate is applied as granules and some of these fell into the dip at the top of the carrot where the leaves meet the stalk (Figure 7.17) so leading to very high residues levels. The residues problem led to the regulations for the use of phorate being changed to reduce the risk to the consumer.

When carrying out risk assessments it is important to remember that the data from different monitoring schemes are very different. NSSs usually give total residues on the food crop (e.g. onions) whereas TDSs give residues in what we actually eat (e.g. a peeled onion, or onions incorporated into a recipe). These differences must be considered when interpreting the results. A good example of this is the risk of OP residues on oranges. The OP residues are mainly on the skin of the oranges, but we don't usually eat the skin and so in this case high residue levels in NSSs are less important than high residues in a TDS.

Take home messages

- Chemical food contaminants include pesticides (e.g. DDT), veterinary medicines (e.g. antibiotics), natural environmental chemicals (e.g. cadmium) and environmental pollutants (e.g. dioxins).
- Most countries carry out food residues monitoring schemes to assess consumer exposure and risk.
- Residues exceeding MRLs and MLs mean that a food cannot be traded.
- Residues exceeding ADIs mean that the risk of consuming the food is unacceptable.
- Residues intake estimates are calculated from residues levels in food and food consumption figures (from National Nutrition Surveys).
- Residue cocktail effects are probably important, but little is known about them toxicologically.

Further reading

Baldi I, Gruber A, Rondeau V, *et al.* (2010) Neurobehavioural effects of long-term exposure to pesticides: results from the 4-year follow-up of the PHYTONER study. *Occupational and Environmental Medicine*, doi: 10.1136/oem.2009.047811.

Bomhard EM, Brendler-Schwaab SJ, Freyberger A, *et al.* (2002) O-phenylphenol and its sodium and potassium salts: a toxicological assessment. *Critical Reviews of Toxicology*, **32**, 551–625.

Botsoglou NA & Fletouris DJ (2001) *Drug Residues in Foods*. Marcel Dekkker, New York.

Carson R (1965) *Silent Spring*. Penguin Books, London.

Committee on Toxicity of Chemicals in Food, Consumer Products and the Environment (1999) *Organophosphates*. Department of Health, London, http://www.doh.gov.uk/cot.htm.

NZFSA (2003/04) *New Zealand Total Diet Survey*. NZFSA, Wellington, http://www.foodsafety.govt.nz/elibrary/industry/2003-04-nztds-full%20report.pdf.

Watson D (ed.) (2001) *Food Chemical Safety; Vol. 1 Contaminants*. Woodhead Publishing, Cambridge. This is a collection of detailed snapshots of food contaminants, including pesticides (Shaw *et al.*), inorganic contaminants (Harrison), chemicals from food packaging (Castle), environmental contaminants (Harrison), veterinary medicines (Dixon) and risk analysis (Tennant).

Chapter 8
Natural Toxins

Introduction

We use many chemicals in growing (e.g. pesticides) and processing (e.g. preservatives) our food; these chemicals are usually covered by legislation that controls their use to maintain the risk to the consumer at an appropriately low level. Plants and animals have evolved to survive in hostile environments and they do this, in part, by producing defence chemicals to protect themselves against others in the environment that see them as food. Many of these chemicals - natural toxins - are toxic to human consumers too and therefore if we consume too much of them they might make us ill. Some plants and animals produce natural toxins that are so toxic that we cannot use them as food; others produce toxins that usually do not exceed human toxicity thresholds, but if the environmental conditions are right (or wrong!) the levels of natural toxins might get to levels that affect human consumers.

In the human food context, natural plant toxins are far more important than animal toxins, but there are a few notable exceptions (e.g. tetrodotoxin in puffer fish used as a Japanese delicacy) which can cause considerable toxicity problems, even death, to human consumers.

Some natural toxins are produced only in response to adverse environmental conditions; others are present in the plant or animal all of the time. Therefore, to some extent, we can control some of the natural toxins by ensuring that our crops are grown, stored or transported in conditions that are not conducive to natural toxin production. But we have no control over most of them. Indeed, we don't even understand why plants (particularly) produce many of their toxins; if indeed there is a reason.

It is not possible for me to cover all of the natural toxins we might be exposed to in our food in this book. In this chapter I will give selected examples of some of the more important, or interesting, natural toxins and explain where they come from, how we are exposed to them, the effects of cooking on the toxins and how exposure can be minimised.

Natural toxins are often monitored by National Surveillance Schemes to keep an eye on consumers' intake and the changing trends in exposure.

Food Safety: The Science of Keeping Food Safe, First Edition. Ian C. Shaw.
© 2013 John Wiley & Sons, Ltd. Published 2013 by John Wiley & Sons, Ltd.

Why produce natural toxins?

Some natural toxins might simply be accidents of the biochemistry of plant or animal cells; others might confer a survival benefit on their producer – they might deter predators by being poisonous or tasting bad.

A good example of a natural toxin that confers a survival advantage is in the insect world. The Monarch butterfly (*Danaus plexippus*; Figure 8.1 and Plate 8.1) is a beautiful orange migratory butterfly with a yellow and black striped caterpillar – they visit my garden. Their bright colours warn would-be predators of the risk they take if they eat a Monarch. Interestingly, some non-poisonous butterflies mimic the Monarch to frighten predators off … the 'sheep in wolfs clothing' approach to not getting eaten! There is a chemical underlying the Monarch's warning colours, but the Monarch caterpillar does not produce the toxin itself, it absorbs it from its food. So, the Monarch puts a toxin from its food to good use.

Monarch caterpillars eat milkweeds (e.g. swan plant – *Asclepias fruticosa*; Figure 8.1) which contain a highly toxic chemical – labriformidin (LD_{50} [i.p. mouse]=3.1 mg/kg body weight) (Figure 8.2). The Monarch caterpillar absorbs

Figure 8.1 A Monarch (*Danaus plexippus*) caterpillar (above) and adult butterfly (below) resting on a swan plant (*Asclepias fruticosa*) in my garden in New Zealand; its bright colours warn of the toxin (labriformidin) within. (Photograph by the author.) (To see a colour version of this figure, see Plate 8.1.)

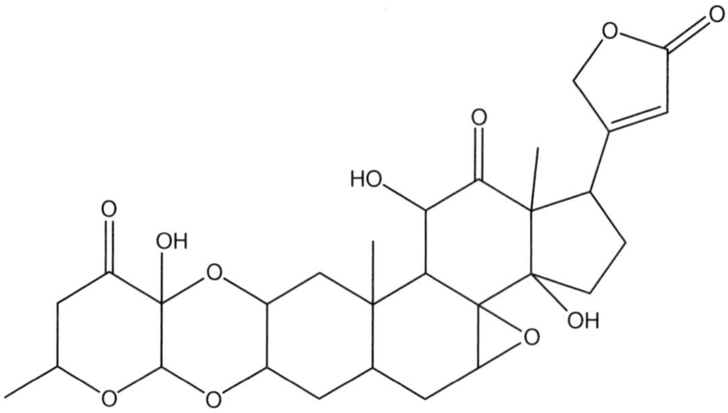

Figure 8.2 Labriformidin – the highly toxic chemical from milkweeds that Monarch butterflies and their caterpillars use to ward off would-be predators.

the labriformidin and becomes poisonous itself. When the caterpillar meta-morphoses into the adult butterfly the toxin residues remain in the butterfly's tissues which is crucial to its survival because the adult only consumes nectar and therefore does not get labriformidin in its diet. From this example it is easy to see the survival advantage of natural toxins and why animals and plants that produce or use them might have evolved.

Labriformidin and the Monarch butterfly illustrate how an animal uses a toxin to survive; plants operate in a very similar way. Some produce toxins to deter animals from eating them (e.g. natural insecticides); others produce natural fungicides or bactericides. If we eat these plants we get a dose of the natural insecticides, fungicides and/or bactericides too.

Natural toxins in the human food chain

Animal toxins

Very few animals we eat contain their *own* toxins; however, there are a sur-prising number of animals that absorb or harbour toxins from animals or plants that they eat. If, in turn, we eat these animals we too will be exposed to the toxins via the food chain – I'll discuss this in more detail later in this chapter.

Tetrodotoxin from fugu

The best example of an animal toxin directly consumed by humans is tetrodo-toxin (Figure 8.3) in fugu – a delicacy used in Japanese sashimi. 'Fugu' is Japanese for puffer fish; several species of puffer fish are eaten as fugu in Japan (see also Chapter 2, *Toxic fugu sashimi – tasty, but potentially lethal*) – every year several people in Japan die of tetrodotoxin poisoning.

Tetrodotoxin is an extreme example of a natural animal toxin that affects people because it is eaten by a very small proportion of the world's population. Other animal toxins are more likely to be consumed; for example, saxitoxin from dinoflagellates – you are probably thinking, but I don't eat dinoflagel-lates! Well let's see …

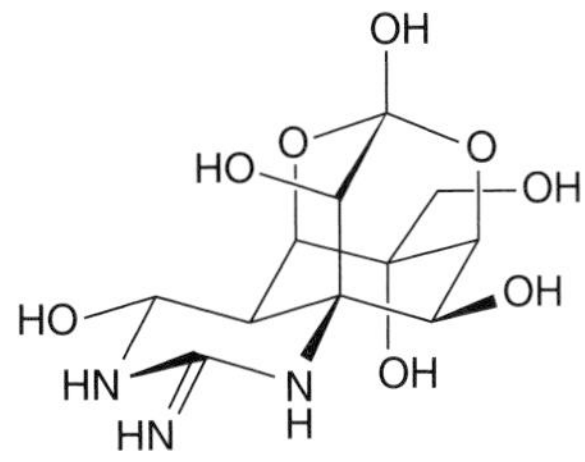

Figure 8.3 Tetrodotoxin from fugu. LD_{50} [mouse] = 334 µg/kg body weight which means that only 25 mg could kill a human.

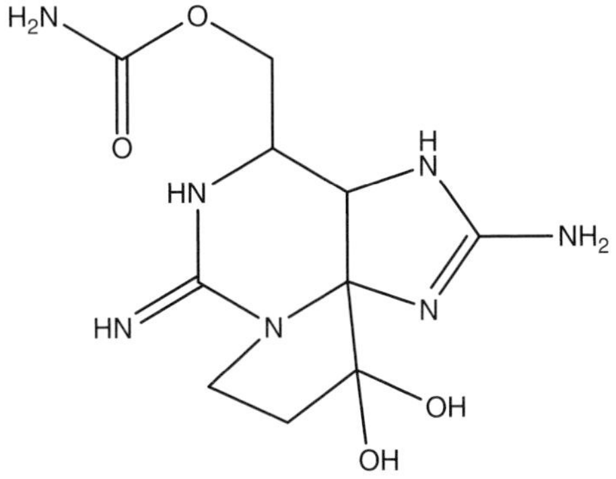

Figure 8.4 Saxitoxin – the cause of paralytic shellfish poisoning. LD_{50} [mouse] = 3 µg/kg body weight; 180 µg could kill a human.

Saxitoxin – paralytic shellfish poisoning

Dinoflagellates are marine protozoa that are consumed by filter-feeding bivalves (e.g. mussels) and so if a human eats a mussel that has eaten dinoflagellates the human will eat the dinoflagellates too. And if the dinoflagellates were synthesising saxitoxin when they were eaten by the mussel the unsuspecting human consumer will get a dose of saxitoxin with his mussels.

Saxitoxin (Figure 8.4) was first isolated from the Butter Clam (*Saxidomus giganteas*), hence its name. It is very toxic (LD_{50} [mouse]=3 µg/kg body weight), and has a mechanism of toxicity that involves blocking Na^+ channels in neurone membranes, so preventing neurotransmission (see Chapter 3, *How botulinum toxins inhibit neurotransmission* for more details of Na^+ channels and neurotransmission) and causing paralysis and sometimes death by respiratory failure. The disease caused by eating saxitoxin-contaminated shellfish is called paralytic shellfish poisoning (PSP).

Most countries where shellfish are eaten and/or exported run environmental monitoring programmes to pick up marine dinoflagellate blooms. When a bloom occurs signs are posted in the vicinity warning people not to harvest shellfish. Some countries also monitor shellfish for saxitoxin *per se* to further minimise human consumer exposure.

Saxitoxin is not destroyed by heat and therefore even if the shellfish are cooked the toxin survives.

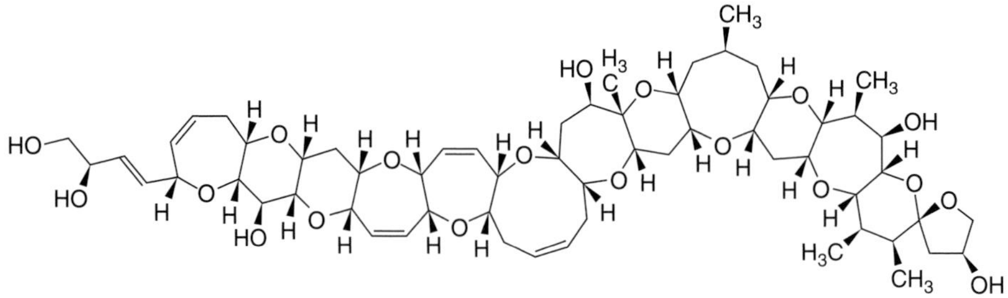

Figure 8.5 Ciguatera toxin from the dinoflagellate *Gambriodiscus toxicus* which is eaten by some tropical reef fish. LD_{50} [i.p*, mouse] = 0.45 µg/kg body weight; 30 µg could kill a human and 0.1 µg causes illness.
*Intraperitoneal – injected into the abdominal cavity; it is considered by toxicologists to be similar to oral dosing.

Ciguatera toxins in tropical reef fish

Ciguatera poisoning is quite common in places where reef fish are eaten; it is estimated that 3% of the populations of the French West Indies and US Virgin Islands in the Caribbean Sea suffer from ciguatera poisoning each year.

Ciguatera toxin (Figure 8.5) is not produced by reef fish, but is absorbed by the fish from a dinoflagellate (*Gambriodiscus toxicus*) that is part of their food chain. Ciguatera toxin is very toxic (LD_{50} [mouse]=0.45 µg/kg body weight). The symptoms of ciguatera poisoning are varied, including vomiting, diarrhoea, stomach cramps, 'pins and needles', and a peculiar (and very characteristic of this poison) reversal of the sensation of hot and cold in the mouth.

Histamine and scombroid fish

This is another example of poisoning by a toxic chemical not produced by the food species itself. In this case, bacteria growing on the surface of some scombroid fish (i.e. mackerel fish; e.g. tuna) can convert the amino acid histidine to histamine (Figure 8.6) – histamine is involved in the body's allergic response and therefore consumption of histamine-contaminated fish results in acute symptom of allergy, including burning and swelling of the mouth, and body rashes. This kind of food poisoning is called pseudoallergic fish poisoning and is treated with antihistamines.

Histamine is a natural biochemical found in most fish, but at concentrations (usually below 1 mg/kg) far below that which would result in exceedance of the toxicity threshold dose (20–50 mg/kg body weight) in humans. The histamine concentration in 'toxic' scombroid fish can reach 100 mg/kg.

Plant toxins

Plant toxins are by far the most commonly encountered natural toxins in our food. There are very many of them, and they have a broad array of molecular structures and mechanisms of toxicity. Many of the natural plant toxins are phytoelexins, i.e. they are produced by the plant in response to stress (e.g. fungal infection) and might protect the plant against the stressors (e.g. fungi) and therefore they are natural pesticides (e.g. fungicides).

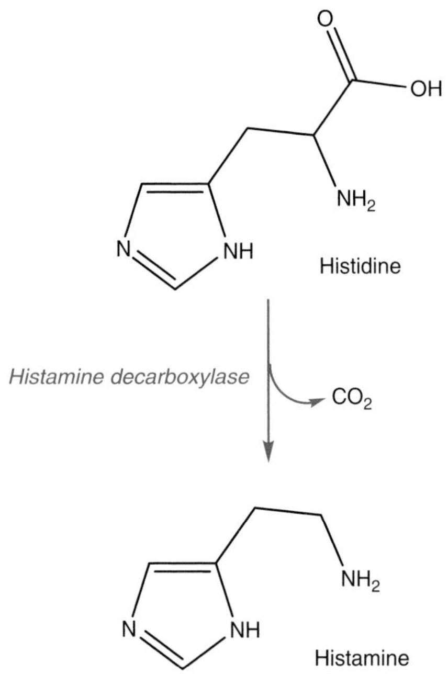

Figure 8.6 Decarboxylation of histidine to form histamine – the reaction that occurs in bacteria living on the surface of some scombroid fish.

I will discuss some examples of the more important food plant toxins below.

Cucurbitacins from the cucumber family

Cucurbitacins (Figure 8.7) is a family of complex toxins thought to be natural fungicides, and perhaps insecticides, produced by members of the cucumber family (Cucurbitaceae), hence their name. They appear to be produced in response to stress. So, for example, if a cucumber plant gets infected with a fungus the stress produced stimulates the plant to produce its own fungicides. This is a very effective way of the plant protecting itself.

There are 17 different cucurbitacins; they all have the same basic molecular structure differing only in the structure of the side chain and groups attached to the pentacyclic nucleus. They are all very toxic, but taste incredibly bitter and so it would be difficult to eat a cucumber with a toxic level of cucurbitacins because it would be unpalatable.

In New Zealand in the summer of 2001, a number of people living in the same geographical region of North Island became ill. Their symptoms were stomach cramps and nausea. They were ill enough to seek medical advice. After detailed investigation the only food they had consumed in common was courgette (zucchini – *Cucurbita pepo*), which is a member of the cucumber family. Many of the affected consumers commented that the courgettes tasted bitter (remember cucurbitacins taste bitter). The summer was warm and humid; just the right conditions for courgette plants to be infected with powdery mildew (a white fungus that grows on the leaves). The courgette plants responded to the fungal infection by synthesising their own fungicides – the cucurbitacins.

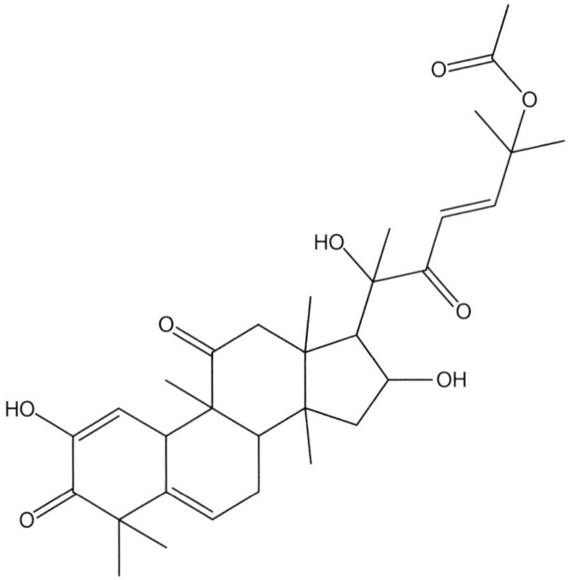

Figure 8.7 Cucurbitacin-B is responsible for the bitter taste of members of the cucumber family. LD_{50} [oral; mouse] = 5 mg/kg body weight; 300 mg could kill a human.

Interestingly, all of the affected people had eaten organic courgettes originating from the same farm. If the farmer had used fungicides to prevent his plants getting powdery mildew perhaps they would not have synthesised their own, very toxic, fungicidal cucurbitacins. This might be an example of the risk of an organic vegetable being greater than its conventional counterpart.

Glycoalkaloids from potatoes

The glycoalkaloids are another group of structurally complex, very similar alkaloids from members of the nightshade family (Solinaceae); the solanines (Figure 8.8) are the principal glycoalkaloids found in potatoes.

The nightshade family is very large and includes species as diverse as potatoes, tomatoes, peppers, aubergine (egg plant), tobacco and nightshades. They all produce glycoalkaloids in greatly varying concentrations (Table 8.1) in the green parts of the plants; we do not eat the green parts of most members of the Solanaceae that are used as food plants (e.g. potatoes and tomatoes). However, if potatoes are left in the light the tuber produces chlorophyll (the green photosynthetic pigment) and begins to photosynthesise and switch on other metabolic pathways; one such pathway is the synthesis of glycoalkaloids. Therefore green potatoes contain glycolalkaloids (Table 8.2), sometimes at high enough concentrations to be toxic. The solanines taste bitter (like most alkaloids) which is why when you eat green potatoes they have rather an unpleasant taste.

The glycoalkaloids are not as toxic (2.8 mg/kg body weight causes toxicity in humans) as many other natural plant toxins, but they can cause mild stomach upset and at high enough concentrations they affect heart rate.

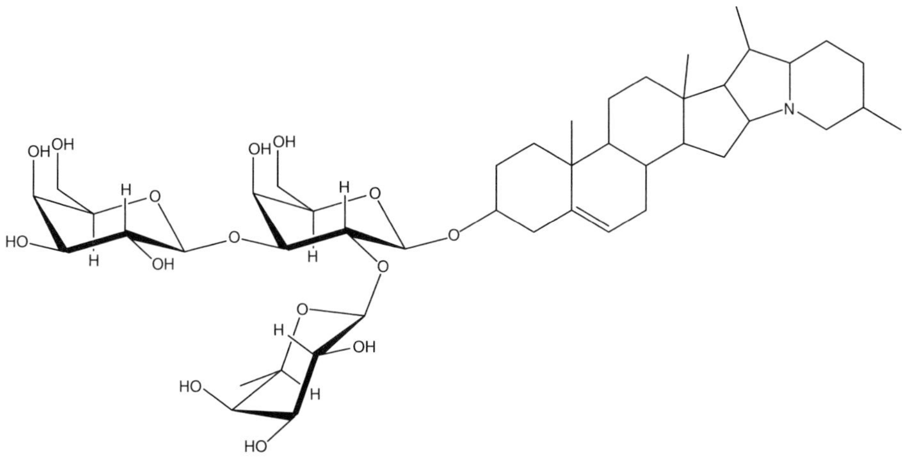

Figure 8.8 α-Solanine, a potato glycoalkaloid. A dose of 2.8 mg/kg body weight (i.e. about 200 mg per person) is known to be toxic to humans, but the lethal dose is very much greater (about 2.5 g).

Table 8.1 Glycoalkaloids in the edible parts of members of the Solanaceae. (Data from Inherent Natural Toxicants in Food, 51st Report of the Steering Group on Chemical Aspects of Food Surveillance (1996), MAFF, The Stationery Office, London, p. 10.)

Member of the Solinaceae	Glycoalkaloid concentration (mg/kg)
Aubergine (egg plant)	65–90[1]
Green pepper	51–117[1]
Red pepper	74–94[1]
Tomato (green)	870[2]
Potato (normal tuber)	75[3]

[1] α-Solanine.
[2] α-Tomatine.
[3] Total glycoalkaloids.

Table 8.2 Distribution of glycoalkaloids in the potato plant. (Data from Inherent Natural Toxicants in Food, 51st Report of the Steering Group on Chemical Aspects of Food Surveillance (1996), MAFF, The Stationery Office, London, p. 10.)

Part of potato plant	Total glycoalkaloid* concentration (mg/kg)
'Normal' tuber	12–20
Green tuber	250–280
Leaves	30–1,000
Sprouts ('eyes')	2,000–4,000
'Normal' skin	300–600
Green skin	1,500–2,200

*Sum of the major potato glycoalkaloids including solanine.

Table 8.3 Glycoalkaloids in different types of potato chips. (Data from Inherent Natural Toxicants in Food, 51st Report of the Steering Group on Chemical Aspects of Food Surveillance (1996), MAFF, The Stationery Office, London, p. 11, paragraph 21.)

Type of potato chips (crisps)	Total glycoalkaloid concentration range (mg/kg)
Peeled	40–150
Skin on ('jackets')	40–720

It would be almost impossible to get a lethal dose from potatoes, but it has happened (see below).

Potatoes are a staple carbohydrate source for a large part of the world and therefore they are consumed in large quantities (e.g. a 35- to 49-year-old male in the UK eats 156 g potatoes/day). For this reason some regulatory authorities issue guidelines for storing potatoes in supermarkets to stop them getting sufficient light to turn green; this minimises consumers' exposure to glycoalkaloid.

There have been cases of severe glycoalkaloid poisoning from potatoes and even fatalities (following excessive consumption of green shoots) in the UK. For example in 1979 a large number of children from a school in London developed gastrointestinal symptoms after consuming green potatoes in their school lunch; several of them needed hospitalisation. More recently (1993) eight people in the north-east of England developed gastrointestinal symptoms after eating potatoes supplied by a farm in Perthshire, Scotland – analysis of the potatoes revealed total glycoalkaloid concentrations of up to 1,750 mg/kg in the potatoes from the Perthshire farm (data from Inherent Natural Toxicants in Food (1996), MAFF, London, pp. 11–12). If someone (mean human weight = 70 kg) ate 156 g (i.e. mean adult potato consumption per day) of these potatoes they would receive a dose of 4.6 mg/kg body weight – which means that a dose <4.6 mg/kg body weight will result in toxicity in humans.

There have been concerns about the toxicity of potato chips (crisps) made from skin-on potatoes. Worst-case scenario calculations show it is possible that a child could consume a toxic dose of glycoalkloids following consumption of only two packets of potato chips with green skins (Table 8.3) – this is clearly unacceptable. For this reason the UK authorities advise against the manufacture of potato chips from skin-on potatoes.

Stability of the glycoalkaloids
Glycoalkaloids are very stable. Cooking, even high-temperature frying, has little or no effect on their concentration in food.

Furocoumarins in parsnips, parsley and celery

Furocoumarins (Figure 8.9) are present in most members of the Umbelliferae (from the Latin *umbella* – sunshade) – a huge family of plants that have umbel-shaped flowers, including carrots, parsnips, parsley (Figure 8.10), coriander (cilantro) and angelica. They are also found in other families and genera (e.g. Moraceae – including figs) (Table 8.4).

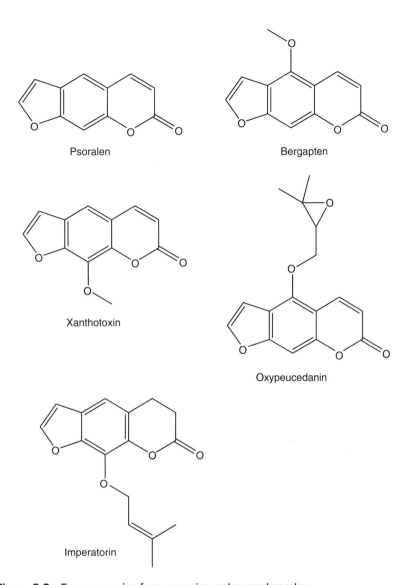

Figure 8.9 Furocoumarins from parsnips, celery and parsley.

The furocoumarins comprise a broad group of tricyclics based on furano-coumarin with varying ring substituents. They are phytoelexins (probably fungicidal) and are photoactivated carcinogens (i.e. they need to be exposed to light before they become carcinogenic) in animals. Clearly we want to min-imise our long-term exposure to furocoumarins via our diet because of their carcinogenicity.

Furocoumarin toxicity

The furocoumarins have extremely low acute toxicity (LD_{50} [rat, oral] >30,000 mg/kg body weight), but as mentioned above they are carcino-genic and therefore long-term exposure is the toxicological issue. In the

Figure 8.10 Parsley (*Apium graveolens*) in flower in my garden. This clearly shows the umbel-shaped flower which is the origin of the name of the family – Umbelliferae – to which parsley belongs. (Photograph taken by the author.)

Table 8.4 Furocoumarin concentrations in fruit and vegetables. (Data from Inherent Natural Toxicants in Food, 51st Report of the Steering Group on Chemical Aspects of Food Surveillance (1996), MAFF, The Stationery Office, London, pp. 24 and 27.)

Plant/part of plant	Furocoumarin	Concentration (mg/kg)
Celery		
Stalk	Psoralen	1.3–46.7
Root	Bergapten	25–100
Seed	Xanthotoxin	0.65
Parsnip		
Root	Bergapten	40–1,740
Carrot		
Root	Total furocoumarins	<0.1–0.9
Parsley		
Leaf (fresh)	Imperatorin	11–112
Leaf (dry)	Oxypeucedanin	300
Fig		
Leaves	Bergapten	480
Sap	Bergapten	620

context of furocoumarins in food, as with all carcinogens, long-term repeat doses are needed to initiate the cellular changes that might lead to cancer. I do not think that most people would eat furocoumarin-containing food (e.g. parsnips) regularly enough to significantly increase their cancer risk. However, some Chinese people like to drink parsley tea regularly, which might be more of a problem in this respect.

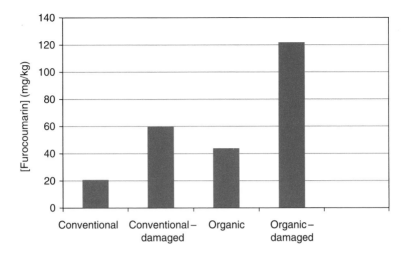

Figure 8.11 Total furocoumarin concentrations in parsnips, showing that organic parsnips have higher levels of furocoumarins. (Data from Inherent Natural Toxicants in Food, 51st Report of the Steering Group on Chemical Aspects of Food Surveillance (1996), MAFF, The Stationery Office, London, p. 28.)

Effects of plant damage on furocoumarin levels

Since the furocoumarins are phytoelexins, if the plant is damaged it responds by producing more furocoumarins to protect itself against the possibility of infection (e.g. by fungi). For this reason damaged furocoumarin-producing plants (e.g. parsnip) have higher concentrations of specific furocoumarins than their intact counterparts (Figure 8.11). This means that the carcinogenic risk is greater if you consume damaged (e.g. bruised) parsnips.

Interestingly, parsnips grown organically often have higher concentrations of the furocoumarin bergapten than those conventionally grown (Figure 8.11). This is likely to be because conventional farmers use pesticides to minimise insect attack (i.e. the stress that stimulates the plant to synthesise furocoumarin) and perhaps fungicides to prevent fungal infection (remember furocoumarins are thought to be fungicidal). So conventional farmers protect their parsnip crop whereas organic crops have to fend for themselves – they do this by synthesising furocoumarins. Continuing this thought process leads us to the conclusion that organic parsnips are more likely to have higher furocoumarin concentrations than conventionally grown parsnips and therefore organic parsnips present a greater cancer risk – this might seem counterintuitive, but it is not the first time we have come to the same conclusion (see *Cucurbitacins from the cucumber family*).

Stability of the furocoumarins

Cooking reduces the levels of furocoumarins in food (e.g. parsnips; Figure 8.12) partly because they degrade at high temperatures and partly because they are water soluble and leach out into the cooking water.

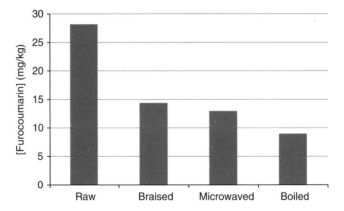

Figure 8.12 Furocoumarin concentrations in raw and cooked parsnips. (Data from Inherent Natural Toxicants in Food, 51st Report of the Steering Group on Chemical Aspects of Food Surveillance (1996), MAFF, The Stationery Office, London, p. 30.)

Phenylhydrazines in mushrooms

Agaritine (Figure 8.13) is a phenylhydrazine present in the mushroom species (*Agaricus bisporus* – hence *agaritine*) we usually buy from the supermarket. It is not particularly toxic *per se*, but it is metabolised in mammals (including humans) to a potent carcinogen.

Toxicity of agaritine

As mentioned above, it is not agaritine that is the toxic problem, but rather its carcinogenic metabolite HMBD (Figure 8.13). However, studies in mice have shown that at doses of 400 µg/kg body weight signs of toxicity, including loss of co-ordination and convulsions, occurred. The effects in humans are unknown, but we must bear in mind the dose that causes effects in mice when we consider the possible toxic effects of agaritine in our diet.

Agaritine levels in cultivated mushrooms are in the range 80–250 (mean 158) mg/kg; using national diet surveys to give an idea of mushroom consumption, it has been calculated that human agaritine intake is in the range 0.6–2.6 mg/person/day. The top of this range equates to a dose of 37 µg/day which is very much lower than the 400 µg/day dose that causes toxicity in mice. This rough calculation suggests that the acute human dose is likely not to be toxicologically significant. However, long-term low level exposure would be necessary to cause cancer. We do not know the human dose of HMBD necessary to cause cancer, but it would need to be regular (e.g. daily) and over a long time (e.g. years). Since mushrooms are unlikely to be consumed daily for long periods the cancer risk is considered to be very low indeed.

Capsaicin, peppers and other flavours

Peppers taste 'hot' because they contain capsaicin (Figure 8.14). Capsaicin binds to a cellular heat sensor protein resulting in a conformational change similar to that caused by high temperatures (Figure 8.15). The signal from the

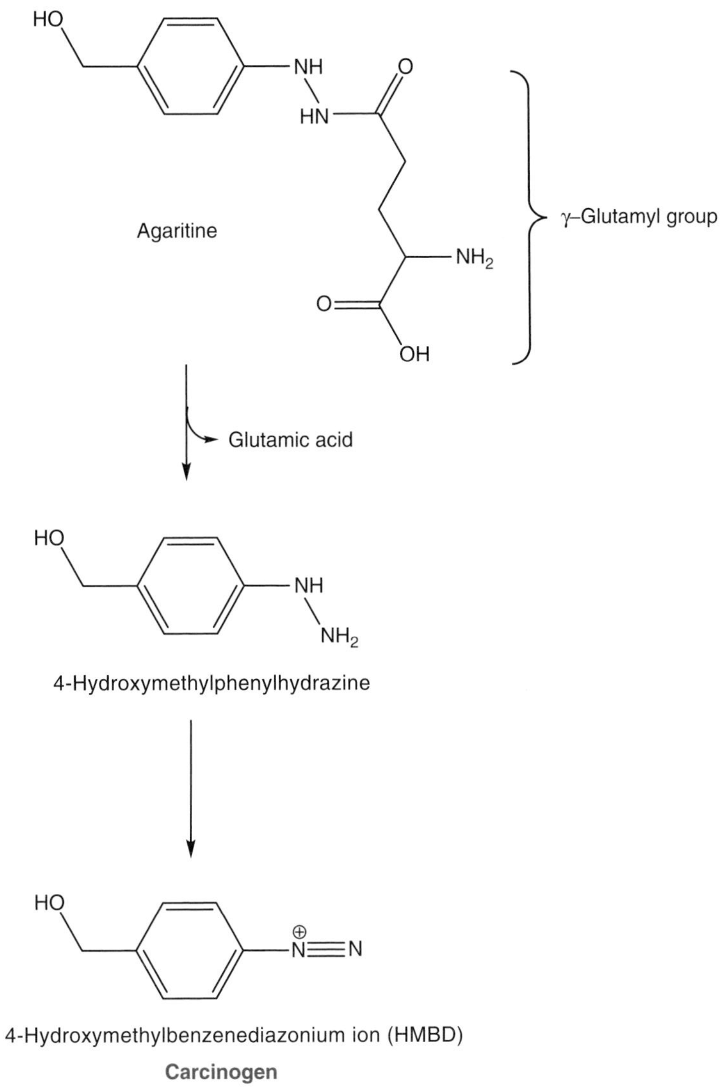

Figure 8.13 The mammalian metabolism of agaritine from mushrooms to form the carcinogen HMBD.

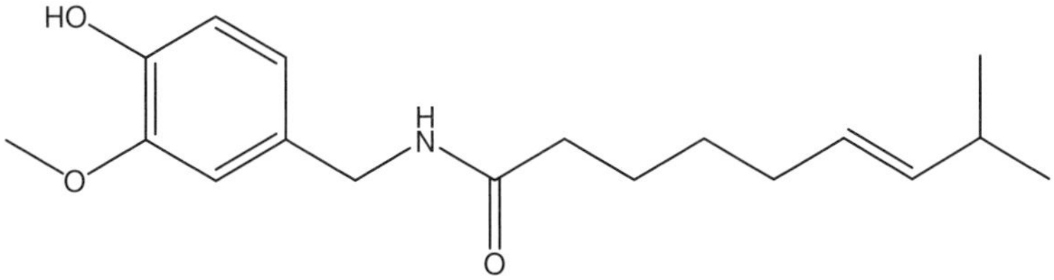

Figure 8.14 Capsaicin – the chemical that makes peppers taste 'hot'.

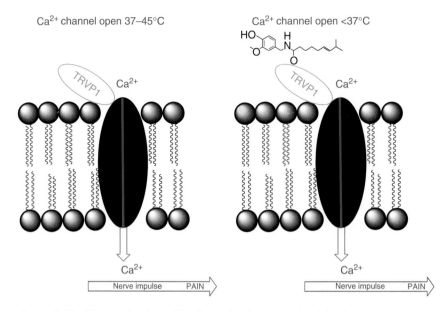

Ca²⁺ channel open 37–45°C Ca²⁺ channel open <37°C

Figure 8.15 The mechanism of heat sensing in mammals. A heat sensor protein – transient receptor potential cation channel V1 (TRVP1) – spans the nerve cell membrane; it changes conformation in response to heat (left) and opens a membrane Ca²⁺ channel which initiates a nerve impulse to 'tell' the brain that something is hot. Capsaicin binds to TRVP1 and causes a similar conformational change (right) which initiates Ca²⁺ influx and the signal 'HOT' is sent to the brain – this is why chilli peppers taste 'hot'.

heat sensor protein initiates neuronal membrane depolarisation and a nerve impulse which signals 'heat' to the brain. This is a highly developed defence response and is the mechanism behind you rapidly moving your finger away from an open flame. It is interesting that a food chemical causes the same response; this is why we describe chilli peppers as 'hot'. The greater the capsaicin concentration in the pepper, the 'hotter' the pepper tastes.

Capsaicin is not absorbed from the gastrointestinal tract and it is not metabolised or degraded during its passage through the gut – this is why the day after a hot curry there is often a burning sensation at the distal end of the alimentary tract! Since capsaicin is not absorbed it is not systemically toxic; however, our gut responds to its presence in the same way that it responds to other toxins - stomach cramps and diarrhoea.

Capsaicin is an example of the many chemicals (Figure 8.16) in our food that are part of the natural flavour chemistry. They are all toxic at high enough dose (remember Paracelsus - 'All things are poisons ...', see Chapter 2, *The factors that contribute to risk*), but most are not sufficiently toxic to cause concern.

Oxalic acid and rhubarb

Oxalic acid is a toxic (LD_{50} [rat, oral]=475 mg/kg body weight) metabolite produced at high concentrations in the leaves of rhubarb, and at much lower concentrations in the petioles (leaf stalks - the part we eat). It is also present in other vegetables, including spinach and lettuce (Figure 8.17), but the weight

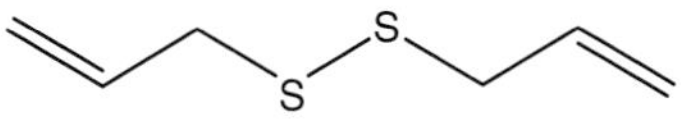

Allicin – garlic

LD_{50} [mouse, sc] = 120 mg/kg body weight

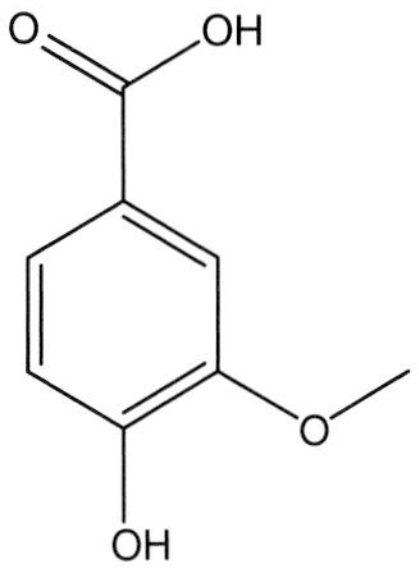

Vanillin – vanilla

LD_{50} [rat, oral] = 1,580 mg/kg body weight

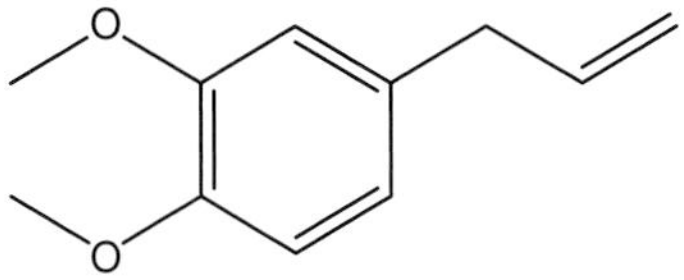

Thymol – thyme

LD_{50} [rat, oral] = 980 mg/kg body weight

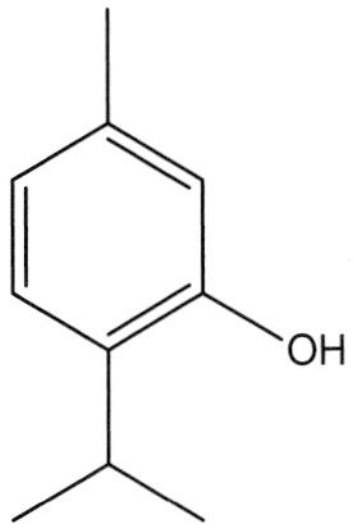

Elimecin – nutmeg

LD_{50} not known

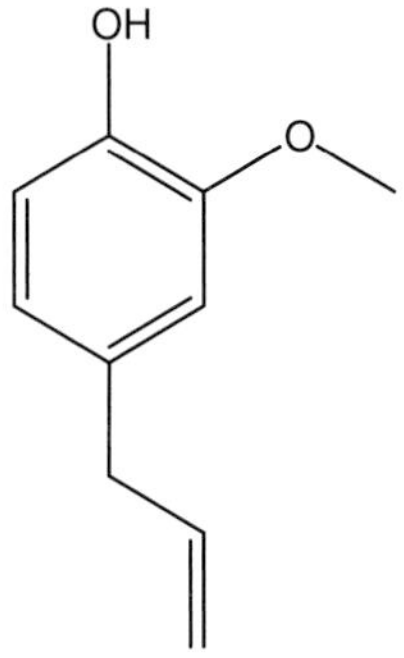

Eugenol – cloves

LD_{50} [rat, oral] = 1,930 mg/kg body weight

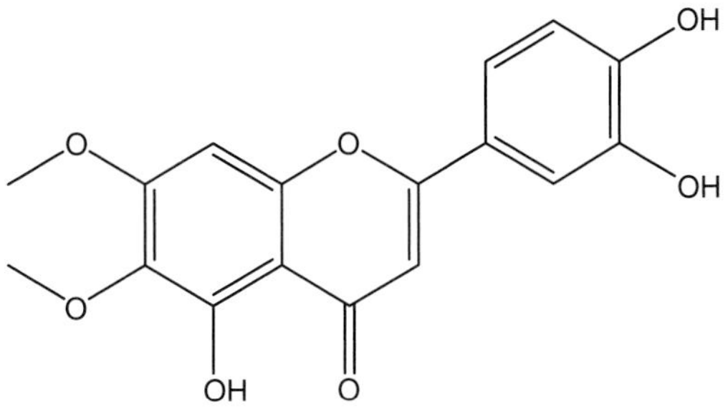

Cirsiliol – sage

LD_{50} [rat, oral] = 3,875 mg/kg body weight

Figure 8.16 Flavour chemicals from herbs and spices and their LD_{50} values. Their high LD_{50}s mean that they are of no toxicological concern to humans when consumed as natural food flavours.

of these vegetables consumed at a single sitting is much lower than rhubarb and therefore even though the levels of oxalic acid might be higher in the plant (e.g. spinach) the dose to the consumer is much lower. You would have to eat many kilograms of rhubarb pie to receive a lethal dose of oxalic acid. In fact to achieve the rat LD_{50} for oxalic acid you would have to eat approximately 6 kg or rhubarb or 3 kg of spinach; clearly this is not possible.

Oxalic acid has several different mechanisms of toxicity, each targeting important cellular processes. It is able to interfere with important mammalian

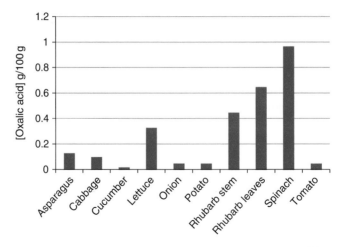

Figure 8.17 Oxalic acid levels in some vegetables. (Data from http://www.nal. usda.gov/fnic/foodcomp/Data/Other/oxalic.html, except rhubarb data which are from Lowry at http://helios.hampshire.edu/~nINS/mompdfs/oxalicacid.pdf.)

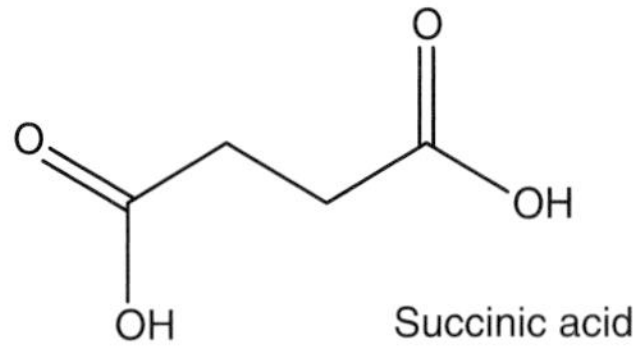

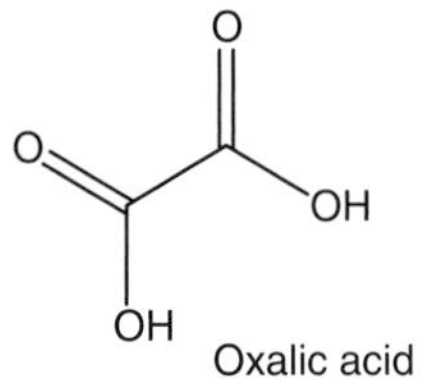

Succinic acid

Oxalic acid

Figure 8.18 The molecular similarities between oxalic acid and succinic acid mean that oxalic acid can occupy the active site of SDH, but not be converted to fumaric acid – this is why oxalic acid inhibits SDH.

metabolic pathways because it resembles the natural substrates (e.g. succinic acid) for key enzymes (e.g. succinate dehydrogenase – SDH) in these pathways (e.g. the tricarboxylic acid cycle). See Figures 8.18 and 8.19.

Oxalate also binds Ca^{2+} (Figure 8.20) and since Ca^{2+} levels in cells are very important in cell regulation, small changes can cause significant Ca^{2+} imbalance which can lead to cell death. In addition, Ca^{2+} is important in neurotransmission; Ca^{2+} is pumped across the neurone cell membrane as part of the process of generating an action potential – oxalic acid removes soluble Ca^{2+} and inhibits this process, and thus oxalic acid is also neurotoxic at high doses.

Oxalic acid is also a kidney toxin because it binds calcium in the kidney; insoluble calcium oxalate (Figure 8.20) is formed which causes kidney damage.

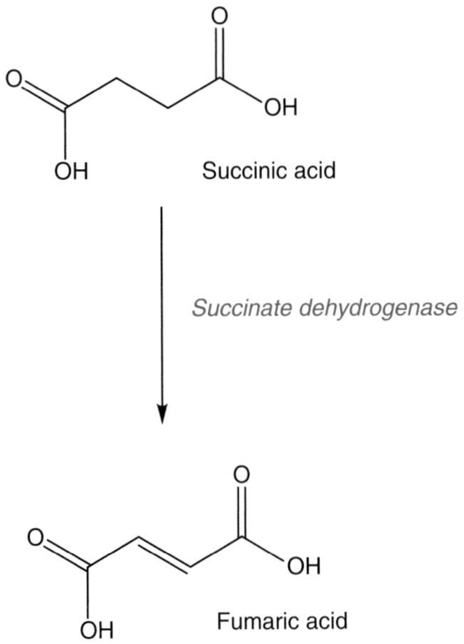

Figure 8.19 The metabolism of succinic acid to fumaric acid catalysed by the enzyme succinate dehydrogenase (SDH) – a key pathway in the tricarboxylic acid cycle.

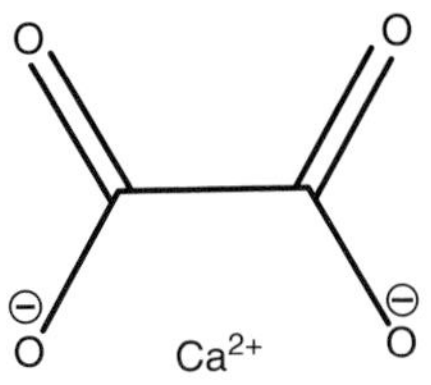

Figure 8.20 The formation of insoluble calcium oxalate changes Ca^{2+} concentration in cells, which affects neurotransmission and kidney physiology.

As an aside - when you eat rhubarb you might notice that your teeth feel rough; this is because the oxalic acid in the rhubarb binds to the calcium in your teeth. If you eat rhubarb with custard made with milk you won't get the same effect because the oxalic acid forms a complex with Ca^{2+} in the milk and is no longer available to bind to your teeth's calcium.

Mycotoxins

Mycotoxins (from the Greek μνκης (*mukes*) meaning mushroom) are produced by fungi. The mycotoxins of interest to food toxicologists are produced by fungi growing on crops or food. There are very many mycotoxins that might contaminate our food, but I will deal with just two here: aflatoxins from mouldy peanuts and patulins from mouldy apples.

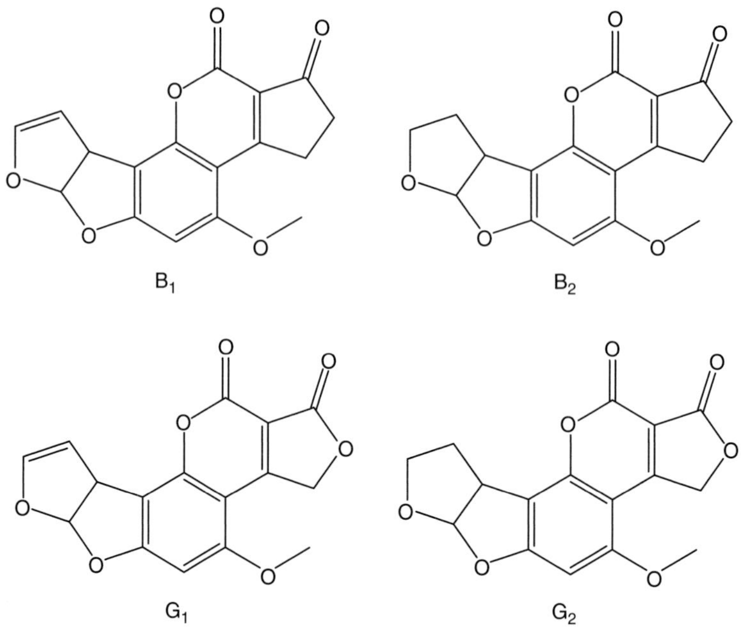

Figure 8.21 The aflatoxins – look carefully; there are only very small differences between the molecules.

Aflatoxins

The most important of the mycotoxins from a human health perspective are the aflatoxins (Figure 8.21) produced by the fungus (mould) *Aspergillus flavus* which grows on stored nuts and grains (particularly peanuts). They are acutely toxic (aflatoxin B_1 LD_{50} [rat, oral]=5 mg/kg body weight) and carcinogenic (cytochrome P_{450} metabolism forms the carcinogenic metabolite aflatoxin M_1; Figure 8.22) at low doses. Aflatoxins are of great concern internationally; so much so that a very low MRL (15 µg/kg total aflatoxins) has been set to ensure that peanuts with unacceptably high concentrations of aflatoxins cannot be traded. In addition, some countries have set MLs for aflatoxins (e.g. in peanut-based foods). We are right to be worried about the cancer risk associated with aflatoxins in food because, of all the food carcinogens we have discussed so far, the aflatoxins are the only ones that humans could receive a carcinogenic dose of - many people eat peanut butter regularly, sometimes every day, and therefore might be exposed to aflatoxins long term. Aflatoxins have been found worldwide in foods containing nuts or grains (Table 8.5).

Fungi growing on animal feed might also produce mycotoxins which in turn might contaminate the animal's meat, milk and/or eggs. Indeed, aflatoxins have been found in milk (Table 8.6) from cows fed contaminated feed.

Patulin

Patulin (Figure 8.23) is produced by the fungus *Penicillium expansum* that can grow on apples and pears causing them to rot. Because rotting apples and pears are unpalatable, patulin is not a problem in fruit directly. However,

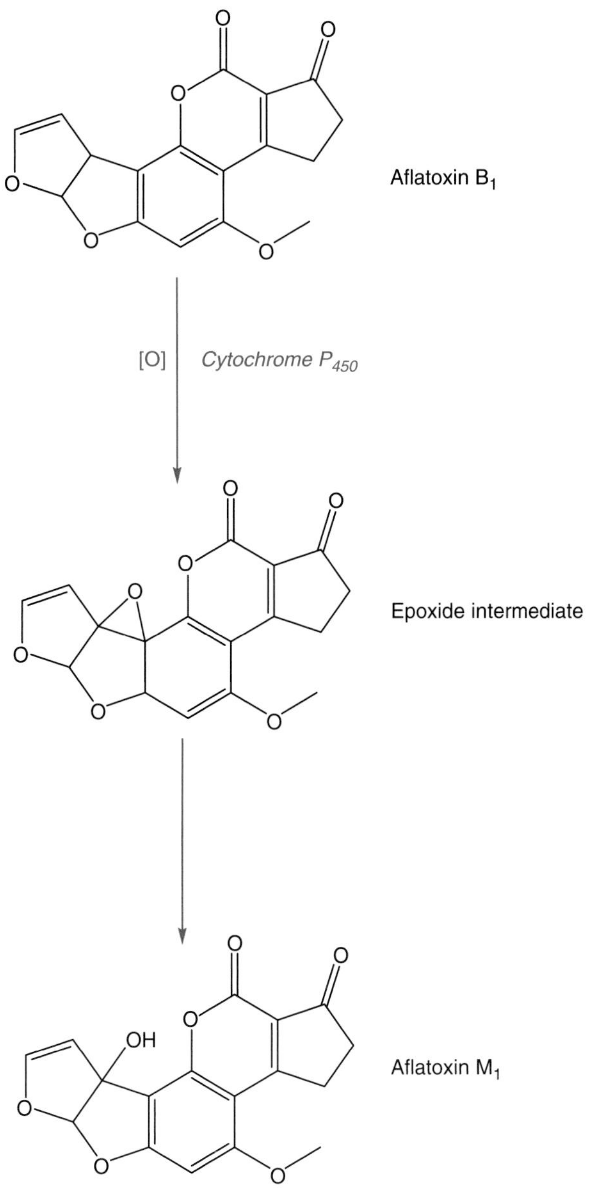

Aflatoxin B₁

[O] | Cytochrome P₄₅₀

Epoxide intermediate

Aflatoxin M₁

Figure 8.22 The mammalian metabolism of aflatoxin B₁ via a very reactive, highly carcinogenic epoxide intermediate to form the hydroxy-metabolite aflatoxin M₁.

sub-standard fruit (i.e. partially rotted) might be used to manufacture fruit juice and thus patulin might be transferred to the fruit juice; indeed, patulin has been found in apple juice at concentrations above 1,000 µg/kg.

Patulin is a suspect carcinogen (although there is dispute about this) and apple juice is often consumed by babies and infants, either directly as fruit juice or indirectly as a natural sweetening agent for baby food products. Children who consume fruit juice or fruit juice products are a particularly

Table 8.5 Aflatoxin levels in foods from around the world. (Data from IARC Monographs (2002), *Aflatoxins*, **82**, 184–185.)

Food	[Aflatoxin] range (μg/kg)	Country of origin of food
Peanut foods	1–1,500	India, Malaysia, Philippines
Nuts and nut products	0.3–128	Japan
Sorghum	0.1–30.3	India, Thailand
Beer	0.0005–0.0831	Japan
Maize	0.11–4,030	China, India, Indonesia, Philippines, Thailand, Vietnam

Table 8.6 Aflatoxin concentrations and percentage of samples positive for aflatoxins in milk from around the world. (Data from IARC Monographs (2002), *Aflatoxins*, **82**, 185.)

Country of origin of milk	Percentage samples positive (n)	[Aflatoxin] range (μg/kg)
Brazil	7.7 (52)	0.05–0.37
Cuba	26 (85)	>0.5
Cyprus	10 (112)	0.01–0.04
France	32 (17,029)	<0.05–0.5
Greece	3.7 (81)	0.05–0.18
India	18 (504)	0.1–3.5
Italy	57 (214)	0.003–0.101
Japan	0 (37)	–
Korea	37.3 (134)	0.05–0.28
Spain	18.7 (155)	0.014–0.04
Thailand	18.7 (310)	0.5–6.6
Europe	4.1 (7,573	≤0.05

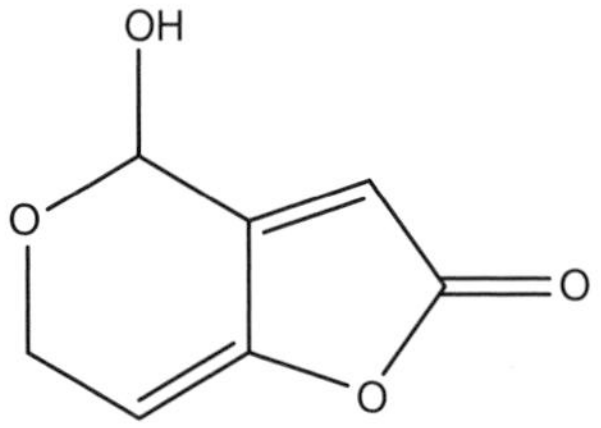

Figure 8.23 Patulin, a mycotoxin produced in rotting apples by the fungus *Penicillium expansum*. LD_{50} [rat, oral] = 27 mg/kg body weight.

worrying at-risk group because they consume large amounts for their body weight (because they are growing) and thus their dose of food contaminants is correspondingly high. A suspect carcinogen contaminant like patulin is therefore taken very seriously, especially when babies and infants are the at-risk consumer group. For these reasons the EU has set MLs of 50 μg/kg for patulin in apple juice and 10 μg/kg for baby foods; the WHO has a similar value of 50 μg/L for apple juice.

Cider is made by fermenting apple juice and so patulin levels in cider give a good indication of the 'freshness' of the apples used for the juice. In a study carried out in the USA, freshly picked apples resulted in cider with no measurable patulin, whereas apples collected from the ground (dropped apples) resulted in cider with patulin levels in the range 40-374 μg/L (data from Jackson *et al.* (2003) *Journal of Food Protection*, **66**, 618-624). The dropped apples had had time to grow patulin-synthesising mould. Therefore, it is easy to control patulin in food – use fresh fruit! Indeed, this is the rule for all mycotoxin contaminations – use fresh produce or produce stored in a way that moulds do not grow.

Phytohaemagglutinins in beans

Phytohaemagglutinins (PHGs) are lectins (proteins) produced by plants. They are important in cell recognition because they are able to recognise and bind to specific combinations of sugars that might be present as glycoprotein (proteins with branched sugar chains attached) on the surface of a cell.

The PHGs' ability to bind to the sugar units of glycoproteins means that they can bind to erythrocytes (red blood cells) because they have glycoproteins on their surface. Indeed, blood group is determined by the specific structure of the erythrocyte cell surface glycoprotein sugar chain (antigen). When PHGs bind to the cell surface proteins of more than one erythrocyte they link erythrocytes together and cause a clot to form – this is the process of haemagglutination (Figure 8.24), hence the toxin's name. The clots block capillaries and might end up in the brain or heart where they can prevent blood flow and lead to death. PHGs are incredibly toxic and are present at lethal concentrations in raw red kidney beans (*Phaseolus vulgaris*).

The PHG of red kidney beans is phasin; its lethal oral dose in humans is about 5 μg/kg – as little as 350 μg could kill a human.

Fortunately, PHGs are destroyed by heat (above 100°C) – they are proteins and they unfold as the temperature rises. This protein denaturation is irreversible so once heated above 100°C red kidney beans are safe to eat. Therefore, cooking above 100°C makes red kidney beans safe to eat.

There are several well-documented reports of serious illness following consumption of red kidney beans that had not been cooked to a high enough temperature. There were seven outbreaks of phasin toxicity in the UK between 1976 and 1979 and two more in 1988. Several were related to the use of slow cookers (crock pots) to cook red kidney bean-containing meals (e.g. chilli con carne). Slow cookers became very popular in the 1970s especially among students (I had one!) because they could add all of the ingredients for chilli con carne (beef, onions, chilli pepper, bell pepper, red kidney beans and stock) to their crock pot, turn it on in the morning and return in the evening to a nice meal. Unfortunately, the cooker did not heat its contents to 100°C (most crock pots only reach 80°C) and so the phasin in the red kidney beans was not destroyed … and the consumer became very ill, very quickly. The problem was not as widespread as it might have been though because most bedsit cooks used canned

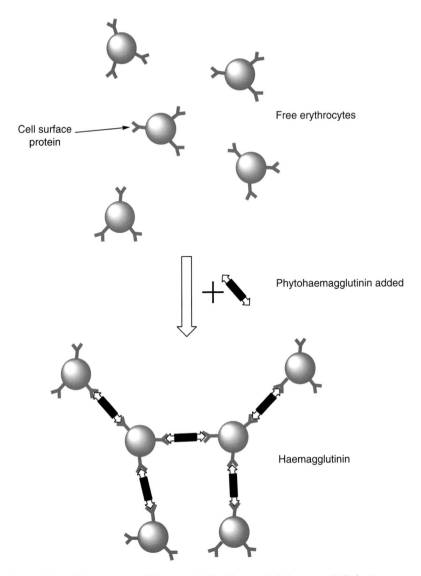

Figure 8.24 The process of haemagglutination – phytohaemagglutinin (e.g. phasin) molecules bind specifically to erythrocyte cell surface proteins and join erythrocytes together to form a clot.

red kidney beans and the canning process attains temperatures >100°C which destroyed the phasin. Interestingly, when red kidney beans are heated to temperatures around 80°C it appears that phasin activity in the beans increases by up to five-fold; so cooking in a crock pot increased the phasin activity!

Amount of phasin in raw and cooked red kidney beans

PHGs (e.g. phasin) are often measured in haemagglutination units (HAUs), i.e. the haemagglutination response is measured rather than the concentration of the PHG itself. The sample suspected of containing a PHG is mixed with erythrocytes and agglutination is measured. The test is simple and quick and gets around the complexities of measuring very low concentrations of a protein.

As discussed above, the amount of phasin is decreased significantly by cooking at or above 100°C. In uncooked red kidney beans the phasin activity (*activity* is used here because haemagglutination, i.e. activity, is measured rather than phasin concentration) is in the range 20,000–70,000 HAUs; after cooking at 100°C for about 10 minutes it drops to 200–400 HAUs (i.e. 99% decrease). This explains very clearly why cooking red kidney beans makes them safe to eat.

Bacterial toxins

Don't forget many food-borne bacteria produce toxins that underlie their pathogenic mechanisms, e.g. botulinum toxin produced by *Clostridium botulinum*. I have already discussed these natural toxins in Chapter 3.

Phytoestrogens

The phytoestrogens are natural molecules produced by many plants including food plants (e.g. soy) and are therefore present in our diet. They have molecular structures that mimic the estrogens and have estrogenic effects on consumers. I will deal with the phytoestrogens in Chapter 9 and I so will not discuss them further here.

Take home messages

- Some animals and plants produce toxins to protect themselves.
- Very few animal toxins are of concern to meat-eating humans.
- Some plant toxins can adversely affect human health and even cause death.
- Some animals (e.g. shellfish) take up toxins from their food (e.g. dinoflagellates) which might have adverse health effects on human consumers.
- Aflatoxins are important natural toxins found in fungus-infected nuts and grains and products made from them. They are of significant health concern because they are potent carcinogens.

Further reading

IARC (2002) Aflatoxins. *IARC Monographs*, **82**, 171–200, http://monographs.iarc.fr/ENG/monographs/vol82/mono82-7A.pdf. This monograph presents a very detailed account of the aflatoxins including their molecular structures, chemical properties, metabolism, toxicity, human exposure and effects.

Krogh P (ed.) (1987) *Mycotoxins in Food*. Academic Press, London.

MAFF (1996) *Inherent Natural Toxicants in Food*. MAFF, London.

Chapter 9
Endocrine Disrupting Chemicals

Introduction

Some chemicals have molecular structures similar to hormones and therefore they are able to mimic the activity of hormones in biological systems – these are the endocrine disrupting chemicals (EDCs). This is a large field of study and I will concentrate here only on chemicals that mimic the estrogens (female sex hormones) – the xenoestrogens (from the greek, ξένος (Xenos) meaning foreign; i.e. foreign estrogens).

There are two types of xenoestrogens, natural and synthetic. The natural xenoestrogens are produced by some plants and are called phytoestrogens – their molecular structures bear a striking resemblance to 'true' estrogens. The synthetic xenoestrogens comprise a vast array of industrial and household chemicals and pharmaceuticals that have molecules with features that allow them to fit estrogen receptors (ERs), but at first sight they might not look particularly like the 'true' estrogens.

Exposure to xenoestrogens is having profound biological effects on animals including humans. I will discuss only the effects on humans in this chapter. As you might expect, the basic effect that exposure to xenoestrogens has is feminisation, usually not overt morphological feminisation, but rather more subtle biochemical feminisation and its attendant physiological effects.

The importance of xenoestrogens in the human food chain is only just being accepted around the world and regulatory authorities are still considering what to do – if anything. However, the importance of the issue is illustrated well by significant studies and risk assessments being commissioned and carried out by the USA and EU regulatory authorities, and the Canadian government's decision to ban a synthetic xenoestrogen, bisphenol A, in babies' bottles ... but more of this later.

Xenoestrogens are going to be a significant issue in the future which is why I have devoted a whole chapter to them.

Food Safety: The Science of Keeping Food Safe, First Edition. Ian C. Shaw.
© 2013 John Wiley & Sons, Ltd. Published 2013 by John Wiley & Sons, Ltd.

The first observations of xenoestrogens' effects

In 1994 an American scientist, Professor Louis Guillette, and his colleagues working at the University of Florida, USA, published a paper (see *Further reading*) that showed that alligators in Lake Apopka in the Florida Everglades had shorter penises than alligators from nearby Lake Woodruff. He concluded that the alligators in Lake Apopka had been exposed to pollutants (e.g. DDT) present in the water, but that Lake Woodruff was cleaner and therefore its alligators had not been exposed to pollutants at such high levels. The finding that Lake Apopka alligators had shorter penises suggested feminising effects of the pollutants. A great deal of fascinating work followed in laboratories around the world and scientists began to uncover what was going on. The pollutants in Lake Apopka were mimicking the molecular shape of the estrogens and fooling the male alligators into 'thinking' they should be female. The first part of the feminisation process was to reduce the physiological expression of maleness, e.g. penis length. But many more things were happening to the biochemistry of the alligators that we now understand very much better (I will return to this later in this chapter).

A year after (1995) Guillette's seminal work, Professor John Sumpter and his colleague Dr Susan Jobling working at Brunel University, UK, reported that male trout caged near the outfalls of sewage treatment plants synthesised the egg protein vitellogenin – you would not expect males to make egg proteins! They concluded that something in the sewage outfall was feminising the male trout. This 'something' was thought to be ethynylestradiol (Figure 9.1) – a synthetic estrogen used in the contraceptive pill – that had been excreted in urine by women taking the Pill and had survived the sewage treatment process.

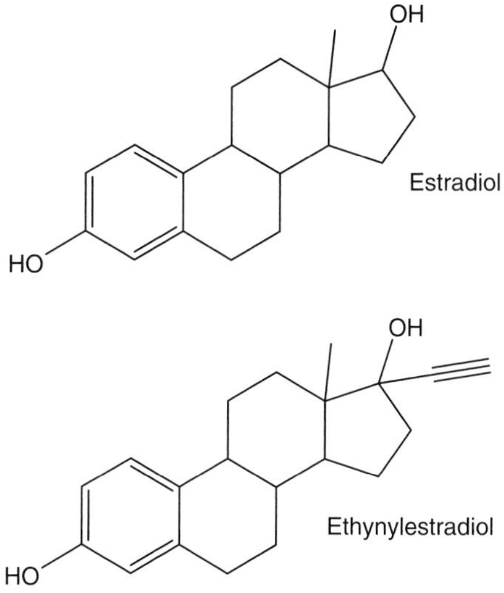

Figure 9.1 The female hormone estradiol and the estrogen analogue ethynylestradiol used in the contraceptive pill – their structural similarities are obvious.

Several years before (1992) Guillette's and Sumpter's papers, work was published (see *Further reading*) by Dr Elizabeth Carlsen and her colleagues at the National University Hospital (Rigshospitalet) and the Panum Institute, Copenhagen, Denmark, which reported that the human semen quality (e.g. number of sperm per millilitre) had been in decline for the past 50 years. The importance of this work was not realised until it was considered in conjunction with Guillette's and Sumpter's findings. Carlsen's observation was the first evidence that pointed to male humans undergoing a degree of feminisation. It was speculated that we too were being exposed to chemicals that mimicked estrogens and that they were affecting us in the same way that Guillette's alligators and Sumpter's trout were affected. Great interest developed amongst many scientists around the world; this has resulted in some very elegant studies which have significantly increased our understanding of exactly how xenoestrogens work. Before I can explain how the xenoestrogens work you need to understand the ERs and how they function.

Estrogen receptors – ERs

I will only give a very brief overview of ERs here – just enough for you to understand how xenoestrogens interfere with their function.

The ERs are large protein molecules (approx. molecular weight=65 kDa; Figure 9.2; Plate 9.1) either present in the nucleus of estrogen-responsive cells or on their cell membranes. The two different types of ER have different functions, but for the purposes of this overview I will not distinguish between them. In addition, there are two main subclasses of ER: ERα and ERβ which are present on different cell types and appear at different times during growth and development. Again, I will not distinguish between them in this overview.

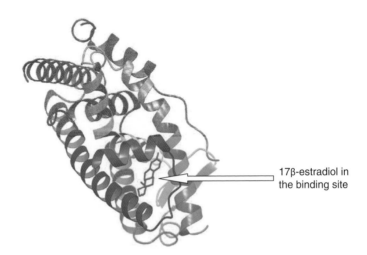

17β-estradiol in the binding site

Figure 9.2 The human ER estrogen binding domain with 17β-estradiol bound. (Created by Lisa Graham in Shroedinger 2008 using published X-ray crystallographic data from Brzozowski *et al.* (1997) *Nature*, **389**, 753–758.) (To see a colour version of this figure, see Plate 9.1.)

ERs have a specific region (binding domain) of their protein structure that binds estrogens, particularly 17β-estradiol (Figure 9.2). They selectively bind estrogens rather like a key fits into a lock. The chemical groups on the surface of estrogen molecules are akin to the 'teeth' on the key that fits a particular lock. If the key fits properly it can open the lock. This analogy can be extended to the estrogen-ER interaction too. If the estrogen key fits the ER lock it can open it and unlock the estrogenic effects.

The amino acids in the binding domain interact specifically with different chemical groups (e.g. –OH) on the estrogen molecule which means that, in theory, only estrogens can bind, but as you will see later other molecules that have similar chemical groups in the same spatial arrangement can also bind to ERs, i.e. they are keys that are not a perfect fit for the ER lock but will still open it. We now understand in great detail exactly how estrogens bind to the ER binding domain.

There are several natural estrogens; they all bind to ERs, but with different binding constants. The ones that bind more strongly have a greater estrogenic effect and those that bind less strongly have a lesser effect. The strength of their binding, and therefore the degree of effect, is related to the spatial arrangement of their chemical groups and how well they fit the binding domain and how well they interact with the amino acids in the binding domain. 17β-estradiol (the β refers to the specific orientation of the hydroxyl group on the 5-membered ring; Figure 9.3) binds most strongly and therefore has the greatest estrogenic effect of the natural estrogens.

Molecular requirements for estrogenicity

To bind to and activate ERs, a molecule must ideally have two –OH groups, one phenolic and one aliphatic, separated by a large hydrophobic region. Optimally the separation of the two –OHs should correspond with the two hydroxyl binding regions of the ER binding domain (Figure 9.4).

When the ER binding domain is occupied by estradiol a conformational change occurs which exposes an amino acid sequence which causes two occupied ERs to bind to form an ER dimer (Figure 9.5; Plate 9.2). The ER dimer undergoes another conformational change that exposes another sequence of amino acid residues that have a great affinity for, and bind to, a specific docking site on DNA. When the ER dimer binds to the DNA docking site it switches on specific genes that code for proteins that result in cellular feminisation (e.g. enzymes of estrogen synthetic pathways) (Figure 9.6).

Estrogens are present in both males and females

Females have high and highly variable (mean 200–900 pg/mL (pg=picogram=10^{-12} g) depending on the point in the estrous cycle) levels of estradiol (Figure 9.7) and many ERs; males have low levels (mean 20 pg/mL in blood) of estradiol and few ERs. So if a male is given estrogens he will develop female sex characteristics (e.g. breasts) after prolonged dosing because he has the molecular apparatus (i.e. ERs) to 'instruct' cells to feminise, but,

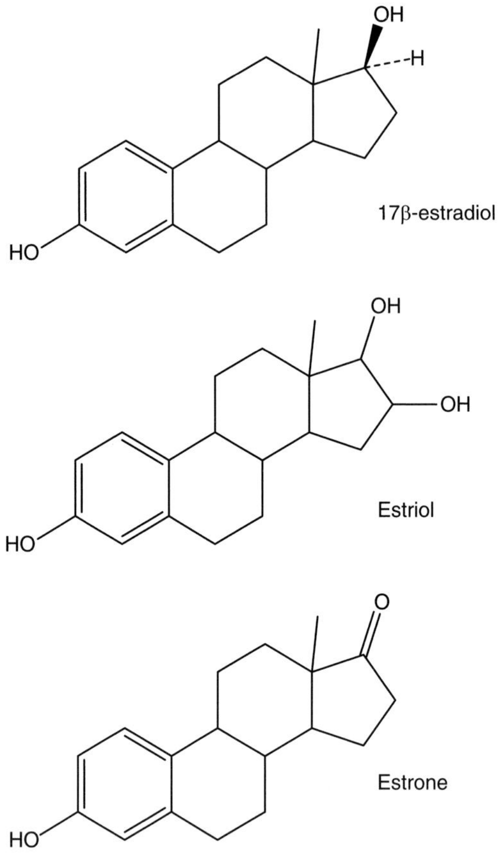

Figure 9.3 Some natural estrogens showing their similar molecular structures. The most estrogenic is 17β-estradiol which is illustrated here to show the specific β-orientation of its 17-hydroxyl group.

under normal circumstance, does not have high enough concentrations of estrogens to stimulate the development of secondary female characteristics.

Xenoestrogens

Xenoestrogens possess some of the molecular attributes of natural estrogens (e.g. estradiol) and therefore fit the ER binding domain and cause the sequence of events that leads to gene up-regulation and cellular feminisation. As discussed above, xenoestrogens can be either natural (e.g. plant estrogens – phytoestrogens; Figure 9.8) or man-made chemicals (e.g. ethinylestradiol; Figure 9.1).

There are many xenoestrogens, but before we look at some examples it is important that you understand how xenoestrogens mimic estrogens in ERs. The important facets of the estradiol molecule that allow it to bind to and activate ERs are:

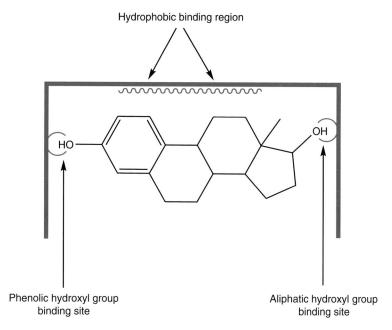

Hydrophobic binding region

HO—

OH

Phenolic hydroxyl group
binding site

Aliphatic hydroxyl group
binding site

Figure 9.4 Schematic representation of estradiol in the ER binding domain (in green) showing the three important estrogen binding areas.

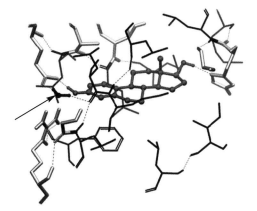

Figure 9.5 17β-estradiol in the binding domain of the human ER. The ER has been simplified so that only the amino acid residues important in the binding of estradiol are shown. The dotted lines are hydrogen bonds and the blue molecule (arrowed) is water which has a key role in the binding of estradiol. You can see that estradiol is bound by hydrogen bonds between its hydroxyl groups and specific amino acid residues in the binding domain. (From Graham and Shaw (2011), SAR QSAR. *Environmental Research*, **22**, 329–350. Reprinted with permission.) (To see a colour version of this figure, see Plate 9.2.)

- two –OH groups the right distance apart;
- one aliphatic –OH;
- one aromatic –OH;
- a hydrophobic region.

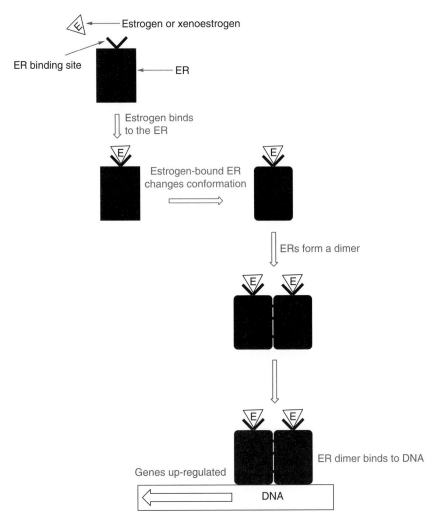

Figure 9.6 Schematic representation of the mechanism of action of estrogens. The estrogen (e.g. 17β-estradiol or a xenoestrogen) binds to the ER; the ER undergoes a conformational change which facilitates ER dimer formation. The ER dimer binds to DNA at a specific binding site and up-regulates specific genes associated with cellular feminisation.

The shape and attributes of a molecule that are necessary to give it particular properties (including biological activity) are termed structure activity relationships (SARs).

An important dietary xenoestrogen is genistein, a phytoestrogen from soy – its molecular structure closely resembles estradiol and therefore it fits and activates ERs.

Genistein

The closer the molecular structure and spatial arrangements of the key groups and regions of a particular ligand to 17β-estradiol the better the ER fit and the

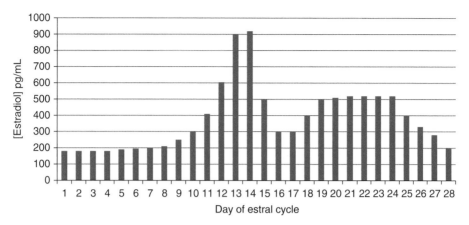

Figure 9.7 Blood levels of estradiol in women showing the large changes during the estral cycle – ovulation occurs on day 13 or 14. (Data from http://commons.wikimedia. org/wiki/File:Estradiol_during_menstrual_cycle.png.)

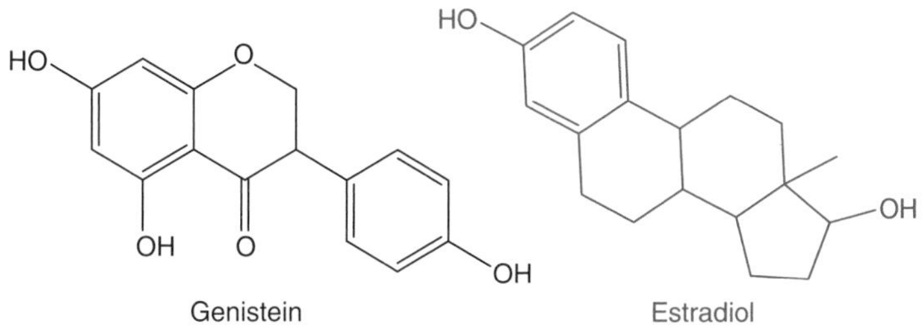

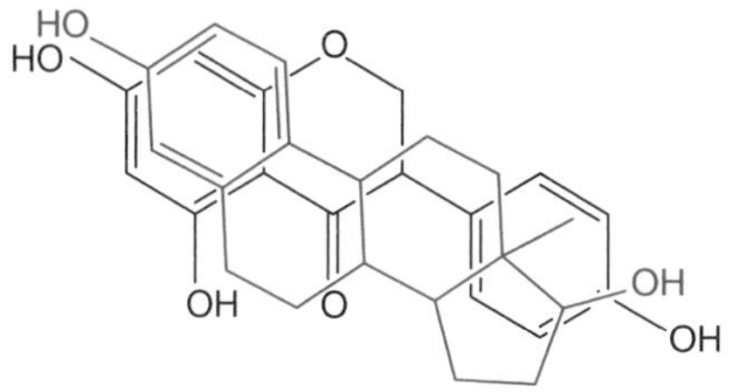

Genistein

Estradiol

Figure 9.8 Genistein and estradiol (top) showing them superimposed (bottom) to emphasise the similarities between their molecular structures and the all-important similarities in spatial arrangements of the key groups (–OHs) and regions (hydrophobic) for ER binding.

greater its estrogenic activity. Genistein fits well (Figure 9.8), but both of its –OH groups are aromatic and therefore it does not interact ideally with ERs. In addition, its hydrophobic region has an oxygen heterocyclic ring; this is not as hydrophobic (because of the electronegativity of the O) as the steroid ring structure of estradiol; this also reduces its ER interaction credentials. For this reason, genistein is estrogenic, but not as estrogenic as estradiol - in fact the estrogenicity of genistein is 10^{-5} that of estradiol so you would need 10,000 times more genistein to have the same effect as estradiol. Estrogenicity is usually expressed relative to 17β-estradiol (i.e. estrogen equivalents).

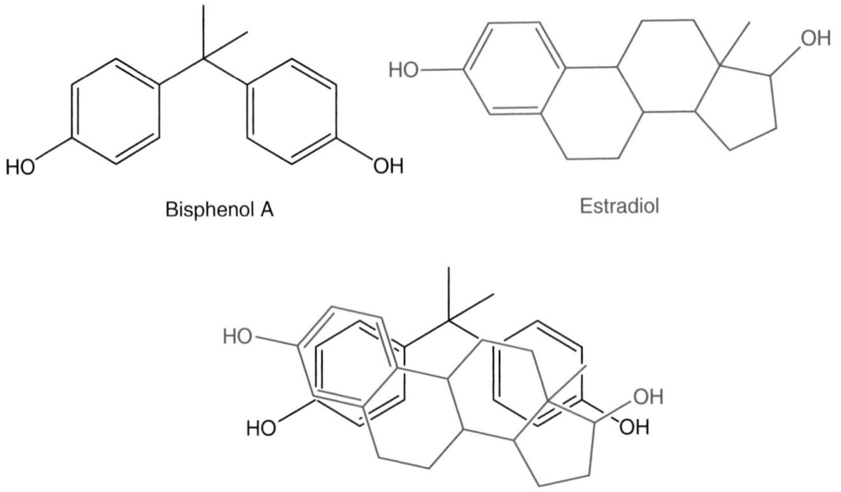

Figure 9.9 Bisphenol A and estradiol (top); BPA has the molecular attributes necessary to bind to ERs as can be seen when estradiol and BPA are superimposed (bottom).

Bisphenol A and 4-nonylphenol

The molecular structure of genistein is clearly very similar to estradiol. Other xenoestrogens, particularly the man-made ones, are often less obvious. Bisphenol A (BPA) is the monomer used to synthesise polycarbonate plastics and is particularly important in food because polycarbonate plastics (e.g. plastic sandwich boxes) and BPA lacquers (e.g. linings of tin cans) are used in food packaging and any unpolymerised monomer can leach into the packaged food (particularly fatty food because BPA is fat soluble). I will discuss BPA in greater detail later, but will use it to illustrate estradiol SARs here (Figure 9.9).

BPA has two –OH groups with approximately the same separation as in estradiol plus a hydrophobic region. These important properties facilitate its binding to ERs. As with genistein it has two aromatic hydroxyls (which is not ideal); this explains why its estrogenicity is only 10^{-5} times that of estradiol.

Finally, some molecules might, at first sight, not appear to possess the estradiol SARs, but on more detailed study they turn out to be xenoestrogens. A good example of this class of xenoestrogens is the non-ionic detergent 4-nonylphenol (4NP) - it looks nothing like estradiol (Figure 9.10), but has 10^{-3} times its estrogenicity.

We now know that 4NP is metabolised both in humans and environmental systems to various hydroxy-4NPs (Figure 9.11) and that these are attracted to the ER binding domain and (amazingly) the chemical microenvironment of the ER binding domain pulls the hydroxy-4NPs into a configuration that resembles estradiol, which, in turn, triggers the sequence of events that leads to estrogenicity. Hydroxy-4NPs have all of the ER binding attributes of an aliphatic and an aromatic hydroxyl with the correct separation plus a hydrophobic region (Figure 9.12); which is why its estrogenicity is 10^{-3} times that of estradiol. This might seem rather far-fetched, but molecular modelling

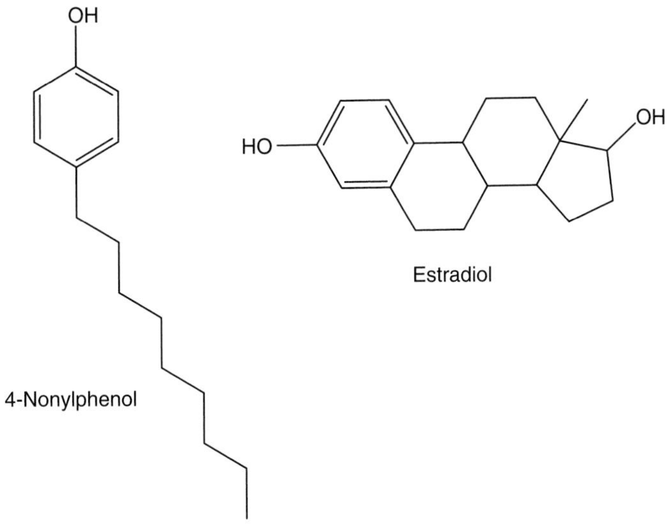

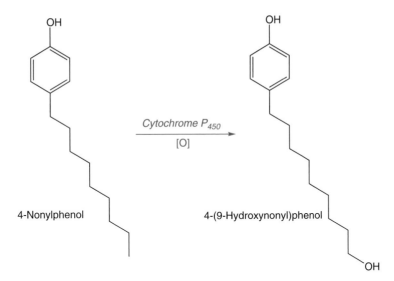

Figure 9.10   Estradiol and the xenoestrogen 4-nonylphenol showing their apparent lack of molecular similarity.

Figure 9.11   Cytochrome P_{450}-catalysed hydroxylation of 4-nonylphenol.

studies in my laboratory have shown that hydroxyl-4NP would refold in the chemical microenvironment of the ER binding domain and when it does it fits like a glove (Figure 9.13; Plate 9.3).

DDT is also a xenoestrogen

Another molecule that does not look like estradiol at first glance is DDT; remember this is where the xenoestrogen story began with alliga-tors' exposure to DDT. However, a close look at the chemistry of the ER

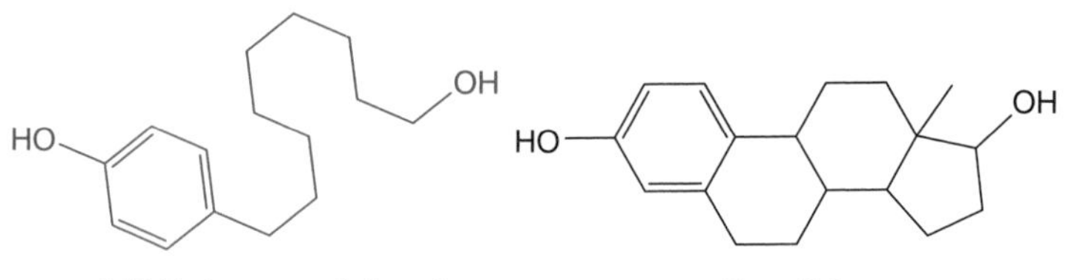

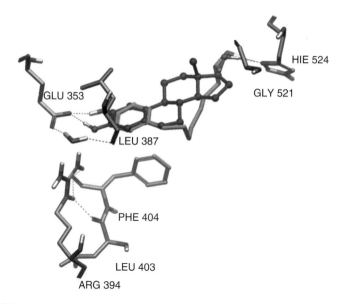

4-(9-Hydroxynonyl)phenol

Estradiol

Figure 9.12 4-(9-Hydroxynonyl)phenol refolded into the shape of estradiol (top) showing that its estradiol SARs line up (bottom).

HIE 524

GLU 353

GLY 521

LEU 387

PHE 404

LEU 403

ARG 394

Figure 9.13 4-(9-Hydroxynonyl)phenol in the ER binding domain with 17β-estradiol superimposed – this shows that 4NP is estrogenic via its refolded hydroxyl-metabolite. (From Graham and Shaw (2011), SAR QSAR. *Environmental Research*, **22**, 329–350. Reprinted with permission.) (To see a colour version of this figure, see Plate 9.3.)

binding domain and the molecular structure of DDT makes it very clear why DDT is estrogenic.

In my summary of the ER binding requirements I listed –OH groups at specific separation and specific orientation. Well, it seems that –OHs are not specifically required, but rather electron withdrawing groups (i.e. electronegative atoms like Cl) that can interact with the –OH hydrogen bonding amino acid residues in the ER binding cleft. Because some electronegative atoms (e.g. Cl) cannot form hydrogen bonds they only have a polar interaction with

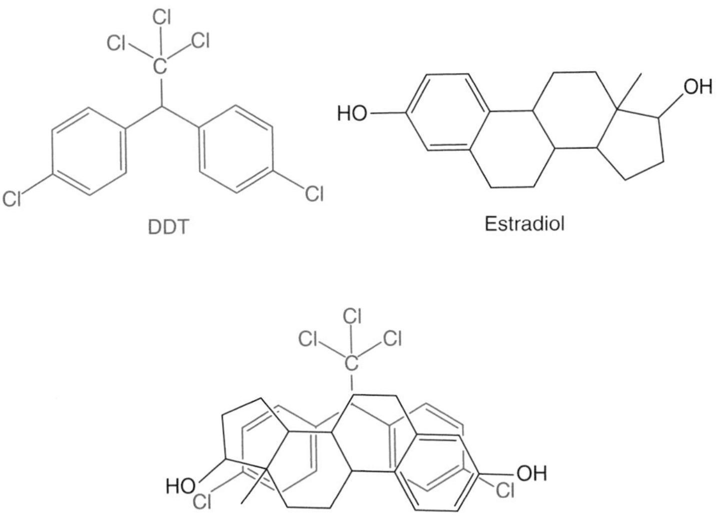

Figure 9.14 DDT and estradiol – superimposing the molecules shows that they have the attributes necessary for ER binding. Interestingly, the ER binding domain has a large hydrophobic cleft that accommodates DDT's trichloromethyl group.

the amino acids rather than forming hydrogen bonds, but this interaction, while not as effective as a hydrogen bond, still induces the conformational change in the ERs that leads to gene up-regulation and an estrogenic response. A close look at DDT's molecule (Figure 9.14) shows that the –Cl groups are in a similar spatial arrangement to the –OHs of estradiol and DDT has a long hydrophobic region between the –Cls ... perfect! DDT has the right shape with –Cls in the right spatial arrangement and it has a hydrophobic region, but it cannot hydrogen bond, therefore it is much less estrogenic than the other xenoestrogens I have already discussed – DDT is 10^{-6} times as estrogenic as 17β-estradiol.

Human exposure to xenoestrogens

It has been said that we live in a sea of estrogens. Our sewage contains the estrogens that we excrete (e.g. ethinylestradiol from the contraceptive pill) which end up in the rivers and sea and can get back into our food chain via food animals and plants, and drinking water. Industrial waste contains xenoestrogens (e.g. 4NP (although this in now banned in some parts of the world)) which also get into our food and water. We eat plants (e.g. soy) that contain phytoestrogens (e.g. genistein) and we package our food in plastics that are made from estrogenic chemicals (e.g. BPA). And these are just the dietary sources. It is outside the scope of this book to discuss other routes of human exposure to xenoestrogens (e.g. modern tooth fillings are made of BPA plastic), but it is important to remember that it does not matter how we are exposed, all of our exposures add up to the total (i.e. combined) estrogenic effects.

The concept of 'adding up to estrogenic effects' is very important because xenoestrogens all work by the same mechanism – they bind to and activate ERs. Therefore, their effects are at least additive. The regulators find this concept difficult, as discussed for the OPs (see Chapter 7, *Pesticide residues cocktails*) because it is difficult to set safety limits for groups of compounds. For this reason, the xenoestrogens for which exposure limits have been set (e.g. BPA; TDI (tolerable daily intake)=10 µg/kg body weight/day) are regulated without any consideration for other xenoestrogens that the consumer might be exposed to. This is a very blinkered approach to risk assessment and risk management.

Xenoestrogens are hormone mimics and hormones work at very low concentrations; therefore xenoestrogens also have effects at very low exposure levels. For example, estradiol manifests its hormone effects at the 10^{-9} ng/L (i.e. nanograms per litre) level and since BPA has 10^{-5} times the estrogenicity of estradiol it would have effects in humans at the 10^{-4} g/L (i.e. 100 µg/L) level in the circulatory system. These are very low levels and could result from low concentrations in food. This is why BPA's TDI is only 10 µg/kg body weight/day.

Humans are exposed to myriad xenoestrogens

It would be impossible to list all of the known xenoestrogens or to speculate about chemicals with estradiol SARS but which have not yet been shown to be estrogenic - there are simply too many. They range from chlorinated pesticides like DDT, chlorinated environmental pollutants like dioxins, plasticisers that used to be used in cling wraps like dibutylphthalate, apparently 'safe' pesticide metabolites like the hydroxy metabolite of the pyrethroid insecticides, and lignans produced by gut bacteria from plant cell wall components ... and these are just a few (Table 9.1).

Human exposure assessments via food

Some xenoestrogens are included in national surveillance schemes and total diet surveys; these data in conjunction with food intake information from national nutrition surveys are used to calculate dietary intakes. It is then possible to compare xenoestrogen dietary intakes with TDIs (based on animal toxicity studies) to determine whether the consumer might succumb to the estrogenic effects of these chemicals - remember all such assessments are carried out on individual compounds not mixtures, but it is the mixtures that are likely to cause the biological effects and therefore the effects are likely to be worse than indicated by such studies.

Phytoestrogens – coumestrol/genistein and soy

Coumestrol and genistein are phytoestrogens found in beans and peas - their concentrations are particularly high in soy beans. Soy is consumed worldwide. It is eaten fermented as tofu or tempeh; it is made into soy 'milk' by homogenising soy beans in water or it is added to flour to increase its protein content (flour improver) for cake and bread making. We all eat far more soy than we might think.

Babies fed soy milk excrete more genistein in their urine than breastfed babies (Table 9.2) which means that they must have absorbed the genistein and it must have interacted with their ERs and probably had a biological effect - I will discuss these biological effects in more detail later.

Estimates of coumestrol intake vary from country to country and reflect different diets. The Western diet contains far less soy than Asian diets and for this reason Asians have high coumestrol intakes and Western people have much lower intakes (Table 9.3).

Table 9.1 Some xenoestrogens that occur in food showing their molecular structures and, if relevant, their estrogenic metabolites.

Compound	Molecular structure	Use/source
Coumestrol		Phytoestrogen/plants
Quercetin		Flavonoid/plants
Dibutylphthalate		Plasticiser – was used in cling wraps
Pyrethroids (e.g. cypermethrin)		Insecticide (see Chapter 7)
Lignans (e.g. enterodiol)		Gut bacterial metabolite of lignin from plant cell walls
Dioxins (e.g. 2,3,7,8-TCDD)		Environmental pollutant (see Chapter 7)

Table 9.2 Genistein in urine of babies fed different types of milk – it is likely that cow's milk contains more genistein than human milk because cattle feed sometimes contains soy bean meal. (Data from Cruz *et al.* (1994) *Pediatric Research*, **35**, 135–140.)

Babies' feeding regime	Urine [genistein] (µg/L)
Breastfed	20
Cow's milk formula	100
Soy milk formula	600

Table 9.3 Dietary exposure to coumestrol in different countries showing that an Asian diet (e.g. Korea) contains higher levels of coumestrol (from soy) than Western diets. (Data from references cited in Thomson (2009) in Shaw (ed.) *Endocrine-Disrupting Chemicals in Food*. CRC Press, New York, p. 218.)

Country	Coumestrol exposure/range* (µg/kg body weight/day)
Korea	6.9–7.1
New Zealand	0.2–0.5
The Netherlands	<0.01
USA	0.2

*Measured in various age group and sex cohorts – see Thomson (2009) (reference in table heading) for full details.

But do these phytoestrogen intakes have any biological effects on the consumer? This question is almost impossible to answer; however, epidemiological studies point firmly to a population level biological effect which has been attributed to our total intake of xenoestrogens (see *Population level effects of exposure to xenoestrogens*) to which genistein and coumestrol are major contributors.

Synthetic xenoestrogens – BPA and plastics

As already discussed, BPA is the monomer used to manufacture polycarbonate plastics. Polycarbonate plastics are widely used; you will almost certainly have used or seen an example today. They can be either hard transparent plastic that can be used to replace glass (e.g. in spectacle lenses) or a softer, but still tough, semi-opaque plastic used for making plastic boxes and containers (e.g. sandwich boxes) or they can be used to make polymer lacquers used to protect metal surfaces (e.g. the lining of food cans). They have many uses in food packaging and therefore food frequently comes into contact with polycarbonate plastics and so it is possible that BPA might leach into the food.

The polymerisation process used to manufacture polycarbonate plastics from BPA (plus catalysts) results in a covalently bonded long polymer (Figure 9.15). It is very unlikely that the BPA units will break off this polymer unless very stringent chemical conditions are applied - it has been suggested

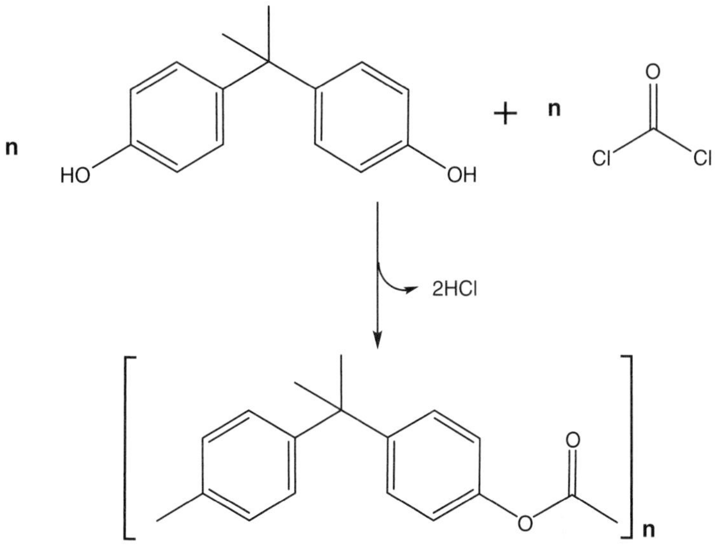

Figure 9.15 The polymerisation of BPA (top left) with phosgene (top right) to form polycarbonate plastic (bottom).

that bacterial action could break the BPA-BPA bonds, but this has not been proved. However, if the polymerisation process is incomplete there will be free BPA molecules amongst the long polycarbonate molecules. It is this BPA that can leach out into food stored in contact with the plastic. The amount of free BPA in a particular sample of polycarbonate plastic depends on the efficiency of the polymerisation process which is controlled in good plastics manufacturing companies by a strict quality assurance (QA) regime. As plastics manufacture is transferred to countries where labour is cheaper and regulations more lax I suspect these QA processes will be compromised.

Can BPA leach from polycarbonate food packaging into food?

As already mentioned, if polycarbonate food packaging with some free BPA monomer comes into contact with food it is possible that the BPA will leach out into the food (Table 9.4). Whether BPA leaches or not depends on several factors, including temperature and the chemical nature of the food. At higher temperatures BPA is more likely to leach for two reasons: the polycarbonate chains will move apart as the temperature rises because the Van der Waals forces that hold them together will be broken by the heat energy input – this will release BPA monomer that was trapped between the polycarbonate chains. And BPA is more soluble in fat than water (in fact it is almost insoluble in water) and so fatty food will 'pull' BPA out of the plastic packaging.

Worst-case scenarios, from the point of a consumer's exposure to BPA, might be microwaving a fatty food (e.g. bacon) in a polycarbonate food container or eating canned fatty food from a BPA-lined can – both would provide the ideal conditions for free BPA to leach into the food.

It is clear from the above discussion that free BPA *can* leach into food, but *does* it? Studies on BPA concentrations in canned food from BPA lacquer-lined cans have shown unequivocally that BPA leaches into food (Table 9.4).

Table 9.4 BPA in some canned foods. The results are very variable because leaching of BPA from polycarbonate lacquer linings depends on many things (e.g. the time the food was in the can), but it is clear that the higher levels tend to be in the fattier foods (e.g. coconut cream). (Data from Thomson and Grounds (2005) *Food Additives and Contaminants*, **22**, 65–73.)

Canned food	Percentage fat	[BPA] (μg/kg)
Fruit salad	0–0.1	<10
Baked beans	0.3–0.4	<10
Tomatoes	0–0.5	<10–21
Peas	0.5	<10
Pineapple	0–0.6	<10
Salmon	6.4–6.5	<20–24
Meat	12–21	<20–98
Coconut cream	17–25	<20–192

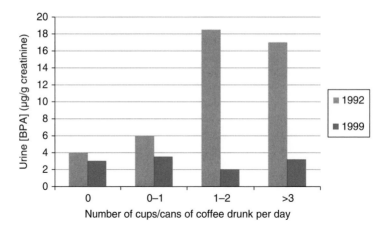

Figure 9.16 BPA in Japanese students' urine – the levels dropped significantly between 1992 and 1999 probably because BPA use in drinks can* linings was reduced voluntarily by the manufacturing industry in 1998. (Data from Matsumoto *et al.* (2003) *Environmental Health Perspectives*, **111**, 101–104.)
* Japanese students like to drink canned coffee.

In addition it has been shown that BPA is present in human urine (Figure 9.16) which means that it must have been absorbed (remember – BPA exposure can also be from non-food sources, e.g. tooth fillings). The next question is can the human dose of BPA cause biological effects? This question has not yet been answered definitively, but most scientists think that BPA in conjunction with other xenoestrogens is responsible for population effects like decreased sperm count – I will discuss this in more detail later (see *Population level effects of exposure to xenoestrogens*).

Population level effects of exposure to xenoestrogens

The biological effects of exposure to xenoestrogens in humans are very difficult to prove unequivocally because they can only be seen at a population rather than an individual level. Other food-borne toxic chemicals and microbes are much easier to study because they have a very obvious effect on the person who consumed the contaminated food. For example, someone eating hazelnut yoghurt (see Chapter 3, *Clostridium botulinum*) contaminated with *Clostridium botulinum* will become very ill, very quickly and the symptoms of their illness will be characteristic of botulinum toxin poisoning. Similarly a fugu sashimi with too much tetrodotoxin (see Chapter 8, *Tetrodotoxin from fugu*) will make its consumer ill quickly – the source of the poisoning would again be obvious.

The situation with the xenoestrogens is quite different because they cause subtle changes that are often not apparent in individual people, but appear as population trends following years of exposure. The issue is proving cause and effect. Does exposure to xenoestrogens cause the effects that we might predict from their known hormone analogies?

First we must determine the effects. Animal studies and some human endocrinology clinical studies clearly show that exposure of males to female hormones causes them to develop female characteristics (e.g. development of breasts). Indeed, this is used to allow transgender males to look more like women – they are given estrogens as implants under their skin. However, to achieve very visible effects like male breast growth, very high estrogen doses are necessary. Our exposure to xenoestrogens via food is very much lower and the activity of the xenoestrogens is considerably lower than estradiol itself (i.e. 10^{-4}–10^{-6} times the activity of 17β-estradiol; see *Human exposure to xenoestrogens*) and so we would expect very much more subtle changes than the overt development of female secondary sex characteristics in males. Carlsen's studies (see *Further reading*) that showed declining sperm count (Figure 9.17) describe the sort of subtle population changes that might be

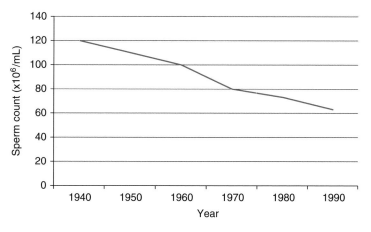

Figure 9.17 Declining human sperm count over 50 years. (Data estimated from Carlsen *et al.* (1992) *British Medical Journal*, **305**, 609–613.)

associated with long-term, low-level male exposure to xenoestrogens (see *The first observations of xenoestrogens' effects*) and indeed the effects described are now attributed to xenoestrogen exposure. There are other population level effects on males that we now believe are due to xenoestrogen exposure, including cryptorchidism (undescended testes) and hypospadias (deformed urethra) in males and early onset puberty in females (I will discuss these disorders in more detail later). Before I go into these effects in more detail we need to understand the difference between males and females with respect to estrogens, their levels and effects.

Estrogens in males and females

As already outlined, males and females both have estrogens. Males have low levels (17β-estradiol = 20 pg/mL) and females have higher and very variable levels (17β-estradiol = 200–900 pg/mL); they go up and down with the estral cycle. Both males and females have ERs – males have fewer than females because they have lower levels of estrogens and less 'need' to express estrogenic effects. It is the estrogens and their interaction with ERs that makes females female.

Males have a constant level of estrogens throughout their lives, but not only do females have variable levels due to estral cycle changes (Figure 9.7), but also the levels also change at different stages in a female's life. At birth, the estrogen levels in a female are very similar to male levels – at this point the female child is hormonally like a male. When she reaches puberty, estrogens are synthesised, her estrogen levels rise and she begins to develop the characteristics of a woman. From puberty to menopause (reproductive phase), the estral cycle governs estrogen levels, but they are always much higher than in males. At menopause, estrogen synthesis subsides and estrogen levels fall. The female begins to revert to hormonal maleness. This is a very simplistic view of sex hormone endocrinology, but I hope it gives you an overview of the hormone situation upon which we can build when we consider the effects of xenoestrogens.

Based on this understanding of estrogens in males and females, it should be clear to you that if a male receives a dose of an estrogen (including xenoestrogens) it is likely to have a greater effect than on a reproductive phase female because male estrogen levels are low to start with and therefore exogenous estrogens will increase the estrogenic activity by a greater proportion than in reproductive phase women. For this reason, males are a target cohort for xenoestrogen exposure for their whole lives.

Females, on the other hand, are very unlikely to be affected by xenoestrogens during their reproductive phase because they have high levels of estrogens to start with and therefore exogenous estrogenicity (e.g. xenoestrogens) will have little impact because it will comprise a smaller proportion of the endogenous estrogen activity than in males. In addition, females are used to highly variable estrogen levels and so small changes are simply tolerated. However, pre-puberty and postmenopausal women are hormonally much more like males and therefore might be affected by the changes in estrogenicity that exposure to xenoestrogens will cause.

This reasoning explains why males might be affected by xenoestrogens for their entire lives, but females are likely only to be affected pre-puberty and postmenopausally. When we look at the population level effects of xenoestrogens in males and females this will become clearer.

Hypospadias

Hypospadias is a developmental disorder of the penis in which the urethra (the urine canal) emerges from the side of the penis rather than its tip. Hypospadias results from an abnormal growth of the penis during the embryo's and fetus' development.

Both male and female sex organs develop in the embryo from the urogenital fold; estrogens cause it to develop into a vagina and androgens (male sex hormones, e.g. testosterone) cause it to develop into a penis. So if the urogenital fold of a male embryo is exposed to estrogens (or xenoestrogens) its development towards the morphology of a penis might be altered. So if a male embryo is exposed to xenoestrogens the resulting sex organ could be part-way between a vagina and a penis, i.e. the urethra does not emerge at the right place.

Male animals (e.g. rats) given estrogens develop hypospadias and male animals given BPA have a higher incidence of hypospadias than controls. This shows that BPA (and perhaps other xenoestrogens) can cause hypospadias.

The incidence of hypospadias in humans is increasing, as is our exposure to xenoestrogens (e.g. BPA). This does not prove cause and effect because there is no direct evidence that the *actual* cases of hypospadias was exposure to xenoestrogens, but it is good evidence, combined with the animal studies showing that exposure to BPA can cause hypospadias.

Cryptorchidism

Cryptorchidism (from the Greek κρνπτος (*kryptos*) meaning hidden; ορχις (*orkis*) meaning testicle) is undescended testes. The testes descend from the abdominal cavity to the scrotum usually before the child is 9 months old. Sometimes the testes descend late or not at all – this is cryptorchidism. Clearly the descent of the testes is an important part of the development of a male and is in part under hormonal control both in the infant and during his life as an embryo and fetus when the canal that the testes will descend via develops. Aberrations in androgen levels or exposure to estrogens (or xenoestrogens) might affect these processes.

The incidence of undescended testes in humans is increasing. Following the same argument for hypospadias above, this could be due to exposure to xenoestrogens, but we cannot prove cause and effect.

Precocious puberty in girls

Puberty in girls is initiated by estrogen synthesis in and release from the ovaries. Its timing is programmed in an individual and is similar for girls with similar lifestyles and ethnic origins. If a pre-pubertal female animal is given estrogens it goes into puberty; a female human would respond similarly. So, if a pre-pubertal girl receives xenoestrogens in her diet at a sufficiently high dose this might fool her body into 'thinking' it's time to begin the process of puberty.

The age of onset of puberty is decreasing in many countries around the world. This could be explained by better diets - it is known that animals go into puberty earlier if they are given a more nutritious diet. On the other hand, it might be explained by exposure to xenoestrogens. Or it might be a combination of both (and perhaps other) mechanisms.

Take home message – are xenoestrogens having an effect on humans?

I have listed a number of effects associated with exposure to estrogens above and have dealt with them in isolation. The conclusion at the end of each example (e.g. hypospadias) is that it is impossible to prove cause and effect in humans. However, when you consider all of the effects together the evidence is very much stronger. And, when you consider that our exposure to xenoestrogens is increasing it becomes a more convincing (but still not proven) cause and effect argument. The argument is so convincing that most scientists now believe that exposure to xenoestrogens is responsible for these, and other, effects.

The Japanese industry voluntarily began reducing the use of BPA in drinks can linings in 1998 and the Canadian government and very recently (July 2012) the US government have banned the use of BPA plastics in babies' bottles - these responses are good evidence that the regulatory world is beginning to believe the scientific evidence even though it can never be conclusive ... it is all about risk.

The positive health effects of xenoestrogens

The old saying 'Every cloud has a silver lining' is true for the xenoestrogens because even though there is concern about their toxic effects on humans (and wildlife), they also have a positive side.

Let's return to the hormone status of a female throughout her life, i.e. male-like → PUBERTY → female → MENOPAUSE → male-like (see *Estrogens in males and females*). At menopause, women often have significant problems overcoming the withdrawal symptoms as their estrogen levels decline; they often suffer unpleasant effects (e.g. 'hot flushes') during this time - their body is responding to unoccupied ERs. In addition, estrogens play an important part in transporting calcium to and from bones. When estrogen levels decline, bones lose calcium and become brittle (osteoporosis) - post-menopausal women are therefore susceptible to osteoporosis.

Estrogens are sometimes prescribed (hormone replacement therapy - HRT) to combat post-menopausal problems. However, some women find that eating phytoestrogen-rich food gives them relief without resort to HRT which can have significant side effects (e.g. breast cancer). This has been picked up by some food (e.g. bread) manufacturers who produce soy-rich products and advertise them as being helpful for post-menopausal women. In this example soy is described as a neutriceutical (a combination of nutrient and pharmaceutical). So, natural xenoestrogens are not all bad.

Take home messages

- Endocrine disrupting chemicals (EDCs) are chemicals that resemble and behave like hormones (i.e. they are hormone mimics).
- Xenoestrogens are EDCs that behave like estrogens.
- Humans are exposed to many xenoestrogens in food, including natural phytoestrogens (e.g. genistein from soy) and man-made chemicals (e.g. bisphenol A from polycarbonate plastics).
- Exposure to cocktails of xenoestrogens is thought to be responsible for population level effects in humans (e.g. reduced sperm count and early-onset puberty in girls).
- Some xenoestrogens have been banned in some countries (e.g. the use of polycarbonate plastics in babies' bottles in Canada).

Further reading

Carlsen E, Giwercman A, Keiding N & Skakkebaek NE (1992) Evidence for decreasing quality of semen during past 50 years. *British Medical Journal*, **305**, 609–613.

Guillette LJ Jr, Gross TS, Masson GR, Matter JM, Percival HF & Woodward AR (1994) Developmental abnormalities of the gonad and abnormal sex hormone concentrations in juvenile alligators from contaminated and control lakes in Florida. *Environmental Health Perspectives*, **102**, 680–688. This is a very important paper that introduced the concept of endocrine disrupting chemicals.

Munck J (2009) Exposure to endocrine disrupting compounds via the food chain: is packaging a relevant source? *Science of the Total Environment*, **407**, 4549–4559.

Shaw IC (ed.) (2009) *Endocrine Disrupting Chemicals in Food*. Woodhead Publishing, Cambridge. This is a detailed research level review of endocrine disruptors in food and their health effects.

Thomson BM, Cressey PJ & Shaw IC (2003) Dietary exposure to xenoestrogens in New Zealand. *Journal of Environmental Monitoring*, **5**, 229–235.

Chapter 10
Genetically Modified Food

Introduction

Genetically modified (GM) food, or genetically engineered (GE) food as it is sometimes called, is the most controversial food issue of our era. There is a vehement debate raging about whether it should be allowed or not. In this chapter, I will discuss the issues and explain why there is such disagreement around the world. Some countries grow and use GM foods (e.g. the USA) while others have legislation that bans the growing, use and sale of GM foods or foods that contain GM components (e.g. New Zealand).

The greatest concern amongst consumers relates to the health risks associated with eating GM products and therefore there is a significant effort by the companies who have vested interests in GM crops to prove that the risks are negligible if not zero. In addition to the possible risks to consumers of GM foods, the risk of growing GM crops on the environment is a point of significant concern to many people including eminent scientists worldwide.

The debate is fuelled by passion for wholesome, natural and safe food, on the one hand, and huge sums of money to be made from the GM industry, on the other. I hope at the end of this chapter you will understand the issues well enough to make your own informed decision and perhaps take part in your own debate with friends and relatives.

The arena for deciding whether or not to allow GM crops to be grown and their products consumed lies at the interface between politics, environment and big business – the most difficult arena to hold any unbiased discussion in.

In essence, GM crops are grown from seeds that have had their genes modified in some way that benefits crop production. For example, a gene can be inserted that codes for resistance to a particular herbicide (e.g. glyphosate) which means that when the crop is being grown the farmer can spray it with the herbicide to kill weeds without harming the crop itself. Or, perhaps, a gene that codes for the production of a particular sugar in a fruit might be inserted into the fruit plant's genome – the crop grown would then be very sweet and command a good price at market.

Food Safety: The Science of Keeping Food Safe, First Edition. Ian C. Shaw.
© 2013 John Wiley & Sons, Ltd. Published 2013 by John Wiley & Sons, Ltd.

GM is not restricted to plants; farm animals can also be modified to give better milk yields, produce milk with particular nutritional components, or synthesise and secrete chemicals (e.g. insulin) useful for the pharmaceuticals industry and the treatment of disease. I will restrict my discussion in this chapter only to food production, but don't forget the GM debate is much broader than just food.

Genetic modification is a very clever technology which is the product of the recent, huge advances in understanding of molecular biology. So, if it is such a good idea what is the problem? These are the issues I will discuss in this chapter – they fall into three main areas of concern:

(1) Inserted genes might be transferred by pollination to other, related, plants and so promulgate whatever the gene codes for (e.g. herbicide resistance) in the natural environment.
(2) The GM crop and the food made from it might contain unexpected chemicals produced by the GM plant or animal – what is the risk of these chemicals to the consumer?
(3) A basic dislike of altering the fundamentals of nature.

At the end of this chapter you should:

- Understand the basic principles of genetic modification of food crops and animals.
- Appreciate the potential effects that adding alien genes to a cell might have on the biochemistry of the cell and set this into a food safety context.
- Be aware of specific examples of GM crops, their advantages for farmers, implications for the environment and effects when fed to animals.
- Be able to make your own mind up about the risk of GM foods to human consumers.

A brief introduction to nucleic acids, genetics and molecular biology

It is an almost insurmountable task to attempt to summarise the biochemistry, molecular biology and genetics that underpin GM food technology, but it is important that you have an appreciation of how alien genes can be inserted into the genetic material of a food crop or animal, the effects they might have on the biochemistry of the receiving cell and the implications for the environment that would arise if such genes could transfer from plant to plant or animal to animal in the natural world.

I'll start with a quick review of the cell's genetic materials – the nucleic acids.

Nucleic acids

The cell nucleus houses and protects the genetic material – deoxyribose nucleic acid (DNA) – on which the cell's blueprint is stored. DNA comprises

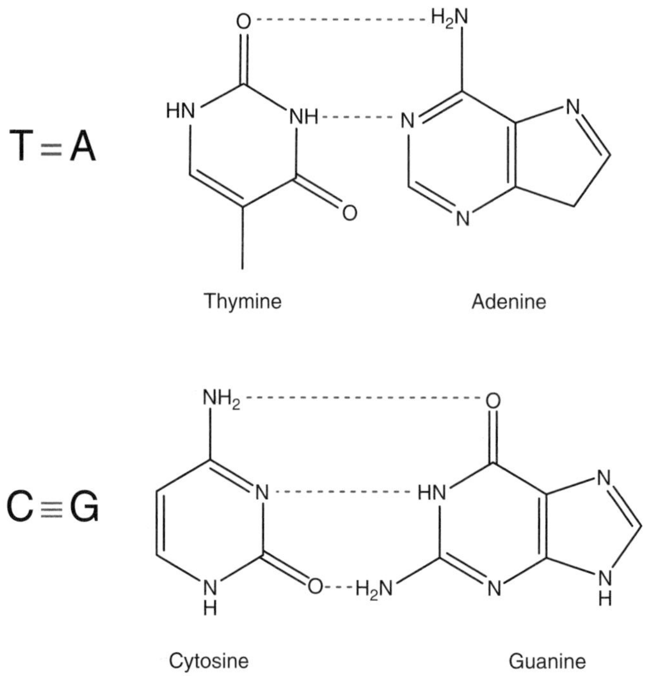

Figure 10.1   The four DNA bases showing how their molecular structures dictate the hydrogen bonding (-------) between them.

two hydrogen bonded single stands made up of a phosphorylated sugar (deoxyribose) backbone with one of four bases (guanine (G), adenine (A), cytosine (C) and thymine (T)) attached. The bases hold the two DNA strands together via hydrogen bonds to form a double helix. The chemistry of the DNA bases only allows bonding between G and C, and A and T (Figure 10.1); G is bonded to C by three hydrogen bonds whereas A is bonded to T by only two hydrogen bonds. If you look at the DNA base molecules it is obvious why this bonding arrangement arises (Figure 10.1). The base pairs' different hydrogen bonding strategies means that A can only bond to T and G to C.

Three bases code for an amino acid and this is how the genetic code is stored. The bases are the code (sets of three bases are called codons) and the code is translated by protein synthesis into a specific sequence of amino acids in a protein molecule. The sequence of amino acids in a protein determines the properties and activity of the protein and indeed the nature and identity of the protein; so the sequence of codons on DNA codes for a particular protein.

Converting the genetic code into a protein

This is a multi-stage process involving a messenger (messenger RNA; see below) to carry the DNA genetic code to the protein synthetic apparatus in the cytoplasm and manufacture of a specific protein from the DNA code.

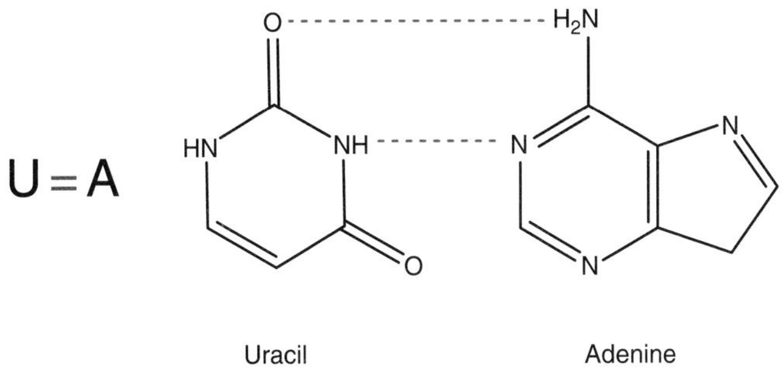

U = A

Uracil   Adenine

Figure 10.2 Uracil replaces thymine in RNA, but can form two hydrogen (----------)
bonds with adenine in an analogous way to the A = T bonds in DNA.

Part I: transcription – transferring the message

The process of converting the genetic code into a specific protein with a
specific activity involves a complex, almost miraculous sequence of events
that has only recently been fully understood – in fact, Venkatraman
Ramakrishnan (Cambridge University, UK), Thomas Stertz (Yale University,
USA) and Ada Yonath (Weizmann Institute of Science, Israel) received the
Nobel Prize in Chemistry for 'Studies on the Structure and Function of the
Ribosome' in 2009 (you can read more about this at http://nobelprize.org/
nobel_prizes/chemistry/laureates/2009/press.html#). The DNA double helix
unzips (think of the hydrogen bonds between bases as the teeth of a zip) to
reveal the reactive base groups (e.g. the $-NH_2$ of guanine with its lone pair of
electrons seeking an acceptor to hydrogen bond with). This unzipped or single
stranded DNA forms the template for the synthesis of a slightly different
nucleic acid – ribose nucleic acid (RNA). DNA and RNA differ in three ways:

(1) DNA has deoxyribose as its sugar backbone – RNA has ribose.
(2) DNA has the four bases, A, T, G and C – RNA has the base uracil (U) in
place of T.
(3) DNA is usually double stranded – RNA is always single stranded.

U can bond to A with two hydrogen bonds just like the A = T bond of DNA
(Figure 10.2).

The RNA synthesised from the DNA template has a sequence of codons that
corresponds to the original DNA sequence. RNA is used as the messenger that
takes the DNA information from the nucleus to the protein synthetic apparatus
in the cytoplasm of the cell. For this reason this type of RNA is called messenger
RNA (mRNA). This part of the protein synthesis process is called transcription.

The mRNA leaves through pores in the membrane that surrounds the
nucleus (nuclear membrane) and goes to the ribosome where proteins are
synthesised – I use the word 'goes' advisedly because it is not an accident that
the mRNA finds the ribosome; it is guided by a complex tramline-like structure termed the cytoskeleton. The ribosome (a huge enzyme complex) is
visible at high magnification under the microscope as two spheres – one large,
one small – attached to each other. The join between the two ribosome
spheres is where the reading apparatus is; here the codon sequence of the

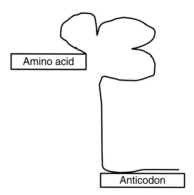

Figure 10.3 A schematic representation of the characteristic shape of tRNA showing its anticodon which corresponds to the mRNA codon and the attached amino acid that also corresponds to the mRNA codon (e.g. UGC = serine) and is added to the protein molecule as it is synthesised at the ribosome.

mRNA is read and converted to an amino acid sequence of a protein, i.e. the protein coded for by the DNA still safely stored in the nucleus.

Part II: translation – making a protein from the DNA message

Converting or translating the DNA code via mRNA to an amino acid sequence that forms a protein with a specific activity is an almost unbelievable molecular process that I still marvel at each time I think of it.

The mRNA codon reading apparatus of the ribosome exposes the mRNA bases in such a way that they are attracted to individual RNA codons (i.e. three base residues in a specific sequence) carried on a highly modified small cytoplasmic RNA molecule with an amino acid attached to it. The amino acid corresponds to the codon sequence of its three exposed bases. This small RNA molecule transfers a specific amino acid to the ribosome – it is called transfer RNA (tRNA; Figure 10.3). The tRNA codon has the opposite sequence to the mRNA codon that codes for the amino acid carried by the tRNA; for this reason it is called an anti-codon.

The tRNA carries amino acids to the ribosome and binds via its anti-codon to the mRNA codon which means that the amino acids are lined up in the sequence dictated by the RNA base sequence, which, of course, was determined by the original DNA sequence. Amazing! The ribosome has enzyme activity that joins the amino acids together by peptide bonds. As the primary sequence of the polypeptide (protein) forms it emerges from the ribosome like knitting from knitting needles and folds into a protein conformation that is dictated by the molecular interactions between its component amino acids. As it folds the protein activity is conferred – active sites of enzymes and the binding sites of receptors are created in this way.

If you would like to know more about the nucleic acids and protein synthesis read *Biochemistry* by Garrett & Grisham, Chapter 10 Nucleotides and Nucleic Acids (see *Further reading*).

Now that I have outlined the principles of the nucleic acids and how they are involved in genetic coding, communicating the code (mRNA – transcription) and expressing this in proteins (protein synthesis – translation) I can discuss the basic principles of GM and how genes are modified in food crops and animals.

The history of GM crops

The GM story began in 1973 when Stanley Cohen and Herbert Boyer at Stamford University in the USA made a DNA molecule outside a cell, then incorporated it into a cell's genome (transformation) and witnessed the cell making the protein coded for by the alien (transformed) DNA. This is the stuff of science fiction, but in 1973 Cohen and Boyer made it scientific fact – they invented genetic modification.

The scene was set and the possibility of modifying a cell's genes became reality, but it was not until 1986 when Kary Mullis of Cetus Corp., Berkeley, USA, modified (engineered) the genes of a prokaryotic cell. This, coupled with Sanger's method for sequencing DNA and the polymerase chain reaction (PCR) method for replicating (amplifying) DNA meant that scientists could identify the sequence of a gene that coded for a particular protein, make the piece of DNA (i.e. the gene) and then make a large number of copies of the gene which could be incorporated into host cells. The host cell's protein synthesis apparatus would then translate the gene into a protein. This is GE or GM.

Over the last 25 years GM technology has advanced to such an extent that it is now commonplace and is used to produce pharmaceuticals in cell culture systems (e.g. *E. coli* expressing the human insulin gene), to study biochemical and molecular mechanisms in cells to help fight disease (e.g. 'knock out' rats with specific genes disabled as models of disease processes) and to produce food more efficiently (e.g. glyphosate-resistant corn), but more of this later.

The first GM crop

The first GM crop to be approved was *Flavr Savr* tomatoes in the USA in 1994. These tomatoes were genetically modified – by inserting the polygalacturonase gene – to ripen more slowly which meant that they built up more sugars and other flavour agents and so tasted better than their conventional fast-ripening counterparts. A plethora of GM crops followed over a short time; most were herbicide (e.g. glyphosate) resistant and included rape, maize and soy. These crops revolutionised farming by allowing farmers to spray their plant crop with herbicides to kill competing weeds, but not kill the crop because it was resistant to the herbicide's effects.

Not everyone regarded GM as good. There were serious concerns about whether the genes (e.g. glyphosate resistance) inserted into food crops could be transferred to related plants via pollen (vertical transmission). Later, concerns developed about a more sinister possible route of gene transmission that involved a gene being 'absorbed' by an unrelated species' cell (horizontal gene transfer) and expressed in that cell and then possibly vertically transmitted later. Horizontal gene transfer has been demonstrated in the laboratory, but never seen in the 'wild'. Vertical gene transfer has been demonstrated in both the laboratory and in the 'wild'. These issues have little or no bearing on food safety, but they are a key part of the debate about whether or not we should use GM technology in food production.

I have only talked about food crops, i.e. plants. GM farm animals are also a possibility, but none is commercially farmed yet. One could envisage GM cows that produce high protein milk, or milk with less fat, or high levels of proteins with health benefits, or more calcium, or low cholesterol, or ... the possibilities are endless and endlessly controversial.

The tools of the genetic engineer

As already discussed in this chapter, GM is now commonplace and the methods easy to carry out; indeed molecular biology supply companies sell kits and equipment to sequence, construct, amplify, transform and transfect genes into cells.

This section is not intended to be a full account of GE/GM processes, but rather a quick run through the techniques used to give you an idea of how a GM crop is created If you would like to know more I have listed some good books in *Further reading* at the end of this chapter.

The gene that codes for a particular desirable trait (e.g. glyphosate resistance – the candidate gene) must first be identified in the species that expresses it; for this specific example the glyphosate resistance gene (GlyR) was identified in the soil bacterium *Agrobacterium tumefaciens* (see *Glyphosate-resistant crops*). The candidate gene is cut out of the genome of the cell that expresses it using restriction endonucleases – these are enzymes that cut nucleic acids between specific bases. When the gene sequence is known it is easy to select the appropriate restriction endonucleases to cut the gene out of the surrounding DNA at just the right place. The excised gene is then amplified by using it as the template for a PCR (a machine is used to achieve this). PCR is a three-stage process:

Stage 1   DNA is denatured ('melted') at 95°C. The hydrogen bonds between DNA strands break and the strands separate.

Stage 2   A primer sequence corresponding to a small section of the beginning of the gene to be copied is added and the temperature reduced to 55°C. The primer hydrogen bonds ('anneals') to the DNA template in the place corresponding to the correct base sequence.

Stage 3   The temperature is raised to 72°C and DNA polymerase is added. The polymerase reaction extends the primer by adding successive nucleotides until the complete gene is replicated.

The process is repeated many times to make many copies of the gene. At the end of the PCR process we have many tens of thousands of copies of the candidate gene (e.g. GlyR). The candidate gene is then incorporated into the genome of the (in this case) plant that we want to express the candidate gene's trait (e.g. glyphosate resistance).

The process (Figure 10.4) involves cleaving the plant genome using a restriction endonuclease and incorporating the candidate gene into the plant's DNA at the cleavage point using a ligase (an enzyme that makes the bonds between the sugar backbone of a cleaved DNA strand). The product is a modified (recombinant) plant genome incorporating the candidate gene. The recombinant gene is put into an appropriate plant cell (e.g. corn) and the

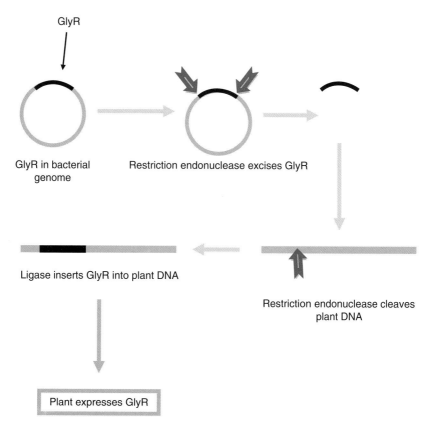

GlyR

GlyR in bacterial genome

Restriction endonuclease excises GlyR

Ligase inserts GlyR into plant DNA

Restriction endonuclease cleaves plant DNA

Plant expresses GlyR

Figure 10.4 Schematic representation of the process of inserting a new gene into a plant. In this example the glyphosate resistance gene (GlyR) from *Agrobacterium tumefaciens* is excised from the bacterial genome using specific restriction endonucleases that cleave the nucleotides at each end of GlyR, so releasing it. The plant DNA is cleaved using a restriction endonuclease (an enzyme that cuts DNA) and the GlyR gene inserted and nucleotide bonds formed using a ligase (an enzyme that joins DNA strands). The resulting plant will be resistant to glyphosate.

cells grown in culture to yield millions of new, identical cells (clones – this is the process of cloning). The cloned cells are grown and allowed to differentiate (i.e. produce different cell types) into a plant and the plant is allowed to mature and reproduce. Its seeds will contain the GlyR gene and can be grown to produce glyphosate-resistant plants (e.g. corn). It does not take long using a concerted programme to produce sufficient glyphosate-resistant seeds to create a commercial product.

GM animals are produced in just the same way as plants except the recombinant DNA is incorporated into an ovum which is fertilised and grown into an embryo in culture (i.e. *in vitro* fertilisation – IVF) and then implanted into the womb of a surrogate mother. GM offspring will be produced.

It is not possible to describe all of the GM crops or animals that have been created here and so I will describe in detail only those most commonly used in food production.

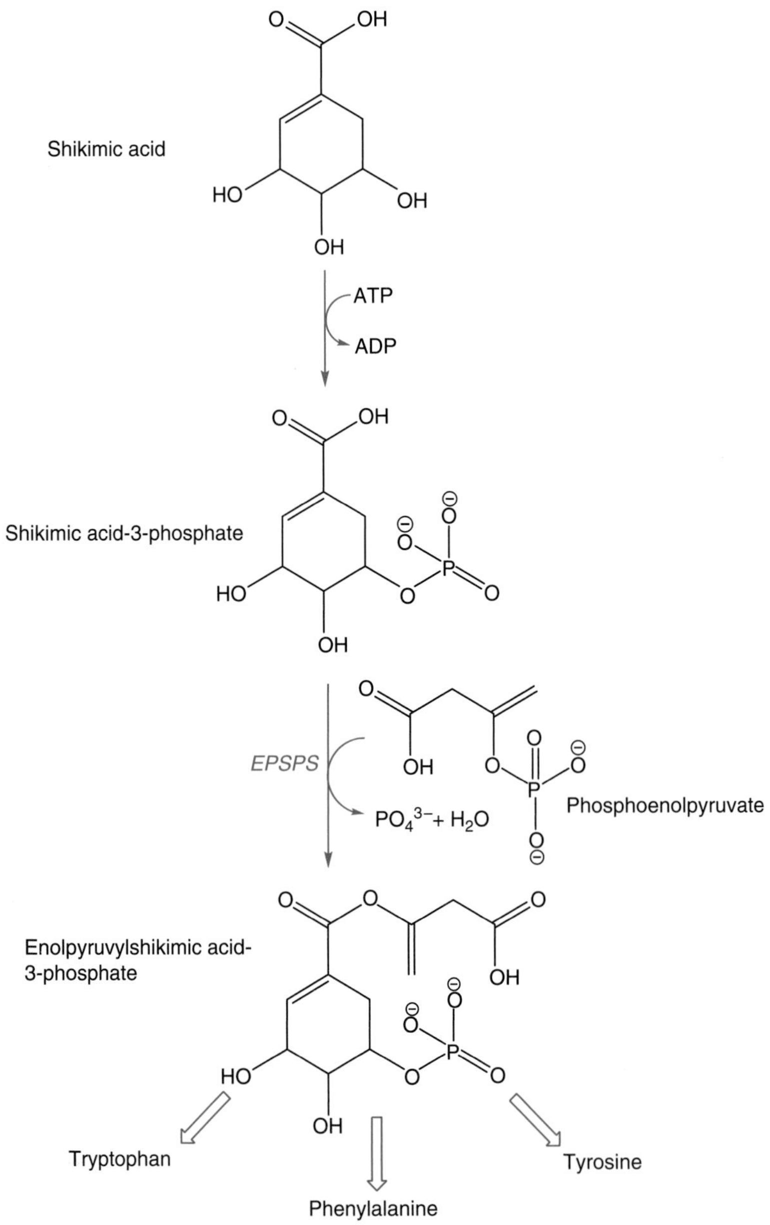

Figure 10.5 The shikimic acid pathway which leads to aromatic amino acid synthesis in plants showing the enzyme EPSPS which is inhibited by the herbicide glyphosate.

Glyphosate-resistant crops

Glyphosate-resistant crops are the most widely utilised GM crops worldwide. The most important food crops are glyphosate-resistant corn (maize), soy and rape (canola).

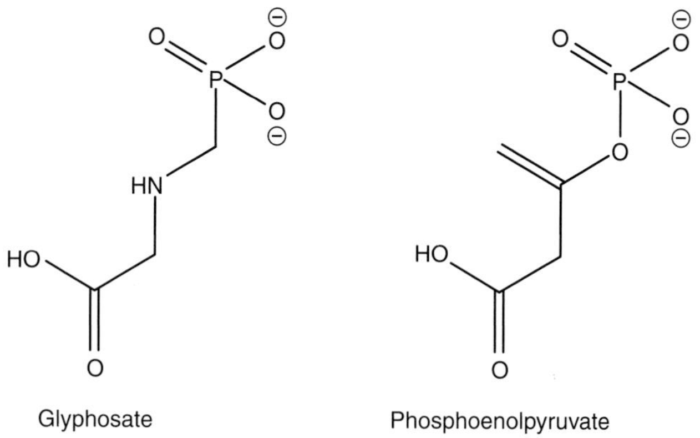

Glyphosate

Phosphoenolpyruvate

Figure 10.6 The herbicide glyphosate is similar to phosphoenolpyruvate, a substrate of EPSPS, which is why glyphosate is an EPSPS inhibitor.

Plants synthesise amino acids in order to make proteins. If they can't synthesise amino acids they can't make proteins and they die. The herbicide glyphosate inhibits a key enzyme, 5-enolpyruvylshikimate-3-phosphate synthetase (EPSPS) involved in the synthesis of aromatic amino acids (e.g. tyrosine) in plants (Figure 10.5). Glyphosate inhibits EPSPS (Figure 10.6), so preventing the plant synthesising aromatic amino acids and thus killing it.

GlyR is the gene that codes for a form of EPSPS in A. *tumefaciens* that is unaffected by glyphosate. Therefore when GlyR is inserted into a plant the plant is able to make its aromatic amino acids even in the presence of glyphosate. Therefore glyphosate-resistant GM crops express A. *tumefaciens* GlyR and so can synthesise aromatic amino acids in the presence of glyphosate. This means that glyphosate can be used to kill weeds amongst GlyR (i.e. glyphosate-resistant) crops.

New GM crops are being developed all the time because they represent huge commercial opportunities for the international agrochemicals companies that engineer and market them. The following are some of the glyphosate-resistant crops available at the time of writing:

- Soy
- Rape (canola)
- Corn
- Sugar beet
- Cotton

Insect-protected crops – BT toxin

Two of the big problems that farmers face are weeds and insect pests. Glyphosate-resistant crops address the former and insertion of genes that code for an insect toxin addresses the insect problem.

The gene used for insect protection is derived from the bacterium *Bacillus thuringiensis*. It codes for a protein toxin (BT toxin) that kills insects by interfering with their digestive process, thus starving them to death. BT toxin is insect specific which means that it does not have major safety implications for the human consumer of BT toxin-expressing crops. The only real safety concern for humans is the possibility of allergy to BT toxin which could occur in sensitive individuals. Corn, potatoes and cotton are all successful GM BT toxin-expressing crops.

It is possible to insert more than one gene into a GM crop, so giving the crop multiple desirable properties. For example, both the BT toxin and glyphosate resistance genes have been inserted into corn which is now grown commercially. The BT/glyphosate-resistant corn is both herbicide resistant and insect protected – a farmer's dream crop!

GM crops with enhanced flavour or nutritional properties

As discussed at the beginning of this chapter, the first GM crop was *Flavr Savr* tomatoes which have the polygalacturonase gene inserted which means that the tomatoes ripen slowly and taste better than conventional crops (at least this is what their marketers claim). Since then there have been numerous gene inserts that code for particular enzymes that produce either flavour molecules or molecules with nutritional value; for example, high oleic acid soy which has enhanced genes that code for enzymes in the oleic acid synthetic pathway, and therefore the soy oil produced is rich in oleic acid. Oleic acid is a commercially important fatty acid (it is a major component of olive oil).

The possibilities are endless and it would be feasible to produce crops that might be able to solve global nutritional problems. For example, imagine rice with a gene for enhanced vitamin B1 (thiamine) synthesis; this could prevent the development of beriberi (neurodegeneration due to lack of thiamine – its name is Sinhalese (Sri Lankan) for 'extreme weakness') in rice-eating undernourished communities. Unfortunately, and cynically, the financial return from such crops might not be great enough to persuade the agrochemical companies to invest the huge sums of money necessary to develop such new GM crops.

Golden Rice

A promising GM rice crop has been developed that could prevent disease due to malnutrition in a large proportion of the world – the crop is Golden Rice. Golden Rice is a GM rice which expresses the *psy* and *lyc* genes from daffodils and *crt1* gene from the bacterium *Erwinia uredovora*. *Psy* codes for phytoene synthetase, *lyc* for lycopene β-cyclase and *crt1* for phytoene desaturase – three enzymes important in carotenoid synthesis (carotenoids make daffodils yellow and carrots red) (see Chapter 11, *The chemistry of food colours*). Rice expressing these three genes synthesises carotenoids and looks yellow (or golden – hence its name), but more importantly it provides its consumer with carotenoids which are the precursors of vitamin A (Figure 10.7). Therefore, Golden Rice prevents vitamin A deficiency which affects a large proportion of the Third World – vitamin A is a key part of the biochemistry of sight and therefore its deficiency affects vision, particularly night vision.

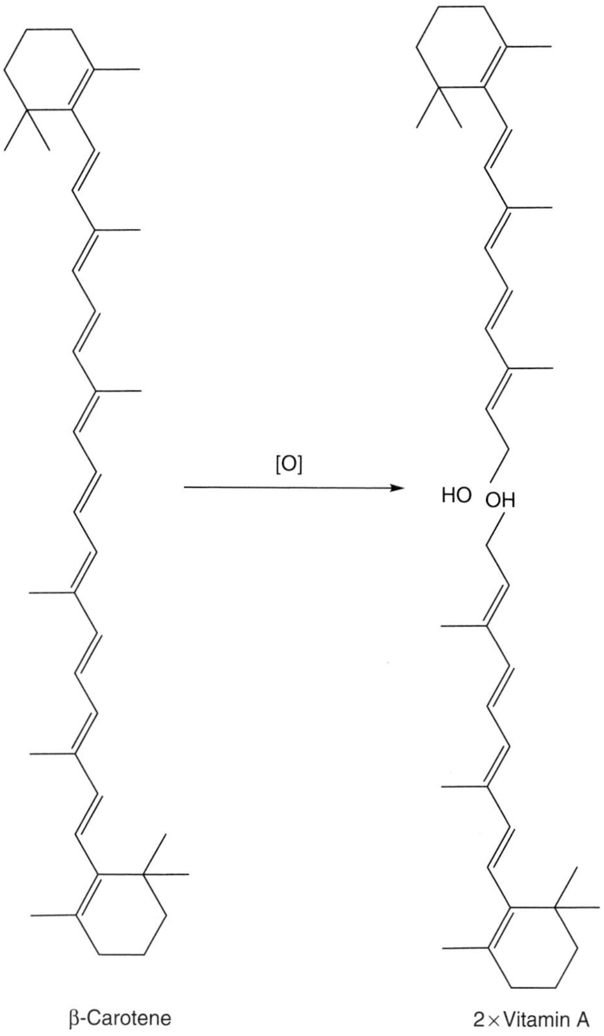

β-Carotene   2 × Vitamin A

Figure 10.7 Oxidative metabolic cleavage of β-carotene to form two vitamin A molecules – this explains why Golden Rice (carotene rich) could be important in preventing vitamin A deficiency in the Third World.

What happens if humans eat GM crops or foods made from them?

There are two distinct food product types made from GM crops – foods containing cells and therefore containing the alien gene (e.g. soy flour from glyphosate-resistant soy), and foods not containing cells and therefore not containing the alien gene (e.g. canola oil from glyphosate-resistant rape). We need to make a wide-ranging risk assessment for the former, but the latter is of less concern because there is no possibility of gene transfer to the human consumer.

The implications of horizontal gene transfer are key considerations when developing and growing GM crops. Remember, horizontal gene transfer is the possibility that genetic material might be transferred from one cell to another simply by growing two cells in close proximity. If this happens it is possible that the cell to which the genetic material was transferred might express it. The main concern in relation to horizontal gene transfer is that if we eat a GM food containing cells (e.g. GM soy flour) the genes in the cells might transfer to the natural bacteria (microflora) in our intestine and then be transferred later to our gut cells. At the present time this is not a significant issue because the modified genes (e.g. GlyR) currently used are of little or no biochemical significance to humans. This might not be the case in the future and therefore we must remain vigilant and consider these possibilities seriously when new GM crops are developed. Horizontal gene transfer from GM crops to human cells has not been demonstrated, but co-culture of GM bacteria with a different bacterial species has been shown to result in horizontal gene transfer to the second species and therefore, in theory, transfer to other cells is possible.

The environmental implications of horizontal gene transfer are significant because the expression of many of the currently inserted genes would have significant environmental impact if they were widely expressed. For example, if the BT toxin gene from a GM crop transferred to wild plants and was expressed it would kill insects indiscriminately and would significantly affect their population and might even result in their extinction. These are serious issues that I cannot develop further here, but they must be taken into account when assessing the safety and acceptability of GM crops.

Changed biochemistry in GM crops

While gene transfer is a significant issue for risk assessors of GM crops, we must also consider the changes in biochemistry that cells undergo when an alien gene is inserted and expressed. For example, insertion of the GlyR gene might not only result in the synthesis of EPSPS, but also interfere with the expression of other genes in the plant's genome. Indeed, it is known that the concentrations of certain metabolites in GM plants differ from their unmodified counterparts. Usually the differences are minor and, usually, relate to concentrations of 'normal' metabolites (e.g. sugars) that would not be expected to have any adverse effects on the consumer, but we must be vigilant in case harmful changes occur in future GM crops. For this reason, GM crops are subjected to extensive testing to make sure that harmful changes have not occurred and therefore the risk is low.

What is the effect of eating DNA and RNA?

It is important to remember that we are constantly eating alien nucleic acid. Every time we eat food containing cells derived from plants or animals we eat DNA and RNA. As far as we know this nucleic acid is broken down by digestive enzymes and absorbed as nucleotides (base + sugar), nucleosides (base + sugar phosphate), bases and/or sugars; there is no evidence that genes from food plants and animals are incorporated into gut microflora or human genomes. This begs the question why would alien genes added to food plants or animals be transferred to human cells following their consumption? Unfortunately we don't know the answer to this question yet.

GM animals

Transgenic or GM animals are still at the experimental stage. Most studies are still utilising small animal models (e.g. mice) to develop the technology, but it will not be long before GM farm animals are a feature of farmyards in some parts of the world. Indeed, in 2002 the first cow to produce human-like milk was developed by inserting genes for the synthesis of specific proteins (e.g. lactoferrin) into the cow's genome. The idea of producing 'human' milk in cows is very interesting because it would solve some of the problems of feeding babies conveniently. Currently cow's milk is used, but it has numerous disadvantages when compared to human milk because it contains proteins (e.g. β-lactoglobulin) that human milk does not contain and some babies develop allergic responses to these proteins which might lead to hyperallogenicity (e.g. asthma) in later life (see Chapter 15).

Transgenic farm animals might also be used to manufacture pharmaceuticals. For example, cows expressing the gene for human insulin excrete insulin in their milk; so it is conceivable that one day farming might become 'pharming' and 'pharmers' will milk their GM human insulin cows and sell the milk to the pharmaceuticals industry where the insulin will be extracted for the treatment of human diabetes. This is far beyond the scope of this book but is an interesting thought that is fast gaining credibility – in just a few years it is likely to be reality.

Take home messages

- Genetic modification involves adding a gene to a crop or animal that confers some positive characteristics (e.g. resistance to herbicides).
- Genetic modification is controversial and there is still significant debate about whether it is acceptable on both safety and environmental grounds.
- Some countries freely allow GM crops providing strict regulations are followed (e.g. the USA). Other countries do not allow GM crops or their products (e.g. New Zealand).
- The commonest GM crops are glyphosate (a herbicide) resistant and BT toxin (an insecticide) expressing.
- GM crops could be used to solve global nutritional problems (e.g. Golden Rice – vitamin A deficiency).

Further reading

Garrett RH & Grisham CM (2010) *Biochemistry*, 4th edn. Chapter 10, Nucleotides and Nucleic Acids. Brooks/Cole, Boston. This is an excellent in-depth account of the chemistry and function of nucleic acids; it provides all of the background you will need to understand the cell biology and biochemistry behind genetic modification.

Kammermeyer K & Clark VL (1989) *Genetic Engineering Fundamentals*. Chapter 7, Recombinant Techniques. Marcel Dekker, New York. This gives a good account of the methods used.

Parekh SR (ed.) (2004) *The GMO Handbook*. Humana Press, Totowa.

Ruse M & Castle D (eds) (2002) *Genetically Modified Foods*. Prometheus Books, Amherst.

Chapter 11

Colours, Flavours and Preservatives

Introduction

The colour, storage and flavour of food have always been important. In prehistoric times food colour was probably important because it warned its consumer of danger, or, perhaps, was associated with nice flavour.

Storing food for the winter when there was less food to be had led early people to ferment, add honey and acids (e.g. acetic acid in aerobically fermented grape juice (vinegar)), and add safe bacteria to prevent harmful bacteria growing (e.g. yoghurt) to preserve food for the lean winter times. In addition, primitive people enhanced the flavour of their food with honey and later the Romans developed a whole economy around salt because it was so highly prized as a flavour enhancer.

So, since time immemorial, human kind has considered colour, preservation and enhancing the flavour of food as important aspects of everyday life. Today, colours, preservatives and flavours have developed into a science in their own right. Without them our tinned peas would be brown, bread would go mouldy very quickly and our Chinese takeaway meal would taste very bland. But there is very much more to colours, preservatives and flavours in the modern world of 'synthetic' food – some foods quite clearly taste of banana or perhaps pineapple, but contain no pineapple or banana, but instead contain synthetic chemicals that taste like banana or pineapple. Our fat-containing foods contain antioxidants to prevent their unsaturated fats oxidising to foul tasting and smelling carboxylic acids. Our cherryade (cherry soda) is red just like cherries, not because it contains cherries, but because it contains a cherry red dye – and for that matter it probably tastes of cherries not because it contains cherries but because it contains chemicals that fool your taste buds into thinking they have experienced cherries. Indeed, it is likely that many cheap cherryades contain no cherries whatsoever.

The use of chemicals to flavour, preserve and colour our food makes it cheaper and more convenient. But there are concerns about the effects that the chemicals we add to food are having on us. The yellow colour (tartrazine) that was added to butter and some soft drinks to make them

Food Safety: The Science of Keeping Food Safe, First Edition. Ian C. Shaw.
© 2013 John Wiley & Sons, Ltd. Published 2013 by John Wiley & Sons, Ltd.

an (apparently) appetising yellow is associated with allergy in children – it is still permitted for use in most countries, but its use is declining because of consumer pressure. Saccharin, the ultra-sweet hydrocarbon, causes cancer under specific conditions in specific strains of mice and was banned in Canada, but not in the USA, which meant that you could step across the Canadian–US border and enter a regime where saccharin was deemed safe or too risky to include in food depending whether you stepped north to south or south to north – this is nonsense ... it is either safe or it isn't!

It is easy to take the 'organic' stance and opt to eat only naturally coloured, preserved and flavoured food, but in these days of mass production, food preservation is an important means of avoiding food-borne illness outbreaks – just imagine a convenience food with no added preservatives growing *Clostridium botulinum* and the health implications of its consumption. So, perhaps, we should accept some of the risks of food preservatives if we set them against the benefit of safe food.

Colours and flavours present a more difficult risk argument, because their benefits are less obvious. One benefit is price – canned peas are much cheaper than fresh peas, but would be a strange unappetising colour if they did not have 'pea green chemicals' added. Moreover, from the manufacturer's perspective they would soon lose their customers if they began to sell brown canned peas. There are many arguments for and against colours, preservatives and flavours, some of which I will discuss in this chapter.

Whether a particular food additive is necessary and safe is a key issue. Indeed, most developed countries have legislation that regulates the addition of colours, preservatives and flavours to foods. Such legislation includes the need for toxicity testing to determine the hazards associated with the chemicals from which risk to the consumer can be assessed. Interestingly, some jurisdictions (e.g. the EU) do not allow additives in baby foods, which is an important risk statement in its own right.

Food manufacturers are aware of more recent consumer desires for 'natural' and will go to great lengths to add only 'natural' chemicals to food to preserve, flavour and colour it. Sometimes these natural chemicals are extracted from nature, while others are synthesised but are identical to the natural chemicals (e.g. natural vanilla versus chemically synthesised vanillin; see Chapter 8, Capsaicin, peppers and other toxic flavours). This gives us a significant risk dilemma – the chemicals are identical to nature's own but at high enough doses might be toxic (remember Paracelsus – 'All things are poisons ...'); we did not worry about them when they were a natural component of our food, but now the manufacturers are adding them we are worried, but why?

As you can see from this preamble, the food safety aspects of colours, preservatives and flavours are not simple. In this chapter I will attempt to unravel some of the science behind colours, flavours and preservatives and at the end you should understand the toxicological issues that have led to the vehement debate about whether or not colours, flavours and preservatives should be added to our food.

I will deal with colours, flavours and preservatives separately below because their risk-benefit assessments are quite different.

Food colours

Colour is very important in all aspects of our lives. Every day we look, see and decide based partly on colour and other visual stimuli (e.g. shape). We choose our food partly on the basis of its colour. A rosy red apple suggests ripe and sweet; the wonderful shiny green and blue mottled hues of a mackerel in the fish market 'shouts' fresh and tasty. On the other hand, a brown apple suggests old and decaying and a cloudy grey-looking mackerel would not be your choice for lunch. So colour is important in making food decisions - whether to buy a particular food or not. The marketers are well aware of this and they want to colour their products in such a way that you want to buy them.

Colour also suggests flavour. I was once involved in a taste panel - we were given different mashed foods to eat; one was blue, one was brown, one was pea-green and one was black. They all tasted different to me. Then we were told that they were all mashed peas and asked which tasted the best, Everyone on the panel said the pea-green one tasted best. Then we were told that they were all from the same batch of peas that had been mashed and different dyes added. The colour had fooled us into thinking one tasted better than the others. And the one that tasted best was the one with the colour we would expect peas to be - pea-green! There is a great deal of psychology in the complex interactions between colour, flavour and desirability. It is far beyond the scope of this book to delve into this interesting subject, but just remember that the colour of food is important for many reasons - not all of them obvious - and that this drives the food industry's use of food colours to make their products more desirable to the consumer.

The chemistry of food colours

All foods have natural colour due to coloured molecules synthesised by the plant or animal from which the food originated. Runner beans are green because they contain chlorophylls - important pigments in photosynthesis - we like our beans to be green because this is what they look like in nature. Beetroot is red because it contains betalains (Figure 11.1) and blueberries, grapes and red cabbage are red or blue because they contain high concentrations of anthocyanins (Figure 11.2) - anthocyanins are found widely in nature and are responsible for so many of the rich colours of plants and flowers. Carrots are orange and peppers and tomatoes are red because they contain carotenes (Figure 11.3). These are all natural colours that we accept because they make our food what we expect it to be.

The psychology of food colour – the introduction of food dyes

Blueberry muffins contain highly coloured blueberries. Even though the flavour of the blueberries is characteristic (and very nice!), sometimes cheaper imitation blueberry muffins are made with small pieces of apple dyed with anthocyanins from grape skins in place of the expensive blueberries. I have eaten these many times and have not noticed the difference at first because the colour was right and therefore my psychology told me that they tasted right. The question is, is there a problem with this? The anthocyanins used to dye the pseudo-blueberries (i.e. apple) are structurally very similar to

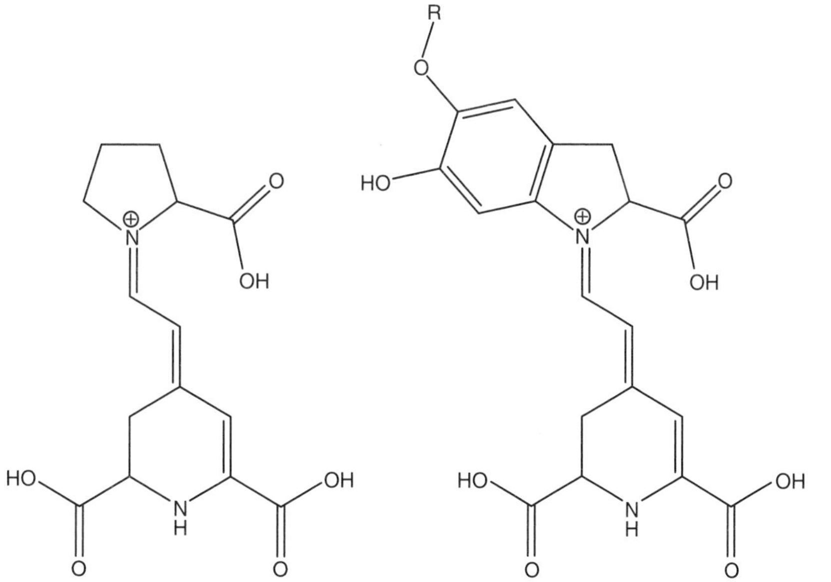

Betaxanthin – orange/red   Betacyanin – blue/red

Figure 11.1 Colour chemicals from beetroot – the molecules have conjugated bonds (i.e. double, single, double, single, etc.) which allows electron delocalisation. The degree of electron delocalisation determines the spectral absorption properties (i.e. colour) of the molecule; betacyanin is more conjugated than betaxanthin which means that it has more electron delocalisation which means that its absorption is at the red end of the spectrum, therefore it reflects blue and looks blue. The combination of blue/red betacyanin and orange/red betaxanthin gives beetroot its intense dark red colour. R, Variable alkyl group.

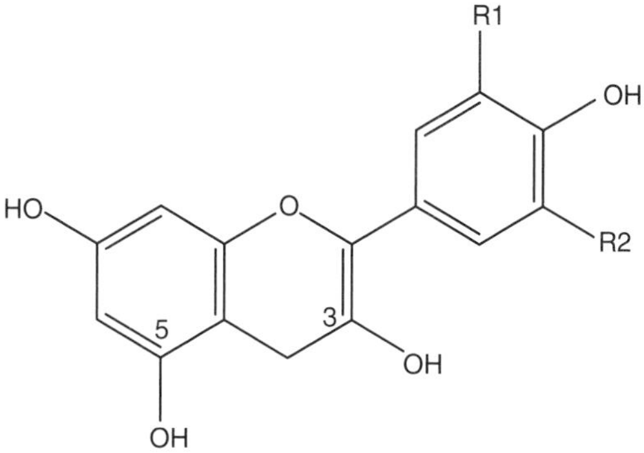

Figure 11.2 The anthocyanins are a group of colours found widely in nature and are based on 5- and/or 3-glycosylated anthocyanidins often named after the plant from which they were first isolated:

	R1	R2
Pelargonidin	H	H
Cyanidin	OH	H
Peonidin	OCH$_3$	H
Delphinidin	OH	OH
Petunidin	OCH$_3$	OH
Malvidin	OCH$_3$	OCH$_3$

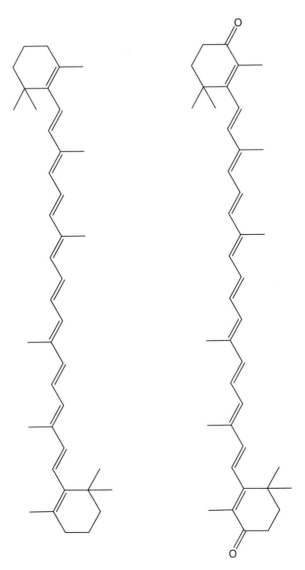

Figure 11.3   β-Carotene (left), the pigment that makes carrots orange, and the red synthetic food colour canthaxanthin (right) – spot the difference! Note the high degree of conjugation of the carotenes and that a small change in conjugation (i.e. addition of two conjugated ring keto groups) makes a great difference to the colour (i.e. orange → red).

blueberry anthocyanins, and they are natural. So what's the problem? I've been fooled, that's the problem. How dare they fool me!

There are many other natural dyes extracted from plants and used to give food the colour we expect it to have. Price very much influences the use of colouring agents. Saffron is a very expensive product used to colour (and flavour; more of this later) food. It is obtained from the anthers of a European crocus (*Crocus sativus*) and its collection and processing

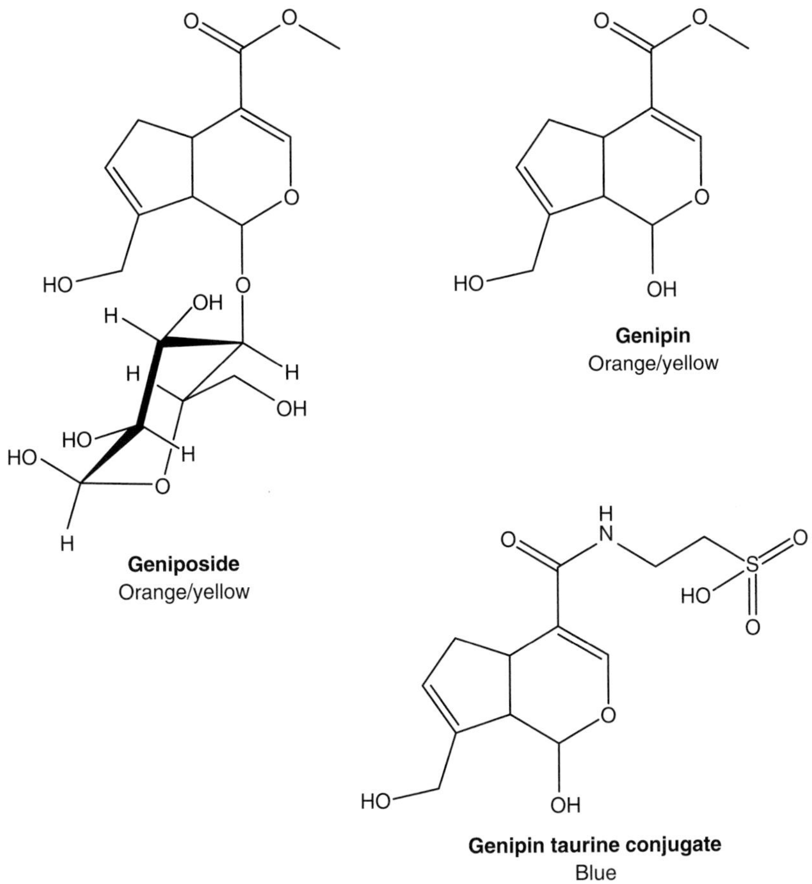

Figure 11.4 Geniposide, the orange/yellow colour from saffron and gardenia, its aglycone (i.e. glucose removed) genipin and the deep blue genipin taurine conjugate.

are labour intensive, hence its high price. Saffron added to Spanish paella gives the rice that wonderful rich yellow colour (and delightful bitterish taste). The molecules that make saffron yellow are the iridoids (Figure 11.4; named after the botanical family *Iridaceae* to which crocuses belong). Other plants also contain iridoids, and some of these plants are much easier to grow and isolate iridoids from, which makes them cheaper; for example, the fruit of Cape Jasmine (*Gardenia jasminoides*) contains a number of iridoids that are used as yellow or orange food colours (Figure 11.4).

Understanding the structural chemistry that results in colour means that chemists can make molecules based on their natural counterparts that have similar colours - this is often cheaper than isolating the chemicals from plants and therefore there is a profit incentive for the industry. Not only do they persuade the consumer (their customer) that the food is something it isn't, but also they do it cheaply.

Synthetic carotenes were the first to be made in 1950; they were followed during the 1950s and early 1960s by carotene methyl and ethyl esters, alkyl substituents, red canthaxanthin and bright orange β-8-carotenol. All became important food colours that made food brighter and more appealing.

In the years that followed many colours were made based on the molecular structures of natural plant colours. The iridoids (see above) produce a wide range of very different colours following simple modifications to their molecules; for example, genipin can be reacted with taurine to produce a blue conjugate (Figure 11.4). Other genipin adducts give greens, deep blues and reds – all are, or have been, used as food colours.

Non-plant natural colours

It seems logical to use plant colours to dye vegetable and fruit food products even if the colours do not originate from the food product itself (e.g. grape-derived anthocyanins are used to colour apple for use in 'blueberry' muffins). However, there are natural dyes from animals that are used to colour foods; perhaps the most famous in carmine.

Carmine

Carmine is a red colour found in several species of scale insect including cochineals (*Dactylopius coccus*). It has been used for a very long time. Records from the Aztec and Mayan people of Central and North America show that they traded in it (mainly as a cloth dye) in the 14th and 15th centuries. The chemical that gives carmine its red colour is carminic acid, which can be extracted from cochineals (and some other insects). The carminic acid (Figure 11.5) gene has been identified and spliced into various bacterial genomes which means that commercial carminic acid can be produced by carminic acid gene-expressing bacteria (e.g. *E. coli*).

Carminic acid has a very low acute toxicity (LD_{50} [rat, oral] = 3,000 mg/kg body weight), but is thought to be allergenic in sensitive individuals, as are many other food colours.

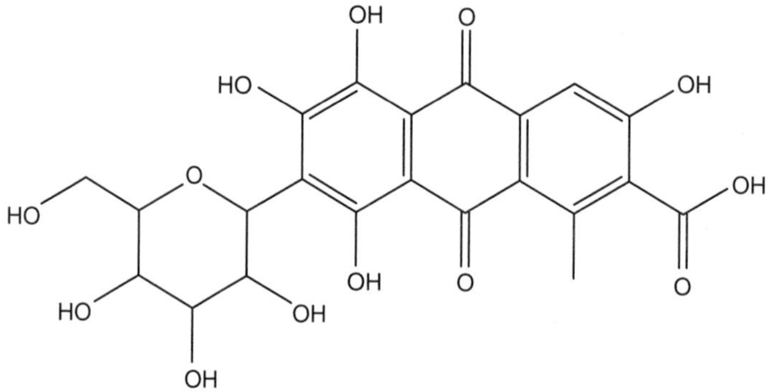

Figure 11.5 Carminic acid – the molecule that makes the cochineal scale insect red and has been used as a food colour for many years.

Table 11.1 The seven food colours permitted in the USA, with their E numbers.

Food colour	Colour	E number
Brilliant Blue FCF	Blue	E133
Indigotine	Indigo	E132
Fast Green	Turquoise	E143
Allura Red	Red	E129
Erythrosine	Pink	E127
Tartrazine	Yellow	E102
Sunset Yellow FCF	Orange	E110

Synthetic food colours

There has been much controversy about the health implications of the synthetic food colours. Some have been linked to attention deficit hyperactivity disorder (ADHD) in children and others are carcinogens or cancer suspect agents. To some extent, it does not matter to the consumer whether a particular dye can be reasonably expected (e.g. it is toxic to animals) to be harmful to humans, it is the fact that some dyes are harmful that leads to wariness of all dyes. For these reasons, many countries have significantly reduced the number of food dyes allowed. For example, the USA FDA permits the use of only seven food colours (Table 11.1 and Figure 11.6).

I will discuss Brilliant Blue FCF, erythrosine and tartrazine in more detail as examples of synthetic food colours.

Brilliant Blue FCF

Brilliant Blue FCF (Figure 11.6) is used widely to colour a multitude of foods (estimated average consumption = 16 mg/person/day); for example, canned peas are often dyed with a mixture of Brilliant Blue FCF (blue) and tartrazine (yellow) to give pea-green (blue + yellow = green). As with most of the synthetic food colours it has a very low acute toxicity (LD_{50} [rat, oral] = 2,000 mg/kg body weight), but has been associated with ADHD in children and is allergenic. The latter two toxicities have led to calls for this and other synthetic colours to be banned.

Erythrosine

Erythrosine (Figure 11.6) is another widely used dye – most cherryades are red because of erythrosine not cherries! Again, it has a very low acute toxicity (LD_{50} [rat, oral] = 1,840 mg/kg/body weight). Erythrosine has four iodine atoms in its molecular structure which is unusual for a food colour chemical. Iodine is a component of the thyroid hormones, T_3, rT_3 and T_4 (thyroxine), and there are some structural analogies between erythrosine and the thyroid hormones (Figure 11.7). Toxicity studies in rats have shown that at very high oral doses (0.4% of diet – approx. 250 mg/kg body weight) of erythrosine the incidence of thyroid tumours increases and levels of thyroid stimulating hormone (TSH – a pituitary hormone that signals to the thyroid to make T_3 and T_4), rT_3 and T_4 also increase, whereas T_3 levels decrease. T_4 is synthesised

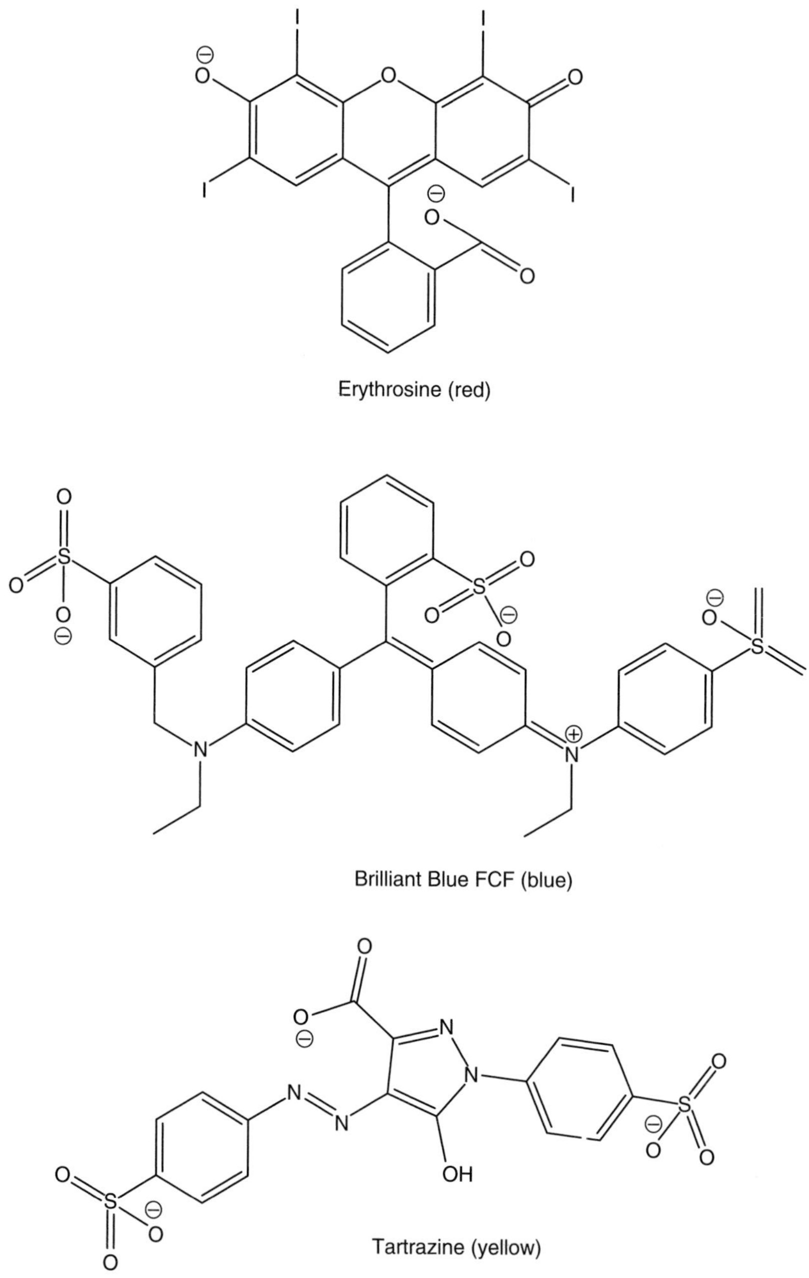

Erythrosine (red)

Brilliant Blue FCF (blue)

Tartrazine (yellow)

Figure 11.6 Three of the seven food colours permitted in the USA – note their highly conjugated structures which give them their bright colours.

in the thyroid from T_3 which is why elevated T_4 results in decreased T_3. These findings led to the ADI for erythrosine being reduced to 0–0.05 mg/kg body weight. The thyroid effects are only seen at very high erythrosine doses and therefore we would not expect effects in humans who consume erythrosine-coloured food from which doses of only picograms per kilogram body weight

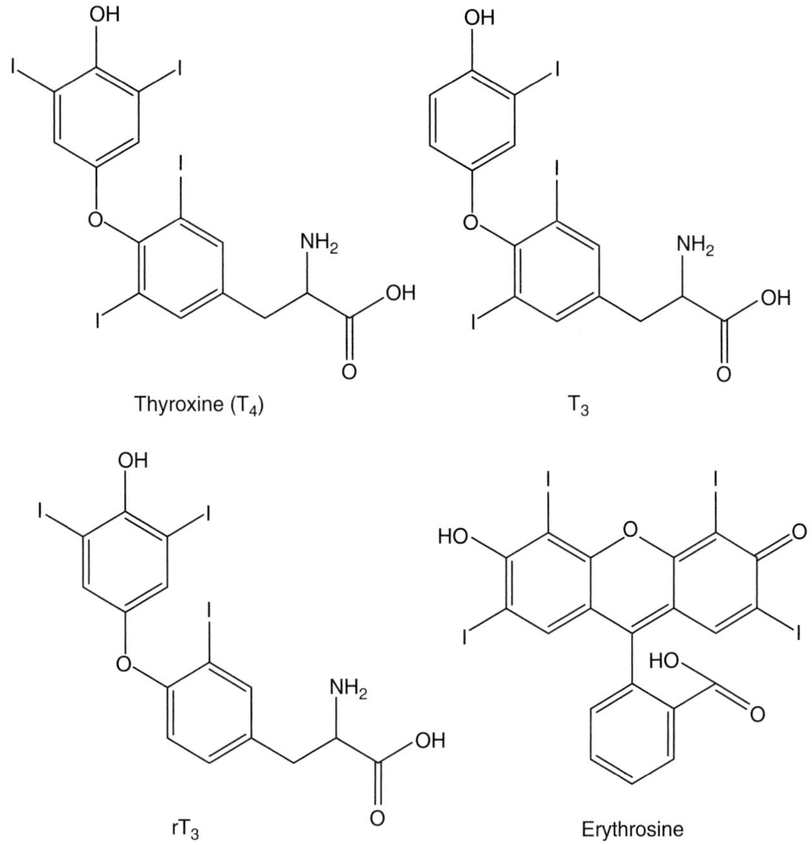

Figure 11.7 The three thyroid hormones, thyroxine (T_4), T_3 and rT_3, and the red food dye erythrosine which interferes with thyroid hormone levels and leads to thyroid cancer. You can see from the structure of erythrosine how it might interfere with either the enzyme synthesis or receptor-based endocrine feedback of the thyroid hormones in animals because its structural similarity to the hormones means that it might inhibit their synthesis by binding to enzyme active sites and thus inhibiting them, or it might fool TSH into 'thinking' that thyroid hormone levels are high by binding to its receptor site and so turning off thyroid hormone synthesis via the TSH/thyroxine feedback mechanism.

would be expected. Nevertheless, thyroid tumours are a significant hazard, albeit at low risk, but the benefit to the consumer is questionable – so should we accept this risk?

Tartrazine

There has been a long-running controversy about the use of tartrazine (Figure 11.6) in food; there is even a specific protest group whose mission it is to get tartrazine banned. So why are some people worried about tartrazine? As with most of the other synthetic food colours, tartrazine has a very low acute toxicity (LD_{50} [mouse, oral] = 12,750 mg/kg/body weight); however, it is allergenic, and its allergenic activity appears to be greater than other food colours – interestingly, its allergenicity is linked to allergy to aspirin. Some countries in Europe have banned tartrazine, but its use *is* allowed under EU legislation.

Take home message – are food colours safe?

Natural food colours in the foods that they naturally colour (e.g. carotene in carrots) are often beneficial because of their antioxidant properties, so they certainly are not problematic from a toxicological point of view. The natural food colours extracted from plants (often food plants, e.g. anthocyanins from grape skins) and used to dye different foods (e.g. apple chunks as pseudo-blueberries for muffins) do not change their properties when they are used in different foods, so, generally, they are, if anything, beneficial too. The modified natural colours (e.g. genipin taurine conjugate – genipin is isolated from gardenia, a non-food plant) are difficult to assess in an all or nothing manner – their toxicity depends upon the chemical modifications to the molecules. On the whole, however, they are of very low toxicity and therefore of little concern. The synthetic colours are quite different. Although their acute toxicities are very low – usually many tens of grams would be necessary to harm a human and they are used in foods at microgram to milligram levels (resulting in doses of picograms per kilogram body weight to the consumer) – many of them are allergenic and some are associated with hyperactivity disorders in children and erythrosine causes thyroid cancer at very high doses (250 mg/kg body weight). Clearly we must treat them with caution and ask the question, does the benefit associated with their use outweigh the risk? Do we really need our soda to be bright yellow (tartrazine) or our canned beans bright green (tartrazine + Brilliant Blue FCF)? I think the answer is *no* on purely allergenicity grounds, but this is a question for society ... and the debate is still raging! What do you think?

Flavours

We expect orange juice to taste of oranges, cherryade to taste of cherries and banana milkshake to taste of bananas ... and they do, but is this because they contain sufficient bananas to make them taste of banana or cherries to taste of cherry? Often the answer is *no* because flavour chemicals are used to enhance our experience and fulfil our expectations. In some cases the product (e.g. orange juice) tastes exactly like oranges, but consumers have developed a different expectation of what orange juice should taste like because they have consumed a coloured, flavoured, reconstituted liquid derived from oranges for so long that they have forgotten what the real thing tastes like. You might like to test this out. Squeeze an orange, chill and drink the juice – it tastes of rather watery oranges, but it is *real* orange juice! Now taste a proprietary boxed or bottled orange juice – it tastes just like you expect orange juice to taste. Now read the label. I had a glass of orange juice with my breakfast this morning – here's the ingredients list from its label:

Reconstituted orange juice (99.9%)
Flavour
Vitamin C

Flavour agents are important to fulfil our expectations of the food we eat and to give us the pleasurable experience we expect from eating, but are their risks worth the benefit? I will explore this important question below, but before I do we need to understand how flavour chemicals achieve their goal - flavour.

How do we sense flavour?

Our tongue is our flavour sense organ and the taste buds on our tongues detect specific taste types (sweet, bitter, sour). Taste buds are complex receptors with binding sites for specific types of molecules. If a molecule is able to bind to a specific taste receptor (e.g. a sweet receptor) it initiates a nerve impulse that sends a 'sweet' message to the brain. Similarly, if a particular molecule fits a bitter receptor, a 'bitter' message is sent to the brain. The brain receives the messages and combines them to give you the complex mixture of individual taste types that make up the flavour of food. We recognise these flavours as particular foods. If I blindfolded you and asked you to drink some orange juice you would probably recognise its flavour. If you saw it, the colour would significantly help your recognition process - flavour is a very complex sensory experience that involves both sight and the sensation of texture too.

It is the shape of a molecule and the chemical groups that are on the molecule's surface that tongue receptors recognise. This is easy to see if we consider sugars (i.e. sweet) - they have very similar molecular structures and it is not difficult to see why they all fit sweet receptors (Figure 11.8). Their

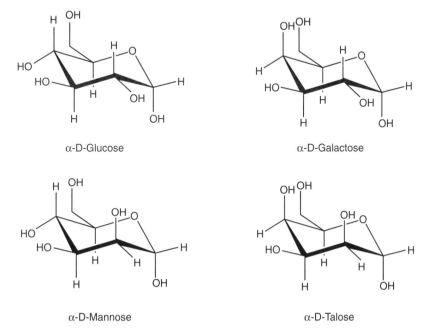

α-D-Glucose α-D-Galactose

α-D-Mannose α-D-Talose

Figure 11.8 Four sugars (pyranose form) showing their molecular similarities – it is not difficult to see why they would all fit sweet receptors and therefore taste sweet.

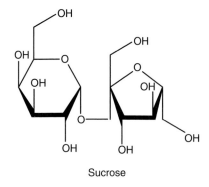

Sucrose

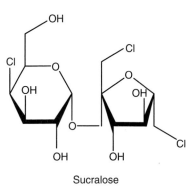

Sucralose

Figure 11.9 The disaccharide sucrose (table sugar) and the artificial sweetener sucralose – the molecular similarities between the two makes it clear why sucralose is sweet.

degree of sweetness is determined by the number of hydroxyl groups and their specific orientations, i.e. how well they fit the receptor.

Knowing the molecular attributes necessary to lead to sweetness means that it might be possible to design a sweet molecule – this is the basis of the non-sugar sweeteners (e.g. sucralose; Figure 11.9). Non-sugar sweeteners are important because they taste sweet, but are not metabolised like sugars and therefore are not fattening (more of this later).

As mentioned earlier, flavour is complex. Many molecules contribute to the flavour of foods and it would be very difficult to recreate this syntheti-cally. It is for this reason that artificially flavoured foods often do not have the depth of flavour of naturally flavoured foods because the synthetic fla-vour usually only has one or two flavour chemicals used to represent a very complex array of chemicals that constitute natural flavour. For example, the natural flavour of raspberries involves at least 13 hydrocarbons, 36 alcohols, 17 aldehydes, 22 ketones and 27 esters (Figure 11.10), and artificial raspberry flavour is just one chemical - 4-(4-hydroxyphenyl)butan-2-one (Figure 11.11).

Many of the natural flavour chemicals occur in many different fruits and vegetables – it is the complex combination and concentration ratios of the chemicals that produce the flavour rather than one molecule alone. Returning to the natural raspberry flavour example; geraniol is present in rose oil and is

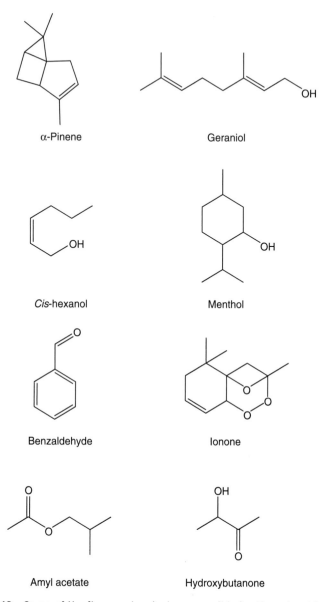

α-Pinene   Geraniol

Cis-hexanol   Menthol

Benzaldehyde   Ionone

Amyl acetate   Hydroxybutanone

Figure 11.10 Some of the flavour chemicals responsible for the natural taste of raspberries.

responsible for the characteristic sweet-scented flavour of rose water; cis-hexanol and menthol are found in mint; amyl acetate is a key component of pear and banana flavours; and ionone is present in quinces. So, raspberry flavour is, in part, a complex coalescence of mint, quince, banana, pear and rose flavours, plus an awful lot more. But if you chop up and mix together mint, quince, banana, pear and rose petals it won't taste like raspberries!

Finally, the spatial arrangements of chemical groups around a molecule are important in imparting flavour because taste receptors are stereospecific; of

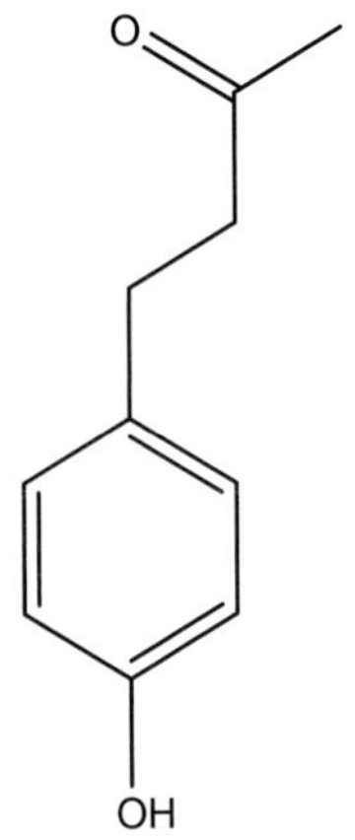

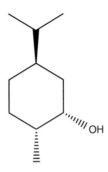

Figure 11.11 Artificial raspberry flavour – 4-(4-hydroxyphenyl)butan-2-one.

Figure 11.12 1R, 2S, 5R-menthol – the specific stereo-isomer of menthol that tastes most minty.

all the stereo-isomers of menthol, 1R, 2S, 5R-menthol has the strongest mint flavour (Figure 11.12).

Flavours used in the food industry can be divided, rather like the food colours, into three groups:

(1) Natural flavours – natural flavour chemicals extracted from fruits or vegetables (e.g. vanilla extract).
(2) Nature identical flavours – chemically synthesised molecules identical to their natural counterparts (e.g. vanillin).
(3) Artificial flavours – chemicals that impart a flavour characteristic of a particular food, but do not resemble in molecular terms the natural flavour molecules (e.g. saccharin; Figure 11.13).

Natural flavours

As with other food additives, consumers prefer natural flavours because they think (i.e. risk perception) they are safer than artificial flavours – that might be true, but it is not based on any toxicological reasoning. Just the feeling

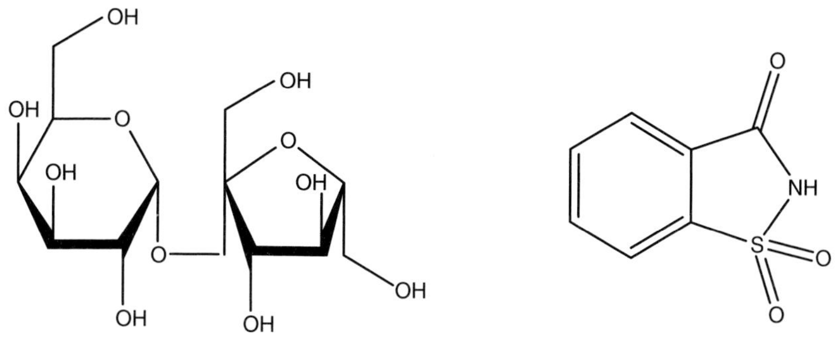

Sucrose   Saccharin

Figure 11.13 The natural sugar sucrose compared with the artificial sweetener saccharin – there is little or no structural resemblance, but they both taste sweet.

that *natural* is good and safe may not be enough (don't forget natural tetrodotoxin is one of the most poisonous chemicals known! See Chapter 8, *Tetrodotoxin from fugu*).

The subject of natural flavours is huge and I cannot do justice to it in the space available. I will just give two examples to illustrate the sort of flavouring agents used and their chemical composition.

Vanilla essence

Vanilla is one of the most widely used flavouring agents worldwide and has been around for a very long time – the Aztecs used it to flavour their chocolate beverages in the 15th century. It is used nowadays in bakery (e.g. cakes) and many other products (e.g. custard). Vanilla flavour chemistry is complex; this complexity can be appreciated when we consider how vanilla flavours are produced.

Perhaps surprisingly the vanilla flavour chemicals are not all present naturally in the vanilla pod (the fruit of two species of climbing orchid – *Vanilla planifolia* (from South America) and *V. tahitensis* (from Oceania)); they are produced during a curing process involving drying and leaving the pods in the sun for long periods after harvest. Some very complex chemistry goes on during the fermentation process which produces the rich flavour of vanilla. There are very many chemicals present in natural vanilla extract, including vanillin (about 3%), vanillic acid (Figure 11.14), sugars, alcohols, aldehydes and esters; they all have a part to play in the flavour of real vanilla. Vanilla essence is an ethanol extract (tincture) of cured vanilla pods.

The toxicity of vanilla is very low (LD_{50} [rat, oral] vanillin = 1,580 mg/kg body weight).

Citrus oils

Citrus oils are expressed from the skins of the fruits of citrus trees (e.g. orange, *Citrus sinensis*). They are widely used as flavouring agents in foods (e.g. orange-flavoured cookies/biscuits) and drinks (e.g. Earl Grey tea is flavoured with oil of bergamot from *C. bergamia*).

Oil of orange is a complex mixture of volatile molecules including limonene, citrals, decylaldehyde, methylanthranilate, linalool and terpineol with a strong orange smell and flavour (Figure 11.15). Some of its components are also

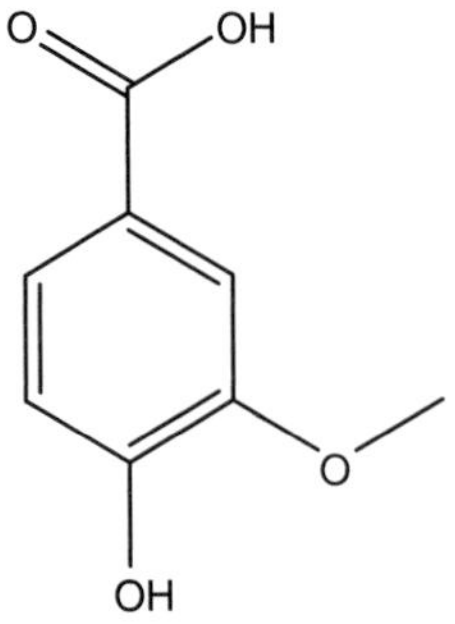

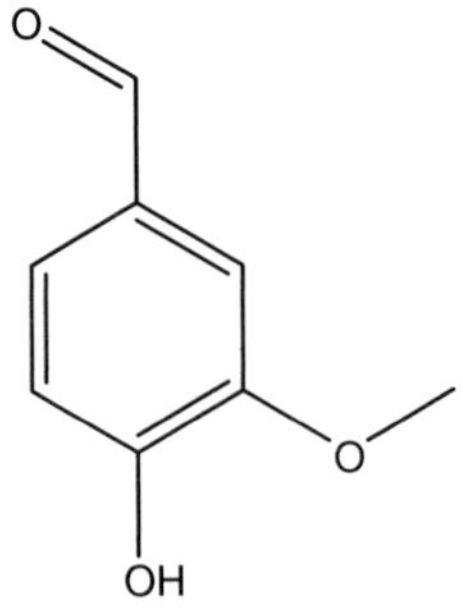

Figure 11.14 Two important flavour components of vanilla extract – vanillic acid does not smell because it is not volatile (MP = 210°C), but has flavour; vanillin has both smell, because it is more volatile (MP = 80°C), and flavour. The two molecules together contribute to vanilla's characteristic smell and taste. Their structural analogies mean that they taste similar because they fit the same taste receptors. MP, Melting point.

present in other citrus oils (e.g. oil of lemon), but it is the complex combination and differential concentrations of each component that gives orange oil (and the other citrus oils) its characteristic smell and flavour.

Nature identical flavours

Vanillin can be made in a chemical factory; it is exactly the same molecule as vanillin from vanilla pods, but it is not natural and therefore it is described as 'nature identical'. Interestingly, vanillin is often made from the much cheaper flavour chemical (eugenol) from cloves - a small change in chemical structure results in a huge change in flavour (Figure 11.16).

Many other components of natural flavour mixtures are synthesised by chemists and used in the food industry.

Artificial flavours

Artificial flavours are not necessarily based on the molecular structures of natural flavour chemicals. They are chemicals that taste like a particular food (e.g. pineapple); presumably they have a similar taste because our tongue taste receptors 'recognise' the flavour molecule, or some part of it, and send a, for example, pineapple message to the brain. If you compare a synthetic flavour with the real thing the difference is usually obvious, but without something to compare the flavour with, artificial flavours are often quite convincing. Indeed, we are exposed so often to artificial flavours that the real thing might taste strange! (Remember our orange juice experiment at the beginning of this chapter.) The artificial flavours also often smell like the food they are mimicking (Table 11.2).

The link between flavour and smell

It is very well known that smell and taste go hand-in-hand. If something smells bad it usually tastes bad too; the opposite is also true. The reason for this

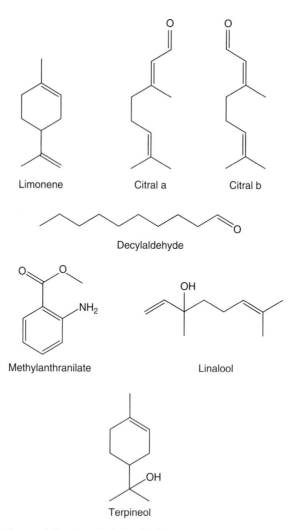

Figure 11.15 Some of the chemicals in oil of orange that are responsible for the characteristic smell and flavour of oranges.

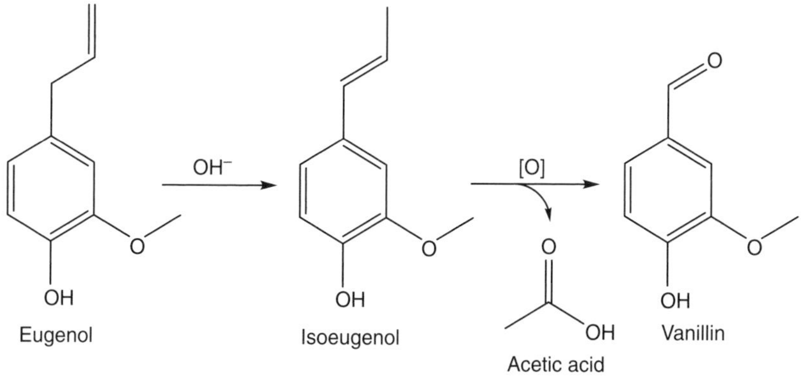

Figure 11.16 The synthesis of vanillin from eugenol – this method was used to synthesise vanillin industrially for many years and shows clearly how a small change in chemical structure leads to a huge change in flavour (i.e. clove → vanilla).

Table 11.2 Some artificial flavours showing that most have have very low acute toxicities. Interestingly, some can be used to represent more than one flavour; e.g. isoamyl acetate can taste like either banana or pear according to the situation – if it is used to flavour banana milkshake it will taste like banana because banana flavour has been suggested to the consumer in the name of the product!

Flavour	Artificial flavour	Molecule	LD_{50} [rat, oral] (mg/kg body weight)
Butter	Diacetyl		1,580
Banana	Isoamyl acetate		16,600
Almond	Benzaldehyde		1,300

Cinnamon	Cinnamic aldehyde		2,220
Grape	Methyl anthranilate		2,910
Pear	Ethyl-2,4-decadieneoate		>5,000
Pineapple	Allyl hexanoate		218

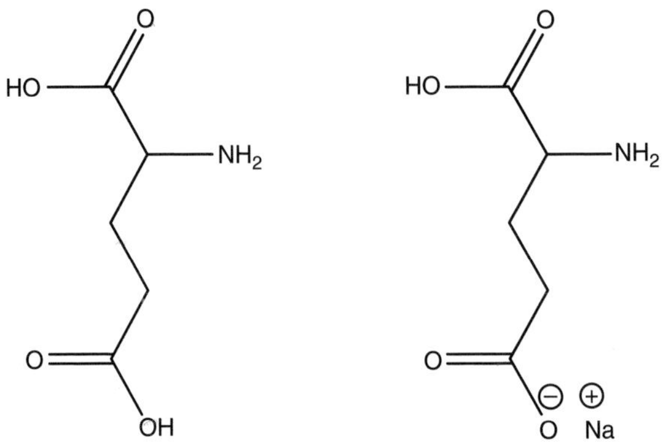

Figure 11.17 The amino acid glutamic acid and its sodium salt monosodium glutamate (MSG) which is used as a flavour enhancer in food.

connection is that our sense of smell and taste are closely linked. A chemical that, for example, smells of pears (e.g. amyl acetate) also tastes of pears. I cannot go into more detail here, but it is important that you realise this connection when thinking about food flavour.

Flavour enhancers

The ubiquitous monosodium glutamate (MSG; Figure 11.17) is the best example of this group of 'flavours'. MSG does not add any particular flavour in its own right, but it enhances the existing savoury flavours. It is the sodium salt of the natural amino acid, glutamic acid, and is present naturally in some foods (e.g. parmesan cheese); indeed in these foods it is responsible for the wonderful burst of flavour one gets when they are eaten (try eating a small piece of parmesan cheese and you will see what I mean). Glutamic acid (and its salts) sometimes crystallises in aged cheeses (e.g. Cheddar) – this gives the cheese its characteristic intense flavour.

Chemically manufactured MSG is added to some foods to give a flavour burst – known as flavour enhancement. It has been used for a very long time in Chinese cuisine; most traditional Chinese cooks would have a jar of MSG and would use a pinch or more in their recipes. MSG has more recently found wider application in many convenience foods (particularly dried foods, e.g. dried soup mixes).

The use of MSG has become controversial because it has been associated with acute effects including 'Chinese restaurant syndrome' (now renamed MSG symptom complex) in some consumers. The symptoms of MSG symptom complex are numbness at the back of the neck, palpitations and weakness; the onset of symptoms is rapid (about 20 minutes after consuming MSG-containing food). Many other symptoms have been reported following consumption of MSG, including headache, nausea, vomiting, mood changes and dizziness. Interestingly, many of the symptoms are central nervous system (CNS) associated (e.g. headache) and in this context it is equally interesting that MSG can be meta-bolised to γ-aminobutanoic acid (GABA; Figure 11.18), an important inhibitory neurotransmitter. GABA occupies a receptor in the CNS – when the receptor is occupied it inhibits the response associated with the receptor's neurones.

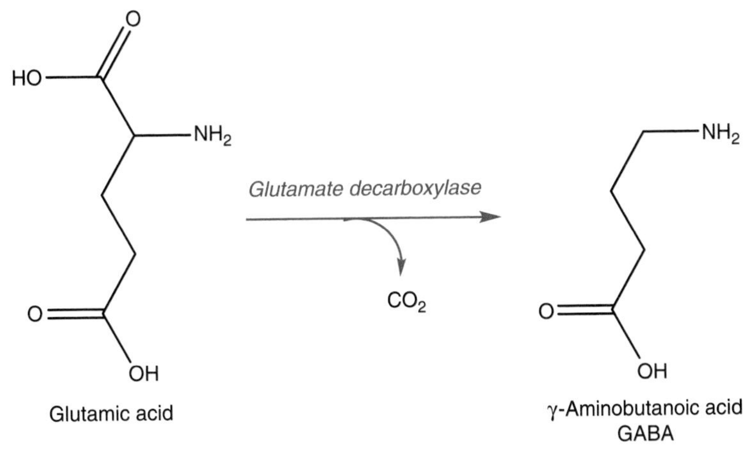

Figure 11.18 The metabolism of glutamic acid to the neurotransmitter GABA.

This debate is controversial and ongoing; even the acceptance of MSG symptom complex is controversial – currently it is not accepted by regulators as an MSG toxic effect (see Walker & Lupien (2000) in *Further reading* for an indepth review of the toxicity of MSG).

Sweeteners

Sweeteners are an important group of food flavours, or perhaps more correctly food flavour enhancers (I have already discussed sugars in the context of flavour perception; see *How do we sense flavour?*). Most humans like sweet foods; chocolates and candies are often regarded as treats; children love them, as do many adults. The problem is that the natural sweetening agents (usually sugars) can have adverse health effects if consumed in excess. The two most important health implications of overconsumption of sugars are increased weight (which can result in obesity with a multitude of knock-on health effects like coronary heart disease and type II diabetes) and tooth decay.

Obesity results from eating too much sugar because sugars can be metabolised to fats and the fats are laid down in tissues. Tooth decay is due to the bacteria that live on our teeth metabolising sugars in food to acids (e.g. lactic acid); the acids dissolve the enamel of our teeth, resulting in dental caries and more extensive decay. The negative health implications of consuming sugars, coupled with our like of sweet foods, has led to the development of artificial sweeteners that do not result in weight gain or dental decay.

Some sweeteners are important for medical reasons. For example, people suffering from diabetes (a group of diseases in which sugar metabolism is compromised) cannot eat too much glucose because they can become ill and even die. Glucose is present naturally in sweet foods (e.g. fruit) and is a component of complex sugars (e.g. sucrose – a disaccharide of glucose and fructose) which are broken down in the body to release glucose (and other sugar components). Diabetics have to avoid foods containing high sugar concentrations. There are other sugars that taste sweet but contain no glucose (e.g. mannitol) which are used in diabetic foods. I will not discuss them further here because they are not widely consumed and have no negative health implications.

Artificial sweeteners

Artificial sweeteners are big business and there is a whole industry involved in their synthesis and marketing for food use. It is not possible to cover all aspects of artificial sweeteners in this book, and so I will discuss just three commonly used artificial sweeteners and the health-related debate that surrounds them.

Saccharin

Saccharin (Figure 11.19) was the first artificial sweetener to be widely used in food and drinks. It was first synthesised in 1878, but did not become an important food and beverage sweetener until the mid-1950s. By the early 1970s, studies had shown that at high doses it caused bladder cancer in rats and by 1985 it was listed as a substance 'reasonably anticipated to be a [human] carcinogen'. This was the beginning of a controversy about the use of this particular artificial sweetener; the controversy continues today.

A huge amount of work has been carried out by scientists to explore saccharin's mechanism of carcinogenesis and it is now thought that it is via a specific protein and that it only occurs in rats and so has little or no implications for human consumers.

Clearly, the risk assessment for saccharin is not simple; on the benefit side of the risk/benefit equation its use reduces obesity and associated diseases, but on the risk side it is a carcinogen, but almost certainly only in rats and only at very high doses. Whether you like the idea of artificial sweeteners or not, there is little or no regulatory reason to prevent the use of saccharin based on a very great deal of excellent toxicology research.

Cyclamate

The intense sweetness of sodium cyclamate (Figure 11.19) was discovered in 1937 in the USA, and by the early 1950s cyclamate was in use as a sweetener for food and drinks. Cyclamate's chequered history is akin to saccharin's and by 1968 it had been declared a carcinogen in animal studies and in 1985 - in the same report as saccharin - it was listed as a substance 'reasonably anticipated to be a [human] carcinogen'.

Saccharin   Sodium cyclamate

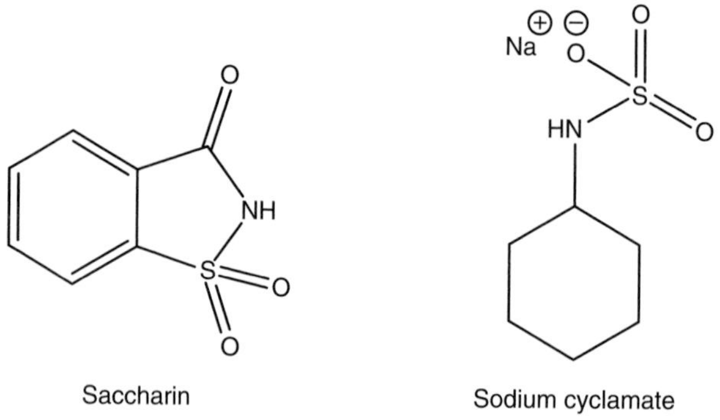

Figure 11.19 Saccharin and sodium cyclamate, two non-calorific artificial sweeteners used in food and drinks.

In parallel with saccharin, its carcinogenicity was found to be species specific and now it is not thought to pose a cancer risk to humans. Unlike saccharin, however, cyclamate was banned in some jurisdictions (e.g. the USA) and allowed in others (e.g. Europe).

Aspartame

Aspartame (trade names Nutrasweet® and Aspartil®) has a very different molecular structure (Figure 11.20) to saccharin and cyclamate – it is a dipeptide ester made up of the amino acids aspartic acid and phenylalanine. It is metabolised in animals (including humans) to its component amino acids and therefore presents no health risk to most people. There is, however, a group

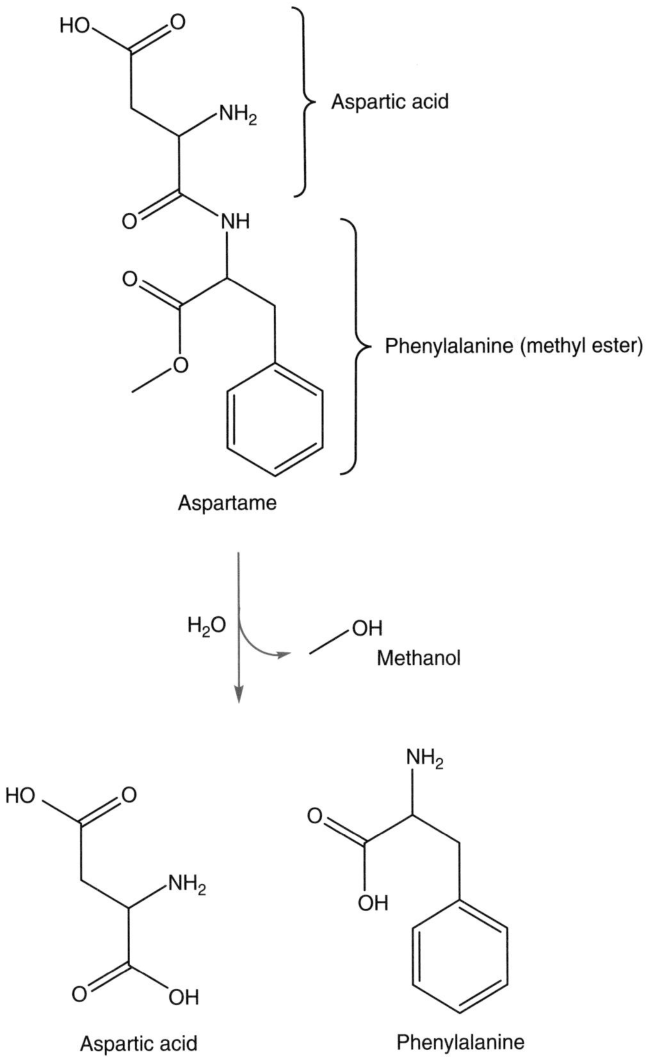

Figure 11.20 The metabolism of aspartame in humans by de-esterification (loss of methanol) followed by peptide bond cleavage to release its two component amino acids, aspartic acid and phenylalanine.

of people who suffer from a rare disease – phenylketonurea (PKU) – who are very susceptible to the effects of aspartame.

PKU is a very rare (1 in 10,000 births) inherited enzyme (phenylalanine hydroxylase) deficiency in which phenylalanine is not metabolised normally (i.e. to another amino acid, tyrosine) and builds up in the blood. The high circulating levels of phenylalanine lead to an unusual metabolic pathway being activated which produces phenylpyruvic acid (a phenylketone – hence the disease's name; Figure 11.21). Phenylpyruvic acid is toxic to the central nervous system, resulting in severe mental retardation. Interestingly, PKU

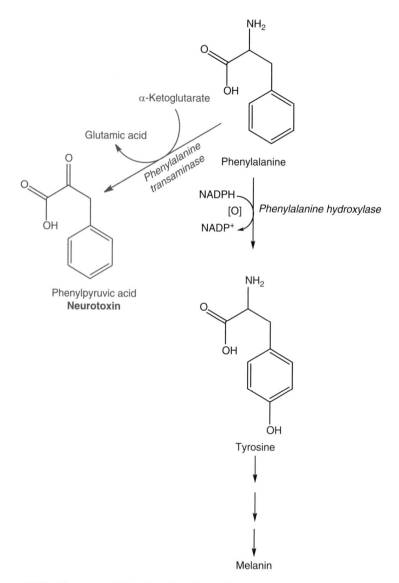

Figure 11.21 The normal (black) metabolism of phenylalanine to form tyrosine which is metabolised by multiple steps to melanin, and the unusual metabolic pathway (green) which forms neurotoxic phenylpyruvate in phenylketonurea (PKU) in which the enzyme phenylalanine hydroxylase is absent.

sufferers are usually blonde haired and blue eyed because tyrosine is the precursor of the black pigment, melanin, which is responsible for hair, skin and eye colour – indeed it is thought that this is the origin of the unfortunate phrase 'dumb blonde'. An important part of the treatment of PKU is not to eat phenylalanine or proteins containing it from birth to adolescence (i.e. when the nervous system is still developing) and to eat a reduced phenylalanine diet in adulthood. Clearly, PKU sufferers would be very severely affected by aspartame and therefore a key part of the aspartame risk assessment is to consider this important minority group. This is dealt with effectively by warnings on aspartame-containing foods and drinks.

Preservatives

Food has been preserved since time immemorial to allow the glut of the harvest to be spread over the year. Ancient food preservation methods involved fermenting foods to produce acids that inhibit microbial growth (e.g. making salami, a fermented sausage), by allowing 'good' bacteria to grow to lower the pH of the food and outgrow bacterial pathogens (e.g. yoghurt), by adding acid directly (e.g. pickling in vinegar (acetic acid – CH_3COOH)), adding sugar (or honey) to increase the sugar concentration above that at which spoilage microorganisms will grow (e.g. jam making), adding high salt concentrations to prevent microbial growth (e.g. salted fish) or cooling to inhibit microbial growth – this was possible in cold countries even before refrigerators were invented. These are just a few of the traditional methods of food preservation that have evolved over thousands of years. More recently canning was introduced during which a food is heated to a high temperature (usually 121°C) in a can (or jar) to kill spoilage microorganisms, followed by sealing the sterile food in the can – the food can be kept for many years after this method of preservation. In addition to the above methods, herbs and spices were sometimes added to foods as preservatives; interestingly many herbs and spices have antimicrobial chemicals as part of their natural chemical makeup. For example, thyme contains thymol (Figure 11.22) which is a powerful antiseptic (and also responsible for the characteristic flavour of thyme – see also Chapter 8, Figure 8.14, *Capsaicin – the chemical that makes peppers taste 'hot'*) that was (and is) used in herb mixtures to help to preserve food. As an aside, thymol is the chemical used in dentist's antiseptic mouth-wash; next time you go to the dentist and she/he asks you to wash your mouth out, think about the flavour of the mouthwash – it tastes of thyme!

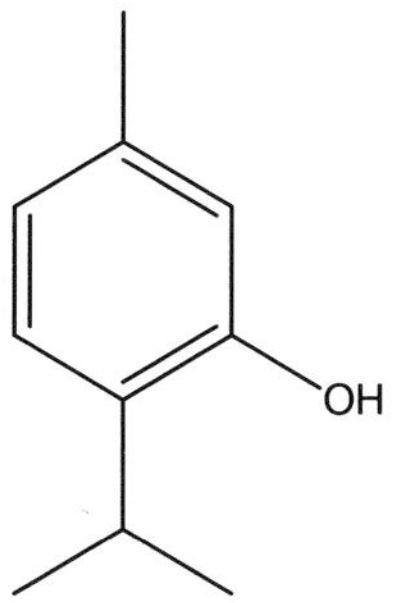

Figure 11.22 Thymol, the natural antiseptic from thyme (*Thymus vulgaris*).

The food preservation methods I have discussed so far all prevent spoilage microorganism growth. Some preservatives don't stop microorganisms growing, but rather prevent the food being chemically modified (e.g. oxidation), which can result in flavour changes. Fats, for example, become rancid when left in air for a long time. Antioxidants have long been added to food to prevent these unpleasant flavour changes – citrus juices are effective antioxidants because they contain the powerful antioxidant vitamin C (ascorbic acid) (see later in this chapter for more detail).

The use of herbs to aid food preservation is actually using the specific chemical components of the herbs (e.g. thymol) as the preservatives. As the technology of food preservation evolved, it became possible by the 1950s to use chemical preservatives added to foods to significantly extend food shelf lives. Despite the introduction of very effective chemical preservatives, some of the ancient preservation techniques survived and are still very commonly used (e.g. the use of bacteria to preserve milk as yoghurt).

Synthetic food preservatives can be divided into two main categories:

(1) Antimicrobial
(2) Antioxidant

Antimicrobial food preservatives

These are chemicals that prevent food spoilage microorganisms (e.g. bacteria) growing; they have a wide range of chemical structures and mechanisms of action. There are very many antimicrobial food preservatives and I will illustrate the group with examples of the most commonly used chemicals in modern food preservation.

Sorbic acid
Sorbic acid (Figure 11.23) is present in members of the rowan (mountain ash – *Sorbus aucuparia*) family of trees. It is interesting that in the past, rowan berries were sometimes incorporated in preserves (e.g. jams)

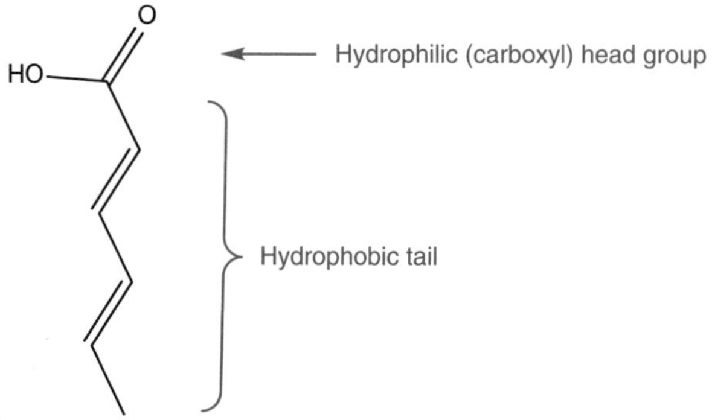

Figure 11.23 Sorbic acid showing its hydrophilic head group and hydrophobic tail – the latter means that it can diffuse across cell membranes, so reducing the cytoplasmic pH and inhibiting the growth of food spoilage microorganisms and pathogens.

possibly as a natural preservative. Chemically synthesised sorbic acid is now commonly used to preserve cheeses, cakes, syrups and dressings. Sorbic acid works by reducing the cytoplasmic (i.e. the cell's aqueous medium) pH of spoilage bacteria and fungi. Reducing the cytoplasmic pH either kills or severely inhibits growth of the microorganisms. Sorbic acid is hydrophobic (lipophilic; Figure 11.23) which means that it can diffuse across microorganisms' cell membranes easily into their cytoplasm (see Chapter 3, *The biology of bacteria*). When in the cytoplasm it reduces pH and prevents their growth.

In effect this is rather like pickling (i.e. preserving food in acid), but instead of reducing the food's pH and so changing its flavour significantly, sorbic acid has a similar effect by altering the pH of the spoilage microorganism or pathogen without altering the food's pH and thus flavour. Sorbic acid has a very low acute toxicity (LD_{50} [rat, oral] = 7,360 mg/kg body weight) and has no known long-term effects, which makes it a safe food additive.

Benzoic acid
Benzoic acid (Figure 11.24) is used to preserve pickles, soft drinks and dressings. Benzoic acid esters (methyl- and propylbenzoates; Figure 11.24) are also commonly used as food preservatives, particularly in marinated fish products. They all work by gaining entry to the microorganism's cytoplasm due to their hydrophobic benzene ring (see *Sorbic acid*). Benzoic acid is able to directly lower the cytoplasmic pH whereas the benzoic acid esters must first be de-esterified by microbial cytoplasmic esterases (enzymes that break

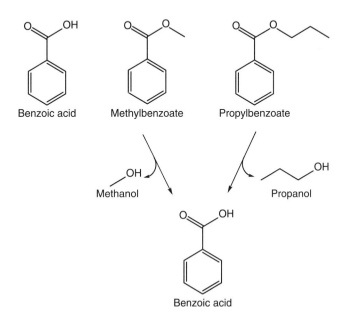

Benzoic acid   Methylbenzoate   Propylbenzoate

Methanol   Propanol

Benzoic acid

Figure 11.24 Benzoic acid and the two benzoic acid ester food preservatives. The benzoic acid esters are de-esterified inside microorganisms to release benzoic acid which can then lower cytoplasmic pH and inhibit microbial growth.

Table 11.3 The benzoate preservatives have very low acute toxicities.

Benzoate	LD_{50} [rat, oral] (mg/kg body weight)
Benzoic acid	2,350
Methylbenzoate	1,177
Propylbenzoate	2,500

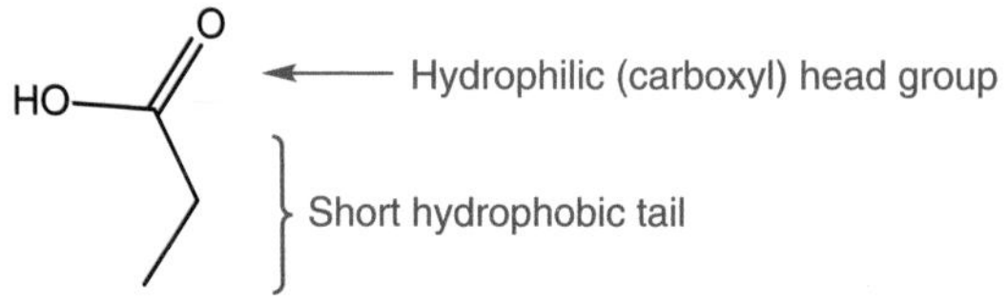

Figure 11.25 Propionic acid – its hydrophobic tail facilitates its uptake by food spoilage microorganisms and pathogens.

ester bonds) to release benzoic acid (Figure 11.24). The benzoic acid esters are more hydrophobic than benzoic acid itself which means that they will traverse the microbial cell membrane more efficiently.

Benzoic acid and its methyl- and propyl esters have very low acute toxicities (Table 11.3).

Propionic acid
Propionic acid (Figure 11.25) works in just the same way as the other hydro-phobic side chain-containing organic acids by being readily taken up by spoil-age microorganisms or pathogens and lowering their internal pH. It is used to preserve bread, cakes, cheese and grain products.

Other organic acids
Acetic acid (vinegar) is an important food preservative that has been used for very many years. It is used to lower the pH of sauces, mayonnaise, dressings, drinks, fruit juices and many other foods. Acetic acid works by changing the pH of the food environment and so inhibiting microbial growth, rather than being taken up by the microorganism and changing its cytoplasmic pH directly.

There are other organic acids used in food preservation; most of them are of natural origin, but are now synthesised chemically and added to foods; for example, lactic (Figure 11.26) and malic (Figure 11.27) acids. Lactic acid is produced by normal anaerobic metabolism and is responsible for fermentation-based preservation (e.g. salami); malic acid is found in apples (from the Latin *malum* meaning apple). They too work by lowering food pH.

This group of organic acid preservatives have very low toxicity indeed and are not associated with any long-term adverse effects. They are safe food additives.

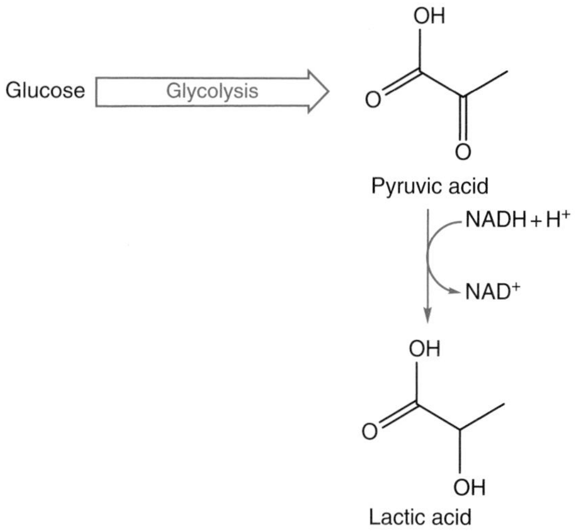

Figure 11.26 The metabolism of glucose to form pyruvic acid via glycolysis followed by its anaerobic metabolism to lactic acid. This is the process of fermentation used to preserve some foods (e.g. salami). Lactic acid is also synthesised chemically and added to some foods as a preservative.

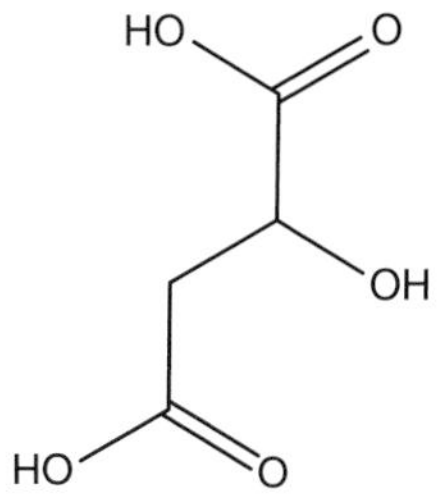

Figure 11.27 Malic acid is an intermediate in cell metabolism (part of the tricarboxylic acid or Kreb's cycle) and is present in apples. It is now synthesised chemically and used as a food preservative.

Sulphur dioxide and sulphites

Sulphur dioxide (SO_2) and the sulphites are powerful cell toxins (Figure 11.28). The most commonly used sulphites in food preservation are listed below:

- Sodium sulphite (Na_2SO_3)
- Sodium metabisulphite ($Na_2S_2O_5$)
- Potassium metabisulphite ($K_2S_2O_5$)
- Sodium bisulphite ($NaHSO_3$)
- Potassium bisulphite ($KHSO_3$)

The exact mechanism of cellular toxicity of the sulphites is not known; indeed it is more likely that they act by a plethora of enzyme and cellular process inhibitions that disable the cell's biochemistry and eventually kill it.

Sulphur dioxide *per se* is rarely used in food preservation (except as a fumigant); instead the sulphites (listed above) are used. The chemistry of sulphites

$$SO_2 + H_2O \rightleftharpoons H_2SO_3 \rightleftharpoons H^+ + HSO_3^- \rightleftharpoons 2H^+ + SO_3^{2-}$$

Sulphur dioxide   Bisulphite   Sulphite

Figure 11.28 The equilibrium between sulphur dioxide, bisulphite (hydrogen sulphite) and sulphite in the aqueous environment of food.

and SO_2 in aqueous solution is complex. It involves an equilibrium between the sulphites and SO_2 which means that, whether bisulphite, sulphite or SO_2 is used, the mechanism of food preservation is probably the same because these ions and molecules are interchangeable in the aqueous environment of food (Figure 11.28).

Sulphites can be used either as antimicrobials in food or to prevent enzyme-mediated degradation of food products (e.g. fruit juices). In addition, the sulphite ion (SO_3^{2-}) is an antioxidant (reducing agent) because it is readily oxidised to sulphate (SO_4^{2-}) and therefore it inhibits the natural oxidation process involved in food deterioration. They are most commonly used in beverages (e.g. wine), fruit juices and dried fruits.

The sulphites (and SO_2) have been associated with toxic effects in consumers, particularly hyperallergenic people (e.g. asthmatics) who have developed severe allergic reactions after eating food preserved with sulphites or SO_2. The sulphites are sensitisers which means that they heighten the immune response; so following exposure to sulphites you are more likely to produce antibodies in response to an antigen. For most people this is not a problem because they do not illicit major immune responses to external antigens (e.g. pollen), but for asthma sufferers this can be serious because their immune system is particularly sensitive to specific allergens (i.e. antigens, e.g. pollen) and the heightened sensitisation stimulated by sulphites can result in a very serious, even life-threatening (anaphylactic), response.

Despite health concerns about the use of sulphites in foods they are still very widely used; in fact it has been estimated that we are exposed to more than 20 mg/person/day in our food alone.

Nitrites

The nitrites (e.g. sodium nitrite – $NaNO_2$), like the sulphites, have been used in food preservation for very many years. A good example of the use of nitrite preservatives is the curing of bacon. This involves 'pickling' pork in a nitrite-containing liquid. This process prevents microbial food spoilage and, because nitrite is a powerful reducing agent (it is readily oxidised to nitrate – NO_3^-), maintains the red colour of the meat by inhibiting the oxidation of bright red haemoglobin to brown methaemoglobin and by forming bright red nitroso-myoglobin from myoglobin (the oxygen-carrying pigment in muscles).

Interestingly, nitrite *per se* was not originally used to cure meats; instead nitrate in the form of 'saltpetre' (potassium nitrate – KNO_3) was used. However, the nitrate is reduced to nitrite by bacteria on the meat and it is the nitrite that is responsible for the preservation of the meat.

The antimicrobial activity of nitrites is akin to that discussed in *Sulphur dioxide and sulphites* above; nitrite inhibits or interferes with very many biochemical and cellular mechanisms and kills cells by a myriad unspecified means. It is a very effective antimicrobial.

Nitrate is associated with a potentially very serious toxicological mechanism (see Chapter 7, *Nitrate – NO_3^-*) involving the formation of carcinogenic nitrosamines in the consumer's intestine. For this reason, the use of nitrate preservatives in food has been subjected to considerable scrutiny over the past 20 years and the discussions continue today. The risk of nitrites as food preservatives must be set in perspective. There are many dietary sources of nitrite including nitrate fertiliser residues (nitrate is reduced to nitrite in the anaerobic environment of the intestine) and 'natural' nitrates contained in leafy vegetables (e.g. spinach). All of these dietary nitrite sources add up to our total nitrite exposure, and, of course, it is the total nitrite that determines the level of risk. Nitrite from cured meats is a small proportion of our total nitrate/nitrite intake and therefore it is inappropriate to 'ban' the use of nitrite as a meat preservative even though it is a carcinogen. Putting this risk argument aside, many people would accept the risk of nitrite toxicity for the benefit of being able to eat ham and bacon; indeed we have the choice whether or not to consume these foods.

Antioxidant food preservatives

It is difficult to strictly categorise the food preservatives because some fall into more than one category; for example, sulphites and nitrites could also be categorised as antioxidants. In this section I will only discuss the food additives solely used as antioxidants.

When food is left in contact with air it slowly oxidises. This process involves the oxidation of a multitude of molecules naturally found in food and can result in colour and/or taste changes. Some of these changes are desirable, e.g. briefly airing a bottle of wine enhances its flavour, but prolonged exposure to oxygen results in the oxidation of ethanol to acetic acid which makes the wine undrinkable. In general, oxidation causes unwanted changes in our food.

The use of antioxidants prevents oxidation and allows food to be stored, while retaining its colour and flavour. Over the past few years the use of modified atmosphere (e.g. nitrogen – N_2) or vaccuum packing has been used to significantly extend food's (particularly meat's) shelf life.

As for the other food additives discussed in this chapter, there are many antioxidants used in foods; I will discuss only those more commonly used.

Vitamin C (ascorbic acid)

Vitamin C is an important micronutrient; without it serious disorders such as scurvy (a skin disease) develop. It is important in the cell because it maintains the oxidation status and is involved in mopping up reactive chemical species (e.g. free radicals) which would cause significant harm to the cell if they were not removed quickly. Without vitamin C, cells and therefore animals cannot survive. The antioxidant properties of vitamin C that are so important to the wellbeing of cells are utilised by food technologists to preserve the colour and flavour of food. Some foods have sufficient vitamin C naturally (e.g. lemon) and therefore do not need antioxidant preservation; in fact traditionally these

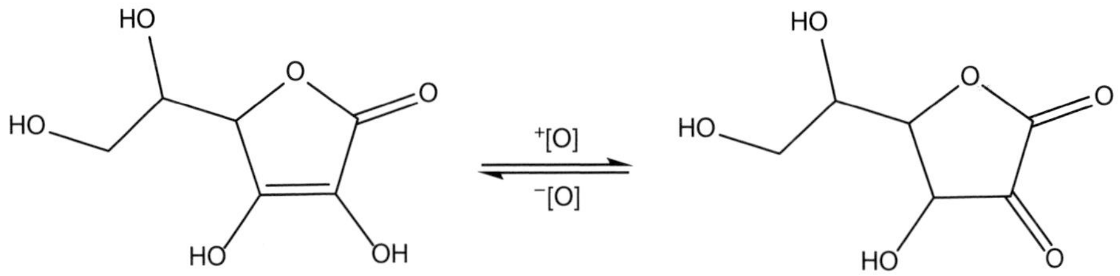

Figure 11.29 The oxido-reduction equilibrium of vitamin C which makes it such a good antioxidant – it readily forms the keto form (right) and in the process mops up oxygen.

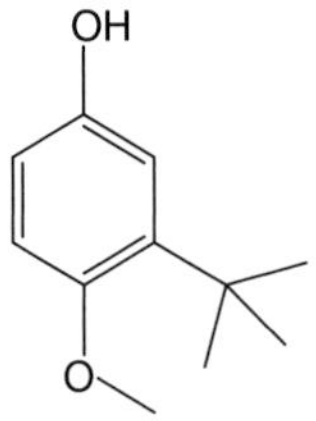

Butylated hydroxyanisole
BHA

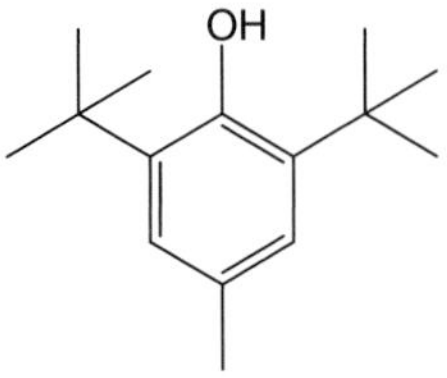

Butylated hydroxytoluene
BHT

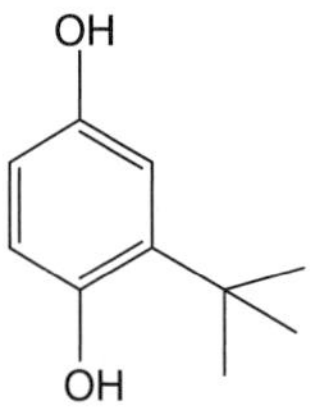

Tertiary-butyl hydroxyquinone
TBHQ

Figure 11.30 BHA, BHT and TBHQ, three food antioxidants commonly used to inhibit the rancidification of fats.

foods are used to preserve other foods – I often sprinkle lemon juice onto cut apples before adding them to a fruit salad to prevent the oxidation reaction that causes them to turn brown.

Vitamin C is a good antioxidant because it is readily oxidisable itself (Figure 11.29) and so is able to both scavenge oxygen from aqueous systems and inhibit oxidation reactions.

Vitamin C's toxicity is extremely low – in fact a dose of 5 g causes no harm in humans. Its use as an antioxidant in food presents negligible or (more likely) no risk; indeed, it is likely to be beneficial to the consumer.

Butylated hydroxyanisole (BHA), butylated hydroxytoluene (BHT) and tertiary-butyl hydroxyquinone (TBHQ)

BHA, BHT and TBHQ (Figure 11.30) are phenolic antioxidants commonly used in fat-based foods – because of their hydrophobic molecules they are fat soluble and therefore protect fats against oxidation particularly well. When fats are oxidised they take on an unpleasant flavour due to the presence of

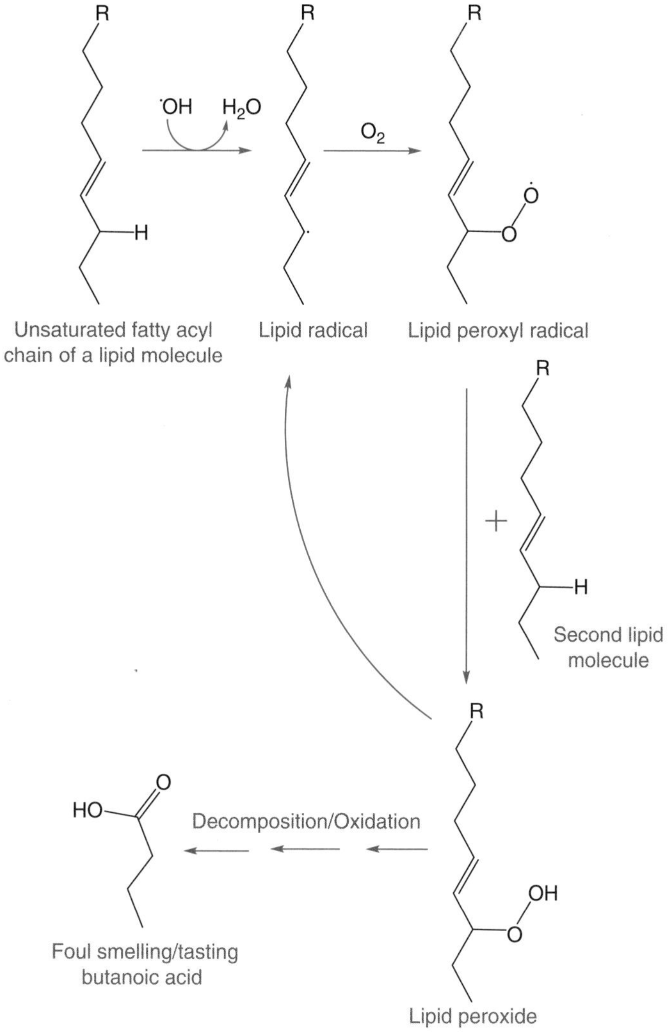

Figure 11.31 The free radical-mediated oxidation of fats to form foul smelling and tasting chemicals that are responsible for rancidity.

lipid peroxides – they are said to be rancid. In addition, BHA and BHC have antimicrobial activity and can prevent food deterioration by suppressing food spoilage microorganisms. I will discuss only their antioxidant/antirancidity activity here.

Before we can understand why BHA and BHT prevent fats becoming rancid we must understand the mechanism of fat rancidification (oxidation) (Figure 11.31). It involves an unstoppable chain reaction which is initiated by a hydroxyl free radical ('OH) derived from water (UV light generates 'OH from water) – a free radical is a reactive species with an unpaired electron. The hydroxyl free radical donates its unpaired electron to one of the lipid's carbon atoms; this reactive lipid species then reacts with oxygen to form a lipid peroxyl radical which forms a lipid peroxide and in so doing donates its spare

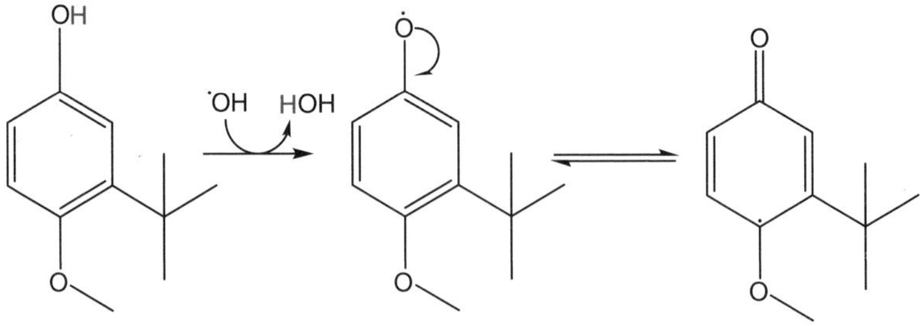

Figure 11.32 BHA (left) 'deactivates' the free radicals (e.g. the hydroxyl radical −·OH) that cause oxidation of fats by scavenging and delocalising their unpaired electron – a harmless water molecule (HOH = H_2O) is formed by the ·OH picking the hydrogen atom off BHA's hydroxyl group (shown in green). BHT and TBHQ work in a very similar way.

electron to another lipid carbon atom, and so on and so on. Lipid peroxides break down and can be further oxidised to form short chain, foul-smelling carboxylic acids and aldehydes (e.g. butanoic acid – $CH_3CH_2CH_2COOH$) which give old fats their characteristically rancid smell and taste (interestingly, it is one of the components of sweaty feet too).

BHA and BHT are free radical scavengers by virtue of their phenolic hydroxyl groups which take the unpaired electron from a free radical and incorporate it into their aromatic delocalised electron resonance *pi*-cloud. BHA and BHT react more quickly with free radicals (e.g. the hydroxyl radical that causes fats to go rancid) than fats do and therefore they scavenge the free radicals that oxidise fats and thus prevent fats going rancid (Figure 11.32).

BHA, BHT and TBHQ health concerns

The three butyl phenols cause tumours in animals and therefore there is concern about their effects in humans. On the other hand, they inhibit the development of tumours in animals exposed to some carcinogens; this is probably because they deactivate free radical intermediates in the mechanisms of toxicity of some carcinogens. So in some circumstances they are carcinogens, but in others they are anticarcinogens. This makes determining their future as food additives on safety grounds very difficult. Currently they are allowed as food additives in most countries with a low ADI (JECFA (Joint Expert Committee on Food Additives) ADI = 0–0.5 mg/kg body weight/day) which reflects the safety concerns. Interestingly, exposure to the butyl phenol antioxidants varies considerably around the world; in Japan an average consumer is exposed to only 1% of the ADI whereas in Australia and New Zealand the average exposure level is 80% of the ADI. These figures reflect the different countries' attitudes to the use of these compounds.

Propyl gallate

Propyl gallate (Figure 11.33) is an antioxidant/free radical scavenger that works in the same way as the butyl phenols. It is hydrophobic and therefore is used as an antioxidant for fats and oils or fatty foods much in the same way

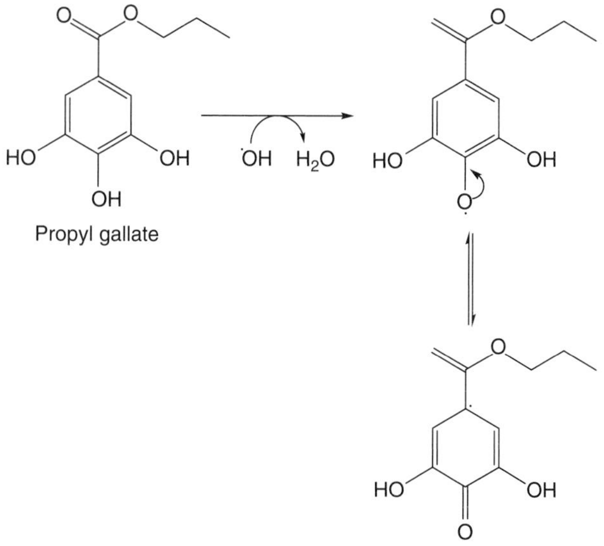

Figure 11.33 The antioxidant propyl gallate showing its mechanism of free radical scavenging.

that the butyl phenols are. Propyl gallate has a very low acute toxicity (LD_{50} [rat, oral] = 3,800 mg/kg body weight) and is not a carcinogen or cancer suspect agent and therefore its use in foods is less of a concern than the butyl phenols.

Ethoxyquin

Ethoxyquin is an antioxidant and free radical scavenger that works via a different mechanism to the aromatic hydroxy compounds discussed above. Instead of the -OH group accepting the unpaired electron from a free radical, ethoxyquin has an N-heterocyclic ring and the nitrogen accepts the unpaired electron and delocalises its charge in the electron *pi*-cloud of the conjugated ring structure (Figure 11.34). The effect is just the same as for propyl gallate and the butyl phenols. By this means free radical-mediated fat rancidification is inhibited very effectively.

Ethoxyquin has a very low acute toxicity (LD_{50} [rat, oral] = 1,920 mg/kg body weight) and is not a carcinogen or cancer suspect agent which means that its use as a food antioxidant is not controversial.

Smoking

Smoking is one of the oldest food preservation methods – it is thought to have originated soon after cooking itself. In addition to preserving food, smoking adds flavour. The flavour is imparted by a complex mixture of hydrocarbons (both aromatic and aliphatic) and phenols produced when wood is burned at low temperature (Table 11.4) – the temperature range of the food smoking process is 43–71°C. These chemicals dissolve in the food's fat component. Many of the same molecules that impart flavour also have

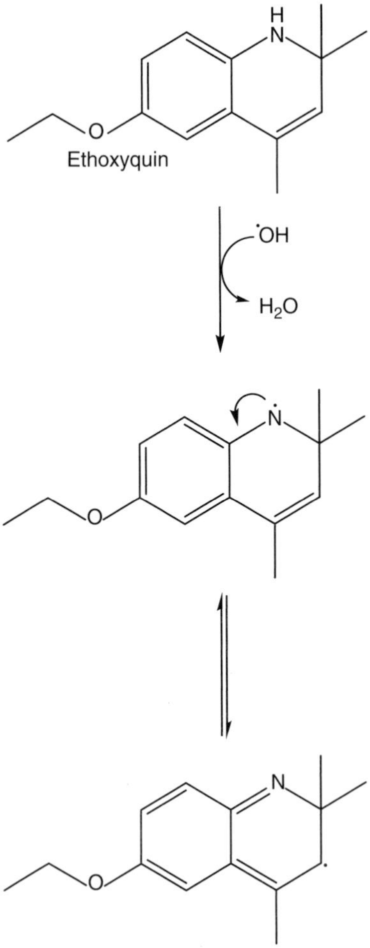

Figure 11.34 The antioxidant ethoxyquin showing its mechanism of free radical scavenging.

antimicrobial and antioxidant properties and therefore preserve the food, e.g. phenols.

A good example of a smoked food that keeps for very much longer than its unprocessed counterpart is smoked salmon. A smoked salmon will keep unrefrigerated for weeks whereas fresh salmon would probably only last for a day or two. Smoking is a very effective method of food preservation.

There is a negative side (i.e. risk) to smoking though. Some of the fat soluble hydrocarbons produced by the smoking process are potent carcinogens (e.g. benzo[a]pyrene; this is also one of the carcinogens present in cigarette smoke; Figure 11.35). Indeed, the incidence of some cancers (e.g. gastric cancer) is higher in countries where smoked food is more commonly consumed (e.g. Korea and Japan; Table 11.5).

Table 11.4 Some polycyclic aromatic hydrocarbons (PAHs) found in smoked fish and their cancer causing status. (Data from Basak *et al.* (2010) *Turkish Journal of Fisheries Aquatic Science*, **10**, 351–355; Simko (1991) *Food Chemistry*, **40**, 293–300.)

Substance	Molecular structure	Carcinogen?
Naphthalene		Yes (high doses in mice)
Acenaphthylene		Equivocal – some animal studies resulted in tumours
Pyrene		Inhalation/skin contact associated with cancer in humans
Benzo(k)fluoranthene		Yes (mouse)
Anthracene		Insufficient data to decide
Benzo(g,h,i)perylene		Insufficient data to decide
Acenaphthene		No
Benzo(a)pyrene		Yes (multiple species)

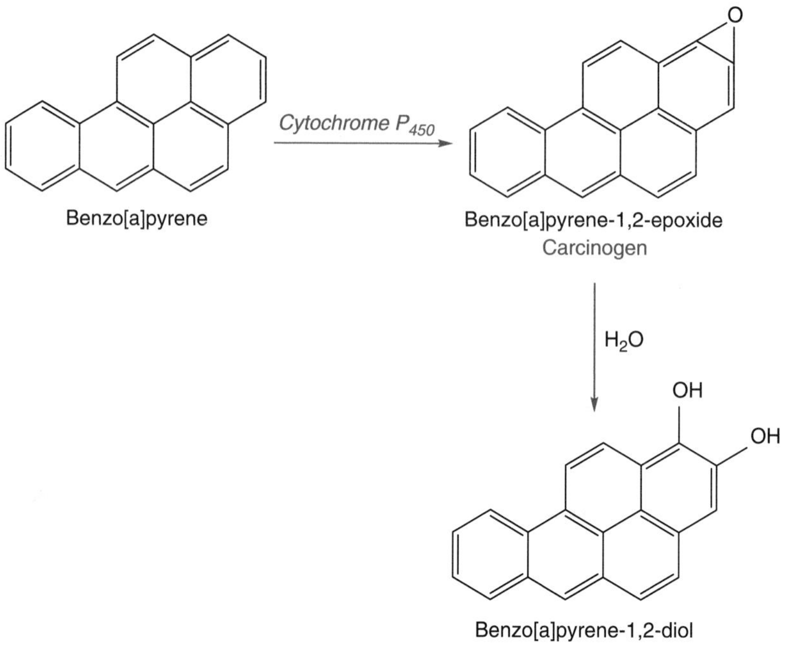

Figure 11.35 The human metabolism of smoke-derived benzo[a]pyrene to a highly reactive, carcinogenic intermediate, benzo[a]pyrene-1,2-epoxide.

Table 11.5 The incidence of gastric cancer around the world. Consumption of smoked food is thought to be, at least in part, to blame for the high incidences in Japan and Korea. (Data from Inouse & Tsugane (2005) *Journal of Postgraduate Medicine*, **81**, 419–424.)

Country	Gastric cancer incidence/ 100,000 population/year	
	Male	Female
Korea	69.7	26.8
Japan	62.0	26.1
Northern Europe	12.4	6.0
Northern America	7.4	3.4
Northern Africa	4.4	2.5

Take home messages

- Colours, flavours and preservatives are extensively used in modern foods to make them look appetising, taste good and have long shelf lives.
- Traditional preservatives (e.g. sugar) have been used for centuries to allow summer produce to be stored for use in winter.
- There are consumer concerns about the health effects of some colours (e.g. tartrazine), flavours (e.g cyclamate) and preservatives (e.g. sulphur dioxide), but when a risk analysis is carried out the risks to the consumer are very low and often restricted to susceptible sub-populations.
- Preservatives offer the greatest benefit because they prevent food spoilage and reduce food-borne illness. Their benefit usually outweighs the risk.
- Colours and flavours are less beneficial and even if the risk is low it is questionable whether they should be used. Some countries restrict the use of artificial colours (e.g. the EU does not permit their use in baby foods).

Further reading

MacDougal DB (ed.) (2002) *Colour in Food*. CRC Press, New York.

Russell NJ & Gould GW (eds) (1991) *Food Preservatives*. Blackie & Sons, London.

Walker R & Lupien JR (2000) The safety evaluation of monosodium glutamate. *Journal of Nutrition*, **130**, 1049S–1052S.

Chapter 12
Food Irradiation

Introduction

The food industry strives to deliver pathogen and spoilage micro-organism-free food to consumers. The technology used to achieve this goal has changed significantly over the years. As discussed in previous chapters, this began with fermentation processes (e.g. yoghurt and salami), pickling at low pHs to inhibit microorganism growth, bottling and canning at high temperatures, embraced the microorganism growth inhibitory low temperatures of refrigeration and freezing, and more recently utilised microorganism growth inhibitory chemicals (preservatives). These pathogen and spoilage microorganism growth inhibition methods developed as our understanding of microbial growth developed and new technologies (e.g. refrigeration) became available. It is therefore understandable that a very recent (in the context of the above history) food preservation technique would rely on one of man's recent (again in the context of the above history) discoveries – radioactivity and its cytotoxic properties.

All cells are killed by high levels of penetrating (e.g. γ) radiation and therefore it is possible to kill microorganisms very effectively by exposure to, for example, γ-rays from a radioactive source (e.g. Cobalt-60 [^{60}Co], a γ-emitting isotope with a relatively long half life (5 years)). Exposure of food to a γ-emitting isotope does not make the food radioactive, just like exposure of food to light (a form of electromagnetic radiation) does not make the food emit light. Despite this, there is significant uncertainty amongst consumers about the safety of food irradiation. This concern arises from ignorance and its association with the evocative word 'radiation'.

Irradiation does not make food radioactive, but it does change the chemistry of the food. This is because bombarding the complex mixture of molecular components that comprise food with a very high energy wave form (e.g. γ-radiation) initiates chemical reactions that alter the chemistry of the food. Many of these reactions are free radical-mediated because high energy electromagnetic radiation can form free radicals from many molecules (e.g. ˙OH from H_2O). These reactive free radicals can initiate further chemical reactions in the food (e.g. oxidation of fats)

Food Safety: The Science of Keeping Food Safe, First Edition. Ian C. Shaw.
© 2013 John Wiley & Sons, Ltd. Published 2013 by John Wiley & Sons, Ltd.

(see Chapter 11, Figure 11.31). Such changes in food chemistry can alter the properties of the food, including its flavour. For this reason, the dose of radioactivity the food is exposed to must be very carefully controlled to minimise unwanted changes in its chemistry. Some consumers are concerned about the chemical changes that irradiation can cause from a health perspective. Free radicals are highly reactive and many interfere with nucleic acids and so are carcinogens, and so it is not difficult to understand their concerns; however, the reactivity of the free radicals mean that they do not survive long in a food chemical environment and thus are very unlikely to lead to adverse health effects because they are very unlikely to be present in the food we eat.

So, food irradiation is a very effective way of making food safe by killing pathogens and spoilage microorganisms, but, often due to ignorance, it is frowned upon by consumers because of its perceived adverse health effects.

In this chapter I will discuss how food irradiation is carried out, how it kills microorganisms in food, and I will explore the conjecture around its use. Before I can discuss the process, effectiveness and controversy of food irradiation, you need to understand a little about radioactive decay and its effects on living cells.

Different types of radioactivity

Whole books have been written on radioactivity and therefore I will only give a very cursory overview of the subject here – you only need to know the basics to understand how food irradiation works.

There are two basic types of radioactivity that are based on: (1) charged particles (e.g. α-radiation) emitted by an unstable element (i.e. a radioisotope) and (2) electromagnetic rays emitted by an unstable element (e.g. γ-radiation) or a chemical reaction, e.g. ultraviolet light (UV-radiation) produced by the sun's chemical reactions. α-Particles only travel in air for relatively short distances and cannot penetrate paper; γ-rays are able to travel great distances in air and some can penetrate concrete.

Particle-based radioactivity

α-Radiation

Some very large (i.e. high atomic weight; e.g. ^{235}Uranium) elements spontaneously decompose and lose fragments of their nuclei to form lower atomic weight elements which are either stable or decompose further (not only via α-radiation; see *β-Radiation* and *γ-Radiation*) to form a stable element – this is a nuclear decay chain. The nuclear fragment lost in α-decay is a helium nucleus (He^{2+}, i.e. a proton and a neutron). α-Particles are heavy (compared to other forms of radioactive decay) and therefore can cause significant damage to cells simply by the force (they are high energy particles; e.g. ^{238}Pu α = 5.5 and 5.46 MeV) of them bombarding the molecular make-up of the cell; however, they don't penetrate far in air because they collide with air gas molecules (e.g. N_2) and are stopped. In addition, they cannot penetrate cell walls and/or membranes. α-Emitting isotopes are very harmful to cells if they are absorbed because the α-particles can irradiate molecules within the cells without having to pass through the cell wall and/or membrane or travel long distances. For these reasons α-radiation is not used for food irradiation.

β-Radiation

I have categorised β-radiation as a particle because it is an electron emitted by an unstable isotope; however, physicists regard it as being more like a photon (i.e. a quantum of light energy) rather than a discrete particle, but for the purposes of this discussion we do not need to go to this level of understanding.

β-Radiation has a spectrum of energies. For example, the β-radiation emitted by ^{32}P has higher energy (1.71 MeV) than that emitted by ^{14}C (0.156 MeV). The higher the β-energy the greater its penetrating properties. Most β-radiation travels at least centimetres in air and will penetrate paper, cell walls and membranes. β-Radiation, therefore, affects cells when they are exposed to an external β-source.

Even though β-radiation penetrates further than α-radiation its energy and penetrating powers are still not great enough to make it useful in food irradiation.

Electromagnetic rays

γ-Radiation

All γ-rays travel long distances in air, penetrate many physical barriers (e.g. concrete) and, therefore, readily pass through cell walls and membranes. For these reasons, γ-emitting isotopes (e.g. ^{60}Co) are used in food irradiation (Figure 12.1).

X-rays

X-rays are produced when elements with an atomic weight greater than 23 (i.e. sodium - Na) are bombarded with high energy electrons. X-rays are very similar (for the purposes of this discussion) to γ-rays. They have similar penetrative and cell killing powers to γ-rays. Their energy is determined by the metal used for their production.

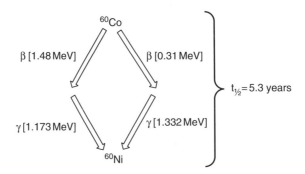

Figure 12.1 The radioactive decay of ^{60}Co – the isotope used in food irradiation – to form stable ^{60}Ni showing the energies of its α- and β-emissions. It is the high energy γ-emissions that are important in killing microorganisms in food.

Ultraviolet radiation

Ultraviolet (UV) radiation is part of the electromagnetic spectrum of light emitted by the sun and, therefore, everything on earth is exposed to 'natural' *uv*-radiation. It is possible to generate *uv*-radiation artificially, e.g. by mercury vapour lamps where an electric current is used to excite mercury atoms [Hg⁰]; when the Hg^0 returns to its ground state it emits energy as *uv*-radiation. *uv*-Radiation is cytotoxic and is used to sterilise surfaces for food preparation; however, its cytotoxic 'power' is not great enough to make it useful in food irradiation because it does not penetrate the surface of matter.

Radioactive half life

The time taken for a radioactive isotope to decay by 50% is its half life ($t_{1/2}$). Half lives vary greatly, from fractions of a second (e.g. ^{214}Po $t_{1/2} = 1.64 \times 10^{-3}$ s) to billions of years (e.g. ^{238}U $t_{1/2} = 4.5 \times 10^9$ years), from isotope to isotope.

The ideal isotope for food irradiation emits high energy γ-radiation and has a $t_{1/2}$ long enough to allow it to be used for a reasonable length of time (i.e. years) before having to be replaced. ^{60}Co is often used for food irradiation because it is a high energy (1.173 and 1.332 MeV) γ-emitter with a $t_{1/2}$ of 5.3 years (Figure 12.1).

How irradiation kills cells

Exposing cells to penetrating (e.g. γ) radiation (this is termed irradiation) leads to chemical changes to some biomolecules. If the irradiation dose is high enough the chemical changes result in cell death. But why does exposure of a cell to radiation harm or kill it? There are many changes that can occur in cells as a result of exposure to, for example, γ-rays; I will give a few examples here and extrapolate the chemical changes to effects on the cell's wellbeing.

When high energy γ-rays pass through water they fragment the water molecule forming reactive free radicals. These free radicals can, in turn, react with important biomolecules changing their chemistry in a way that means they can no longer function normally within the cell. For example, the hydroxyl radical ($^\bullet$OH) generated by γ-irradiation of water is very reactive and reacts with many important biomolecules, including nucleic acid bases, thus interfering with the functions of DNA and RNA. For example, if an aqueous solution of the nucleic acid base, guanine, is γ-irradiated, 8-hydroxyguanine is formed (Figure 12.2). Similarly, if a cell is γ-irradiated, some of its DNA and RNA guanine residues will be oxidised to 8-hydroxyguanine which will interfere with normal DNA/RNA base pairing, thus affecting protein synthesis, DNA replication and mRNA synthesis (see Chapter 10, *A brief introduction to nucleic acids, genetics and molecular biology*). These effects can be so severe that the cell can no longer function and it dies.

Similarly, γ-irradiation of lipids results in their oxidation to lipid peroxides followed by breakdown to short chain carboxylic acids (e.g. butanoic acid) (see Chapter 11, *Butylated hydroxyanisole (BHA), butylated hydroxytoluene (BHT) and tertiary-butyl hydroxyquinone (TBHQ)*). Lipids are important components of cell membranes and the membranes that surround cell organelles (e.g. mitochondria). If lipid structures are changed significantly by oxidation,

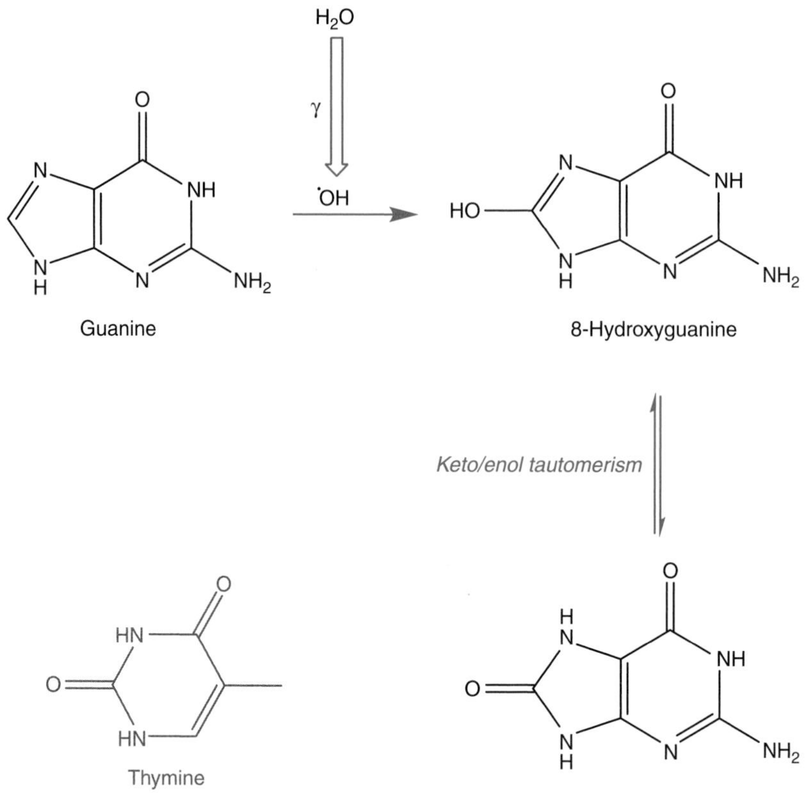

Figure 12.2 Hydroxyl radicals (˙OH) formed when water is γ-irradiated hydroxylate the nucleic acid base guanine which in its keto form resembles thymine. This means that nucleic acid base pairing is affected because what was originally a guanine residue that base paired with cytosine now resembles thymine and thus pairs with adenine; this in turn leads to misreading of codons and the synthesis of functionless proteins.

resulting in chain length reduction, the properties of the membranes will change markedly. Since membranes are pivotal in maintaining cell homoeostasis, such changes cause irreversible damage to cells which quickly leads to their death.

I have given just two examples of the effects of γ-irradiation on cells; there are many more, but I think it is clear that the changes are so severe that a cell cannot survive a high dose of γ-rays. Of course, if the cell is a food spoilage microorganism or a pathogen, γ-irradiation results in death of these unwanted food inhabitants and makes the food safer.

The history of food irradiation

Wilhelm Röntgen (1845-1934) discovered X-rays (a form of electromagnetic radiation - like γ-rays) in 1895 and the next year Henri Becquerel (1852-1908) discovered radioactivity. Marie (1867-1934) and Pierre (1859-1906)

Curie followed Becquerel's studies with a great deal of experimental work which shaped our understanding of the radioactive elements. Interestingly, the same year that radioactivity was discovered (1896) it was suggested that it could be used to kill microorganisms in food, but it was not until 1921 that B. Schwartz obtained a US patent to use X-rays to kill the parasite *Trichinella spiralis* in meat (see Chapter 5, *Trichinella sp.*). Thus the concept of food irradiation was born.

Much work on food irradiation was carried out in the UK and USA in the 1930s and 1940s which led to the effectiveness and technology of food irradiation being perfected. The International Food Irradiation Project (IFIP) was launched in 1970; this was a 19 nation fund to finance research into studying and developing food irradiation procedures. The IFIP was completed in 1982 and was replaced by a collaboration between the WHO, the Food and Agriculture Organisation (FAO) and the International Atomic Energy Authority (IAEA) called the International Consultative Group for Food Irradiation (ICGFI). Joint FAO/IAE/WHO expert committees were set up, including one on the Wholesomeness of Irradiated Foods, and the Joint Expert Committee on Food Irradiation (JECFI) to explore irradiation procedures and safety. Their findings were adopted by the Codex Alimentarius Commission (CAC) and became the general Standard for Irradiated Foods and later the Recommended International Code of Practice for the Operation of Radiation Facilities for the Treatment of Foods.

The latter lays down the rules for food irradiation, including the dose of radiation to which a food should be exposed to ensure that microorganisms are killed – that dose is an average of 10 kGy. Gy is the abbreviation for Gray, named after the British physicist Louis Gray (1905–1965); it is a measure of radiation energy absorbed (1 Gy = 1 joule/kg absorbed).

The effect of radiation on microorganisms

It has been known since 1896 that exposure to radiation can kill microorganisms. In a food safety context it is important that the most resilient pathogens are killed. Sporulating bacteria are the most problematic because they form spores in adverse conditions and the spores can 'germinate' later to release active bacteria. For this reason, the most meaningful test of the effectiveness of radiation in a food safety context is to investigate its effects on bacterial spores. Studies on the spores from *Bacillus pumilis* have been used as a model system to investigate the efficacy of irradiation. *B. pumilis* is not a food pathogen, but it forms highly resistant spores and therefore it represents a worst case model system to test irradiation against. A γ-radiation dose of 4 kGy (i.e. 0.4×10^4 Gy on the graph in Figure 12.3) kills 99% (i.e. $\log_{10} = -2$ from the graph in Figure 12.3; this means the survival fraction = 10^{-2} [or 1/100]) of exposed *B. pumilis* spores; thus γ-irradiation is an excellent means of killing even the most resistant microorganisms. The γ-radiation dose used in food irradiation is 10 kGy which is more than twice the dose that kills 99% of *B. pumilis* spores which means that food irradiation is even more effective than in the experimental system used to assess irradiation efficacy. Clearly food irradiation is very effective.

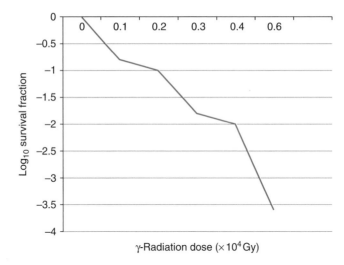

Figure 12.3 The effect of exposure to γ-radiation on the survival of *Bacillus pumilis* spores. (Data from Tabei *et al.* (1984) *Journal of Antibacterial and Antifungal Agents*, **12**, 611–618.)

How is food irradiated?

There are three ways in which food irradiation is carried out. The first uses γ-rays emitted from a radioactive element – ^{60}Co is usually used, but ^{137}Cs is sometimes used. The food is taken on a conveyor belt past a sealed radioactive source (e.g. ^{60}Co) so that it is in the γ-rays for sufficient time to ensure an average dose of 10 kGy which kills the food microorganisms. The second irradiation method uses X-rays which are very similar to γ-rays but are generated by an X-ray machine, and the third uses high velocity electrons generated by an electron accelerator. The high speed electrons are akin to very high energy β-radiation (remember β-particles are electrons) and are able to deliver greater than 10^{10} Gy/s. This huge dose means that the food need only be irradiated for fractions of a second to kill contaminating microorganisms. The sealed γ-source method, electron accelerator method and X-rays achieve the same goal and are carried out in a similar facility which includes the radiation source (i.e. electron accelerator, X-ray machine or γ-emitting element) and a conveyor belt to transport the food at the right speed to achieve a microorganism-lethal dose.

Food (e.g. raw meat) is usually packaged before being irradiated so that the sterile product is not recontaminated during the packaging process.

The effects of irradiation on food chemistry

I have already discussed the effects of γ-irradiation on molecules important to cell function (see *How irradiation kills cells*). γ-Irradiation can also affect molecules that might not be very important in assuring cell survival but might have other important properties. If a flavouring chemical is modified by irradiation the flavour of the irradiated food will be different; if a structural chemical is affected the texture of the food might change, and if a food pigment is affected the colour might change. All of these things can happen as a result of irradiation and they are all undesirable. In addition, high energy electromagnetic radiation increases the temperature of food and so the dose of radiation must be carefully controlled so that the microorganisms are killed but the food is not heated sufficiently to partly cook it, so changing its flavour and texture. The dose of radiation used in food irradiation (i.e. 10 kGy) does not cause a significant increase in the food's temperature.

When food is irradiated, the water molecules in the food undergo ionisation to form ˙OH, which in turn can react with food components (e.g. lipids) to generate many different reaction products. Also, the radiation can interact directly with food molecules (e.g. carbohydrates) to form modified molecules. The chemistry of these (and many more) reactions is complex, extensive and not fully understood. For example, if maize starch (a glucose polymer) is irradiated in the presence of oxygen and water, at least 28 different products are formed (Table 12.1) – this illustrates the complexity of the food chemistry that exposure to ionising radiation can initiate.

Different food molecules are affected differently by different radiation doses. For example, unsaturated fat molecules are more susceptible to the effects of irradiation (they form peroxides) than saturated fats (Figure 12.4). Indeed at the 10 kGy doses used in food irradiation there is little effect on saturated fats, but some effect on unsaturated fats that leads to changes in both the nutritional value of the unsaturated fats and the flavour of the food – one of the products of unsaturated fat peroxidation is butanoic acid which has a foul taste.

I mentioned change in the nutritional value of unsaturated lipid food components above; this is just one example of the potential effects of irradiation on food's nutritional value – there are many others. Vitamins are important food components that are known to be affected by exposure to high energy ionising radiation, are present at low concentrations in foods, and thus their dietary benefit might be reduced sufficiently following irradiation; this might be a cause for concern from a human health point of view. For this reason, there have been many studies that have looked at the effects of irradiation on vitamin levels in food.

The effects of irradiation on vitamins

The fact that irradiation of food results in changes in vitamins has been known for a long time. As long ago as 1919 Kanematsu Suguira and Stanley Benedict from the Roosevelt Hospital in New York, USA, showed that irradiating yeast with radium resulted in rats fed the irradiated yeast not growing as fast as control rats. This was attributed to 'growth promoting factors' in the yeast – these factors were the B-vitamins, although this was not realised at the time.

Studies in the late 1950s on the γ-irradiation of vitamin B_1 (thiamine) demonstrated the release of ammonia; this suggested that the 6-amino group of the pyrimidine ring is lost. Later studies have suggested that the radiolysis (i.e. radiation-induced breakdown) of thiamine is mediated by ˙OH liberated from water, resulting in the formation of oxythiamine (Figure 12.5). Other molecules (e.g. glucose) interfere with γ-induced thiamine radiolysis and to some extent protect thiamine from the effects of irradiation; since such molecules (e.g. glucose) are

Table 12.1 Products formed by γ-irradiating (dose = 10 kGy) starch in the presence of oxygen and water. This illustrates the complex food chemistry that ionising radiation can initiate. (Data from Urbain WM (1986) *Food Irradiation*. Academic Press, London, p. 40.)

Radiolytic product	Concentration (μg/g)
Formol	20
Acetaldehyde	40[1]
Acetone	2.1[2]
Malonaldehyde	2
Glycoaldehyde	9
Glyoxal	3.5
Glyceraldehyde/dihydroxyacetone	4.5
Hydroxymethylfurfural	1
Methylglyoxal	<0.25
Diacetyl	<0.1
Acetoin	<0.1
Furfural	<0.4
Formic acid	100
Acetic acid	<1.8
Glyoxylic acid	<0.5
Pyruvic acid	<0.2
Glycolic acid	<0.6
Malic acid	<1.3
Oxalic acid	<1.4
Methyl formate	Trace
Glucose	5.8
Methanol	2.8
Maltose	9.8
Mannose	0.1
Ribose	0.6
Xylose	0.4
Erythrose	1.2
Hydrogen peroxide	6.6[3]

[1] Dose = 8 kGy
[2] Dose => 20 kGy
[3] Dose = 1–4 kGy

present in food and the radiolysis products of thiamine have never been isolated from irradiated food, there is still considerable conjecture about the importance of thiamine radiolysis following food irradiation. This is a very complex issue that needs a great deal more research to shed light on its significance in a human nutrition context; I have discussed it briefly here to illustrate the conjecture about radiation effects on the wholesomeness of food.

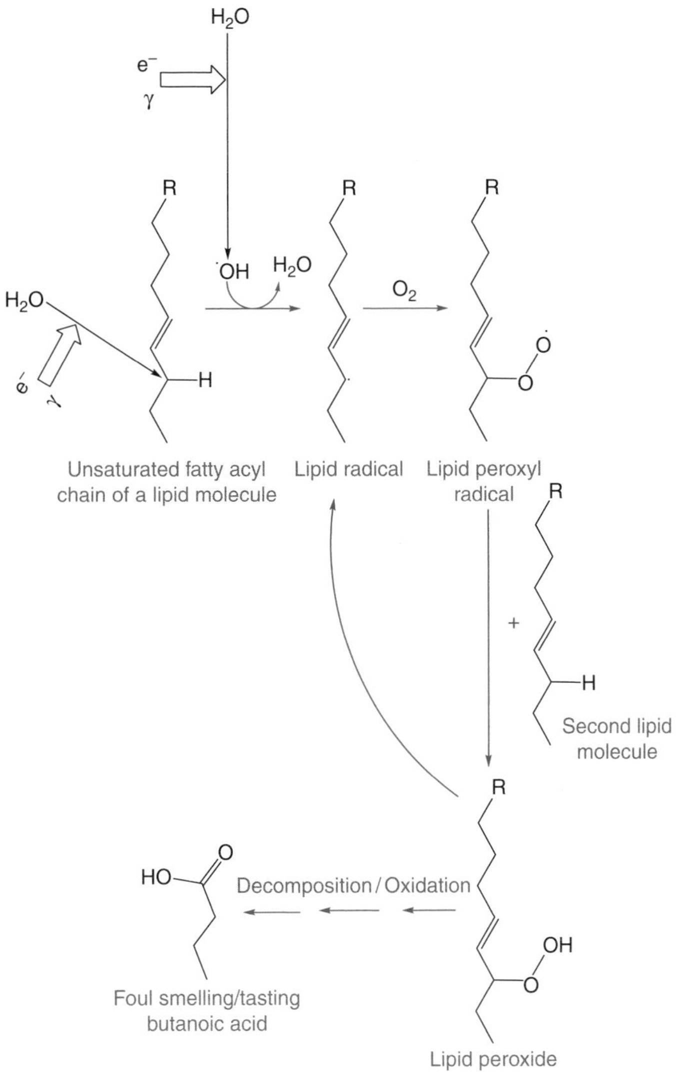

Figure 12.4 The irradiation of food oxidises and breaks down unsaturated fats to form foul smelling and tasting molecules (e.g. butanoic acid). The chemistry of this process either involves liberation of a reactive ˙OH radical from water which in turn reacts with the lipid C–H adjacent to the C=C bond, or involves direct interaction of the high energy γ-rays (e.g. from ^{60}Co source) or e– (from an electron accelerator) to form a carbon radical on the fatty acyl carbon adjacent to the C=C. Both processes lead to the same end point, the reduction in nutritional values of the unsaturated lipid and the generation of foul tasting butanoic acid (and other carboxylic acids).

When assessing the effect of food irradiation on vitamin (e.g. thiamine) levels in food it is important to compare irradiation with other methods of food preservation. Interestingly, freezing and heat sterilisation also cause significant decreases in vitamin levels (Table 12.2) ... so perhaps irradiation is not so bad after all!

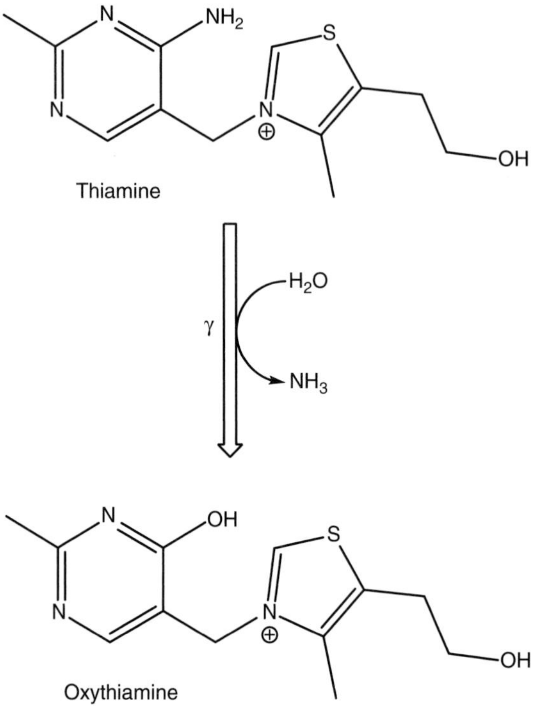

Figure 12.5 The formation of ammonia and oxythiamine following γ-irradiation of thiamine (vitamin B$_1$).

Table 12.2 The effects of different preservation methods on vitamin levels in beef. This shows that freezing, heating and irradiation all cause changes in vitamin levels (some even result in increases). (Data from Josephson et al. (1978) *Journal of Food Processing and Preservation*, **2**, 299–313. Used with permission.)

Vitamin	Storage time (months)	[Vitamin] after different preservation methods (mg/kg)			
		Freezing	Heat	^{60}Co γ-irradiation	Electrons
Thiamine	0	0.97	0.63	0.83	0.77
(vitamin B$_1$)	15	0.68	0.14	0.21	0.26
	Change	−30%	−78%	−75%	−66%
Riboflavin	0	2.8	2.63	2.83	2.6
(vitamin B$_2$)	15	1.69	2.6	2.6	1.46
	Change	−37%	−1%	−8%	−44%
Niacin	0	48.6	48.1	48.8	46.8
	15	57.2	54.9	50.1	44.5
	Change	+18%	+14%	+3%	−5%
Pyridoxine	0	2.5	2.13	3.93	5.2
(vitamin B$_6$)	15	0.97	0.57	0.35	0.42
	Change	−61%	−73%	−91%	−92%

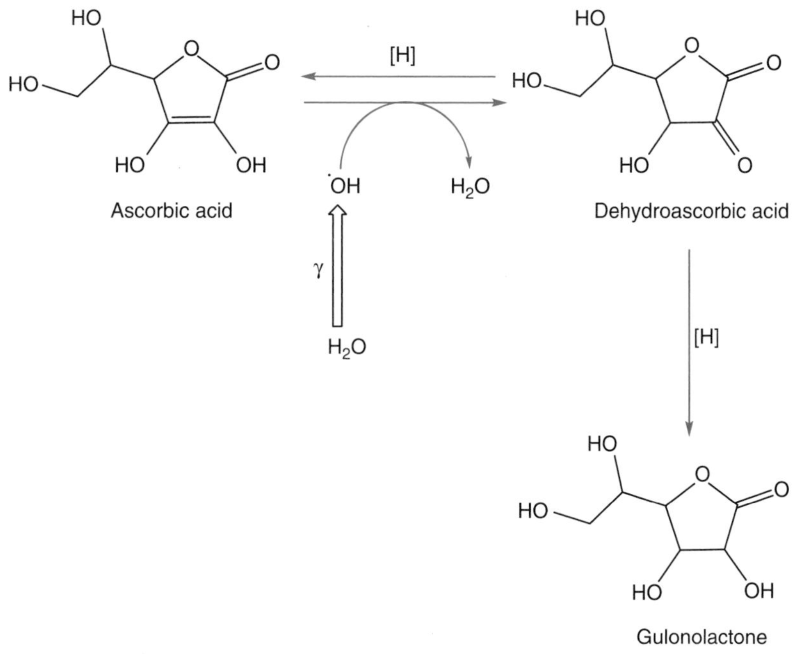

Figure 12.6 Postulated γ-induced (via ·OH) oxidation of ascorbic acid (vitamin C) to dehydroascorbic acid (which is in equilibrium with ascorbic acid) and its breakdown to form gulonolactone.

Other vitamins have also been studied and similar conclusions drawn. For example, vitamin C (ascorbic acid) is oxidised by water-derived ·OH following γ-irradiation to form dehydroascorbic acid followed by a complex series of reactions which result in formation of reactive ascorbate radicals which break down in water (Figure 12.6). This, of course, would reduce the concentration of vitamin C in irradiated food and so reduce its micronutrient value.

Clearly, we have much to learn about the effects of irradiation on the nutritional value of food; this might be reason enough not to move too fast in endorsing irradiation as a means of reducing food pathogens, so making food safer. This is a good example of a risk versus benefit assessment (see Chapter 2). On this occasion the benefit is clear (i.e. reduction of food pathogens), but the risks (e.g. reduction of nutritional value) are uncertain.

Radiation dose

The greater the radiation dose the greater the effect, but if the dose is too great it might significantly affect the nutritional value, flavour and perhaps texture of the food (as discussed above), and therefore the optimum dose must be determined at which the desired outcome (e.g. killing food pathogens) is maximised and the unwanted effects (e.g. changing the food's flavour) are minimised.

Table 12.3 Doses of radiation used for different food applications. (Data from Roberts (1997) *Food Irradiation*, Royal Society of New Zealand Alpha Series: 94. Royal Society of New Zealand, Wellington.)

Use	Radiation dose range (kGy)	Foods
Inhibit vegetable sprouting	0.05–0.15	Potatoes, onions, garlic
Delay ripening of fruit	0.05–0.15	Tropical fruits
Reduce parasites	0.1–0.3	Pork
Kill insects	0.1–1.0	Grain, rice, fruit, vegetables
Delay spoilage (for ambient temperature storage)	0.5–5.0	Strawberries
Delay spoilage (for refrigerated storage)	0.5–10.0	Meat, poultry, fish
Kill pathogens	2.0–10.0	Meat, poultry, seafood, dried foods
Sterilisation	10.0–30.0	Herbs, spices, special diets (i.e. for people with medical conditions)

Irradiation is used not only to make food safe by killing pathogens, but also to kill insects, prevent stored vegetables sprouting and delay fruit ripening. In this book I am only concerned with the food safety uses of irradiation, but it is important to know that this is only one use of the technique and that different radiation doses achieve different effects – for food safety applications the dose is usually approximately 10 kGy (Table 12.3).

Does irradiation make food radioactive?

The simple and definitive answer to this question is NO! The source of γ-rays used in food irradiation (e.g. ^{60}Co) is sealed and the food being irradiated does not come into contact with it. And, as we have learned above, the effect of passing γ-rays through food is purely chemical. There is no way that γ-rays can make other atoms radioactive.

Surprisingly, there is a popular misconception held by some consumers that irradiated food *is* radioactive; this might be one of the reasons that there is staunch opposition to food irradiation from some people … but they are wrong!

Health effects of food irradiation

The strict regulations around food irradiation are primarily to protect operators from the ionising radiation used in the process. However, as discussed above, irradiation of food changes the chemistry of the food in two main ways – initially by the generation of free radicals (e.g. ˙OH) from water and secondarily the reaction of these highly reactive species with other molecules (e.g. lipids) in the food to form different molecules (e.g. lipid peroxides → short chain fatty acids). The free radicals and their food molecule reaction products might be toxic, and, therefore, the effects of the complex chemistry resulting from food irradiation on the safety of the food must be carefully considered.

Arguably the most worrying issue in the context of consumer health is the toxicity of free radicals in irradiated food. Free radicals are very reactive and are able to alter molecules (e.g. nucleic acids – see *How irradiation kills cells*) in consumers' cells, potentially causing harm (e.g. mutations leading to cancer). However, the extreme reactivity of free radicals means that they have very short half lives in food (because they quickly react with other food molecules) and therefore are very unlikely to be present when the food is consumed.

The myriad reactions that irradiation-derived free radicals undergo in food must generate an astounding number of reaction products. It would be very difficult, if not impossible, to assess the toxicity of each of these in turn. For this reason irradiated whole food toxicity studies are used to assess the toxicity of the combined irradiation-induced changes in the food.

A great deal of work on the safety of irradiated foods has been carried out. This work involved feeding animals with irradiated food and studying the effects. The UK Committee on Toxicity (COT) concluded that 'there was no evidence to suggest that any toxicological hazard to human health would arise from the consumption of irradiated food up to an average dose of 10 kGy'.

The food safety issues relating to food irradiation extend beyond creating toxic molecules in food as a result of irradiation. As discussed above, the nutritional value of food might be changed by depleting important nutritional components (e.g. vitamins). This does not, strictly, make the food unsafe (in a toxicological context), but it would make it less efficacious.

In conclusion, irradiated food is safe. In fact it is safer than non-irradiated food because it has no pathogens, but it might have a lower nutritional value which is cause for concern.

The use of food irradiation around the world

Unfortunately the WHO and the UK government's Advisory Committee on Irradiation reported that irradiated food is safe just a few days before the Chernobyl disaster in 1986. This led to much public misunderstanding because they confused the very negative implications of radiation following the Chernobyl disaster with the beneficial use of radiation to make food safe. This confusion remains and has led to some countries (e.g. New Zealand) not allowing food irradiation or the importation of irradiated foods – their concerns have no food safety basis whatsoever. It might be that these countries simply disagree with the use of radioactive sources for environmental reasons and so their objections to food irradiation reflect their general antithesis to radioactivity generally; in this context it is difficult to understand why they will not condone the use of electron accelerators since they are not permanent radiation sources. This rather simplistic argument does not take account of the nutritional implications of food irradiation and perhaps countries that do not allow food irradiation are worried about their citizens' nutrition. This debate is far beyond the remit of this book, but it is worth some serious thought, I believe. What do you think?

Take home messages

- Food pathogens and spoilage microorganisms are killed by irradiation with γ-radiation, X-rays or high energy electrons from an electron accelerator.
- Food irradiation DOES NOT make the food radioactive.
- Food irradiation results in a great deal of complex chemistry that might alter the food's constituent molecules (e.g. vitamins).
- There is no evidence that food irradiation makes food harmful to the consumer.
- There is considerable misunderstanding amongst consumers about food irradiation which has led to much unjustified concern.

Further reading

Advisory Committee on Irradiated and Novel Foods (1986) *Report on the Safety and Wholesomeness of Irradiated Foods*. HMSO, London.

Hartley NH (1996) Toxic effects of radiation and radioactive materials. In: Klaassen CD (ed.) *Casarett & Doull's Toxicology – The Basic Science of Poisons*. McGraw Hill, New York. This chapter gives a detailed account of the effects of radiation on biological systems.

Thorne S (ed.) (1991) *Food Irradiation*. Elsevier Applied Science, London & New York. Chapter 1, An Introduction to the Irradiation Processing of Foods by S. Hackwood is an excellent overview.

Urbain WM (1986) *Food Irradiation*. Academic Press, London.

Chapter 13
Food Safety and the Unborn Child

Introduction

This is a short chapter, not because I don't want to write much, but because there is not very much known to write about. I have included the chapter because food-related developmental effects is currently a field of great research interest and it is likely to assume much more importance as we understand more about how food chemicals affect growth and development *in utero* and influence the health and well-being from birth to death.

Most of what I will discuss in this chapter is based on current research (some of it my own) and thus is far from being definitive - many of the data and ideas are still the subject of much conjecture and debate among scientists. This is very different to the other chapters in this book which ostensibly include well-established facts that have been through the conjecture and debate stage and have emerged unscathed.

I will address the question 'Can chemicals in food affect growth and development of the embryo and/or fetus?' The answer is *yes*, but that the evidence for some chemical exposures and their postulated effects is uncertain and controversial.

'You are what your mother ate'

It has been known for a long time that the mother's diet during pregnancy has a great influence on the health of her child. It is obvious that if mum's diet is poor, since she is indirectly providing her developing child's nutrition, the child will be nutritionally lacking too. There is nowhere that this is more painfully obvious than in the poor parts of the world where food is in short supply and babies are often born significantly malnourished with a much lower chance of survival and resistance to disease than their counterparts from affluent parts of the world. For example, if the mother is deficient in folate (sometimes called vitamin M) in her diet, her baby is more likely to have a

Food Safety: The Science of Keeping Food Safe, First Edition. Ian C. Shaw.
© 2013 John Wiley & Sons, Ltd. Published 2013 by John Wiley & Sons, Ltd.

neural tube defect (e.g. spina bifida) than a baby born to a mother with plenty of folate in her diet. This is simple nutrition; the developing embryo/fetus needs folate to produce its neural tube – if it is not provided with enough folate the neural tube will be defective – simple (I will discuss folate in more detail later in this chapter).

Exposure of the developing child to some food-borne pathogens can have devastating effects on development. Some bacteria (e.g. *Listeria monocytogenes* – see Chapter 3, *Listeria*) can result in abortion or malformation of the child; the specific malformation depends on the time during development that the exposure occurs.

As we understand more about the importance of diet in pregnancy, it is becoming clear that not all dietary components have simple nutritional effects on the development of the child. Some have far-reaching effects on the regulation of genes that can have equally far-reaching effects on the child for the rest of his or her life. For example, an overweight mother is likely to have a baby who will become overweight in later life. This might simply reflect the dietary environment of the growing child; its mother ate too much fat (which is why she is overweight) and therefore her *in utero* child received more fat via its mother, and was born into a family that eats too much fat. This is simple nutrition and very easy to understand – too much fat makes you fat. However, experiments in rats have shown that some overweight rats have overweight offspring (pups) even if the mother is fed a low-fat diet during pregnancy and the pups are fed low-fat diets when they are weaned. This shows that 'overweightness' might be under some degree of genetic control. This again is not difficult to understand because we know that bodily characteristics are encoded on our DNA. But consider a rat experiment where a 'thin' pregnant rat was fed a high-fat diet during pregnancy and lactation. When weaned her pups were fed a 'normal' fat diet, but they still became overweight as they grew up. Clearly this was not because the pups were getting too much fat in their diet once they were weaned, so it cannot be explained by simple 'over' nutrition. It is thought that the high-fat diet during pregnancy changed the pups' genetic expression so that they metabolised fat differently and became fat themselves. This experiment led to new thinking about diet in pregnancy and its effects on the developing child. The New Zealand scientist Professor Sir Peter Gluckman summed up this new thinking when he said, 'You are what your mother ate'.

In order to understand the effects of diet on development, we must first explore how the embryo/fetus gets its nutrition from its mother and how its mother controls what she delivers to her baby. The placenta is the key to this delivery system. The placenta allows transfer of some, but not all, of the components of the mother's circulatory system to her *in utero* child. But what does it allow across the placental barrier and what is barred?

Our discussion so far has focused on normal dietary components (e.g. folate, fat), but there are many other chemicals in the mother's circulatory system that originated from environmental exposures (e.g. chlorinated hydrocarbons from a newly painted room) and her food (e.g. pesticides). Do these cross the placental barrier? If so, what effect do they have on the developing fetus?

Finally, we must explore the mechanism of the effects of both dietary components (e.g. fat in the above experiment) and chemical contaminants (e.g. chlorinated hydrocarbons). Some of these effects are simply nutritional (as discussed above), but others might involve gene regulation. As more research in this area is published – this is a very active area of research at present – it is becoming clearer that the mechanism of gene regulation does not involve chemical modification (i.e. mutation) of the DNA that comprises genes, but appears to involve indirect effects that result in modulation of gene expression (i.e. switching on or off of specific genes). This is termed epigenetics and it is one of the most eagerly researched subjects at the moment; we are only just beginning to understand the power of epigenetics and its implications for the developing child and its life ahead, and perhaps the lives of its children and their children.

Growth and development of the embryo and fetus

Before it is possible to understand how nutrition and chemical and microbiological exposures might affect the developing fetus it is necessary to understand the process and timescale of growth and development. In simple terms, exposures to chemicals affect whatever is developing at the time of the exposure. So if we know what is developing we can understand what might be affected if exposure to a chemical or a microorganism occurs during that period. Dietary deficiencies work in exactly the same way; they affect what is developing at the time of the deficiency.

Darwin said, 'Embryology recapitulates phylogeny'. By this he meant that as the embryo develops it goes through stages that resemble the evolutionary development of the species. So the human embryo goes through a fish-like, an amphibian-like and then a more recognisably mammalian stage. During these stages, specific cell types develop and the cells grow into organs which eventually begin to function. For example, in the early stages of human embryological development a few cardiac cells develop from a very basic cell type (i.e. stem cells) from which all other cells can be made. There are only a very few cardiac cells at first and so they cannot function as a heart. But as they divide and arrange into the morphology of the heart, chambers appear and eventually the signals that cause the heart to beat can be generated and received. At this point the embryo has a functioning, beating heart which is very clear if the mother's abdomen is scanned (by ultrasound). If the embryo is exposed to a toxic chemical during the very early stages of the heart's development, it is possible that this will have a significant impact on the future development of the heart and that a malformed heart that does not function properly (if at all) might result. On the other hand, if the chemical exposure occurred when the heart was fully formed it is far less likely that it would significantly adversely affect the heart because it could not interfere with the growth and development of the cells that formed the heart because exposure was too late.

A chemical that adversely affects the developing embryo/fetus is termed a teratogen (from the Greek τερας [teras] meaning monster; γένεσις [genesis] meaning generation or creation). The drug thalidomide is a potent teratogen; some mothers who were prescribed thalidomide (to treat morning sickness) in the 1950s and early 1960s gave birth to children with severely malformed limbs.

The stages of development (Figure 13.1)

Pregnancy is divided into three trimesters:

- 1st trimester: 0-3 months (weeks 1-13)
- 2nd trimester: 4-6 months (weeks 14-26)
- 3rd trimester: 7-9 months (weeks 27-40)

Birth occurs on average in week 40, but a viable child can be delivered prematurely as early as week 24 and still have a 50% chance of survival. Equally birth may be delayed to week 42 or even beyond (the child is then termed 'postmature').

First trimester
The first trimester is the time that the embryo develops, and represents active growth and development of the body organs from stem cells. This is the most vulnerable time for interference with development because if cells are damaged during the early phase of development of an organ the cells produced from them might also be damaged. This is the most vulnerable time if the mother is exposed to a teratogen. The first trimester ends when the embryo

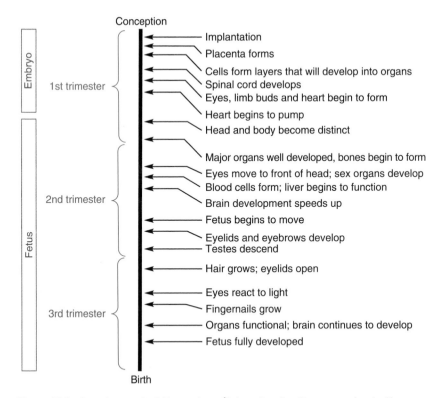

Figure 13.1 Development of the embryo/fetus showing the approximate times that major developments occur. This shows clearly that the first trimester is the time that initiation of organ development begins and thus at this time the embryo is most vulnerable to teratogens.

has a recognisable human form – it is then called the fetus (from the Latin *fetus*; in the UK it is spelled 'fœtus' which is strictly speaking incorrect because the Latin from which the word is derived does not have a diphthong (i.e. œ)).

Second trimester

The fetus has transparent skin at the beginning of trimester 2 and its blood vessels can clearly be seen. As the trimester progresses the skin becomes opaque. The second trimester fetus has active organs; its heart is functioning and its kidney and bladder are working and it can even suck its thumb.

Third trimester

The fetus's hair grows (including its eyelashes) and it grows fingernails. Its eyes respond to light and it begins to change its position ready to be born. At 9 months a fully formed human being is delivered.

Throughout its development the embryo (from about the end of the first week of gestation) and then the fetus are nourished by the mother via the placenta and the umbilical cord.

The placenta

Five to 7 days after fertilisation of the ovum, the zygote (i.e. fertilised ovum) implants into the uterus wall. The *corpus luteum* secretes the hormone progesterone which causes thickening of the endometrium (i.e. the uterus wall). The zygote develops a yolk sack, an amniotic sack, and the tissues from which the placenta and umbilical cord will begin to develop over the following week or so and continue to grow for the entire gestation period.

The amniotic sack encloses the embryo and fetus for their entire development period and contains amniotic fluid in which the developing child is suspended – I will discuss amniotic fluid in the context of fetal exposure to chemicals from the mother's diet later.

The placenta (Figure 13.2) begins as a small growth from the early stage embryo and grows to an organ 15% of the weight (approx. 500 g) of a full-term fetus. It initially sits on the surface of the endometrium (the lining of the uterus) and develops projections (villi) that extend into the uterus wall. As it develops, a distinct physical barrier (the septum) forms between the embyo/fetal side and the maternal side. The embryo/fetal blood vessels penetrate the placenta as far as the septum on the embryo/fetal side and the mother's blood vessels infiltrate the modified outer region of the endometrium (myometrium) that develops at the point of contact with the placenta. Blood gasses (i.e. oxygen from the mother; CO_2 from the fetus) and nutrients (e.g. glucose from the mother) and metabolic waste products (e.g. urea from the fetus) are exchanged as the means of providing for the fetus's needs. During the third trimester the fetus excretes some of its own metabolic waste products (via its active kidneys) into the amniotic fluid rather than relying only on the placenta as its sole disposal service.

The placental barrier

Not only does the placenta provide the developing fetus with its nutrients and dispose of its waste, but also it protects the fetus against the numerous toxic

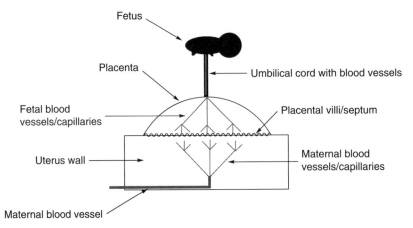

Figure 13.2 Schematic representation of the placenta attached to the uterus wall. The septum presents a barrier which only selected molecules can cross – waste products from the fetus diffuse into the maternal circulatory system for excretion via the mother's kidneys (e.g. urea) into her urine or via her lungs (e.g. CO_2) into her expired air, and nutrients (e.g. glucose) cross from mother to fetus. The septum controls what enters and leaves the fetal side – this is termed the placental barrier.

chemicals present in its mother's circulatory system. These toxic chemicals might be endogenous maternal metabolites (e.g. urea), toxic food components (e.g. natural toxins from food plants) or environmental contaminants (e.g. xenoestrogens) that the mother has absorbed. The placenta acts as a barrier to the fetus and prevents the passage of many of these compounds from the maternal to the fetal circulation. In addition, the placenta has active detoxification systems that metabolise toxic components from the mother's circulatory system, so preventing them crossing the 'placental barrier'.

We do not understand fully how the placental barrier works, but recent studies have clearly shown that some compounds cross the placenta into the fetus whereas others don't. This is an important mechanism by which the mother protects her child from the myriad chemicals she absorbs from her food and environment every day.

As mentioned above, the placenta also has detoxification systems to metabolise components of the maternal blood, thus either reducing their toxicity to the fetus or making them very polar so that they cannot traverse the placental barrier and so remain to be excreted (e.g. in urine) by the mother. For example, the placenta has cytochrome P_{450} which carries out Phase I of toxic compound metabolism, and conjugating enzymes (e.g. glucuronyl transferases) which carry out Phase II metabolism (see Chapter 1, *Evolution of cellular protection mechanisms*). Both Phase I and Phase II metabolism increase the polarity of compounds, so preventing them crossing the placental barrier.

Some interesting work carried out by my colleagues at the National Research Centre for Growth & Development, New Zealand, has investigated the transfer of chemicals across the placenta from the maternal to the fetal side; their experiments showed clearly that some chemicals (e.g. 17β-estradiol) are not allowed to cross freely, whereas others (e.g. bisphenol A (BPA)) pass freely. This is very interesting because high levels of 17β-estradiol in the fetus would interfere with sexual development (especially in males) and therefore it is understandable why the placenta would restrict its transfer from mother to child. The control of transfer mechanisms for endogenous compounds (e.g. the female hormone 17β-estradiol) have evolved to make sure that the child's development proceeds 'as planned'. However, exogenous chemicals (e.g. environmental pollutants) are relatively recent introductions (i.e. post Industrial Revolution – the late 1800s) and therefore placental protection mechanisms for many of them have not (perhaps yet) evolved; for example, the xenoestrogen BPA (see Chapter 9, *Xenoestrogens*) freely crosses the placenta from mother to child (Figure 13.3) and therefore would be expected to exert its feminising effect on the developing fetus of a BPA-exposed mother.

Effects of nutrients

As discussed above, the placenta also delivers nutrients and other important biochemicals to the developing embryo/fetus. This is very important because the child is growing and developing fast and needs a constant supply of the basic biochemical building blocks (e.g. amino acids) from which it can manufacture the components (e.g. proteins) of its cells. In addition to the biochemical building blocks, the child also needs many micronutrients (e.g. vitamins) that are important for normal development and, of course, the biochemicals (e.g. glucose) used to generate energy to drive metabolic processes. Deficiencies in the biochemical building blocks, energy sources or micronutrients can lead to malformations in the child. As discussed above, the time that the deficiency occurs determines the effect; for example, if a key nutrient is deficient during the development of the embryo's limb buds it is very likely that the child will have malformed limbs.

All of these important biochemicals are either derived directly from the mother's food or manufactured metabolically by the mother from components of her food. It is beyond the scope of this book to delve in detail into the effects of nutritional deficiencies on the developing child, but I will give two examples to illustrate the importance of maternal provision of key nutrients to her developing child; both have been outlined already but are covered in greater detail below.

Folate

Folate or folic acid (these terms are often used interchangeably – folate is the deprotonated form of folic acid and is the form found in the body; Figure 13.4) is a B vitamin that is important in the development of the central nervous system (CNS). It is found in many foods, including green leafy vegetable (e.g. spinach), liver, white fish and grains (e.g. rice). It is an essential dietary component because humans are unable to synthesise it; therefore, to be healthy we must have a small amount (>1mg) in our diet every day. Interestingly, our gut bacteria can synthesise folate and therefore they provide a proportion of our daily folate requirement.

Folate deficiency causes anaemia (particularly megaloblastic anaemia) because folate is a key player in the development of some blood cells (e.g. leukocytes). Megaloblastic anaemia occurs particularly in parts of the world

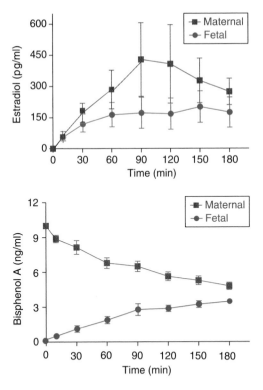

Figure 13.3 Results (*n*=7; mean±SD) of an experiment in which BPA was added to the perfusion medium of an isolated human placenta* and 17β-estradiol (top) or BPA (bottom) levels measured on the fetal and maternal sides. This shows that there is much more 17β-estradiol on the maternal side which indicates that the placenta restricts its trasnsfer to the fetus, but that BPA passes across the placental barrier freely. (Experimental data kindly provided by Dr B. Balakrishnan, National Research Centre for Growth & Development, Auckland, New Zealand; see Balakrishnan *et al.* (2010) *American Journal of Obstetrics and Gynecology*, **202**, 393.e1–393.e7, for the BPA data. Reprinted with permission.)
* A placenta obtained following a child's delivery. The placenta was kept alive by pumping synthetic blood through it containing oxygen and important nutrients (e.g. glucose) and chemicals were added to the perfusion medium and measured in the liquid that came out on the fetal side (i.e. equivalent to the fetal blood supply) and the perfusion medium on the maternal side at time intervals.

where diets are poor and polished (i.e. white) rice is consumed. Polished rice has the husk removed and the husk contains folate.

Folate also plays a key role in the development of components of the CNS, particularly the neural tube which is the early (early 1st trimester) embryolog-ical form of the CNS. The neural tube is a tube-like structure that encases the developing brain and spinal cord and around which the spine develops. Therefore, mothers deficient in dietary folate have a greater risk of their child developing CNS developmental disorders, in particular spina bifida (from Latin *spina* meaning spine and *bifida* meaning split or divided). Spina bifida is a disorder in which the neural tube fails to close during embryological development resulting in a hole in the spine (i.e. split) which exposes the

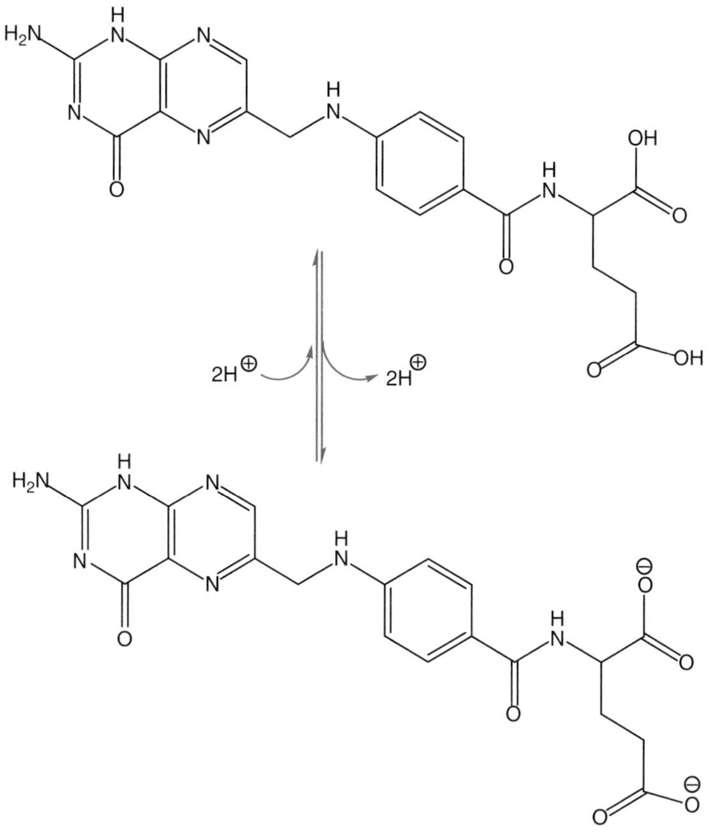

Figure 13.4 The molecular structures of folic acid and folate (deprotonated folic acid).

spinal cord and leads to varying, sometimes severe, degrees of disability (from bladder problems to paralysis). If the defect is not repaired surgically, the spinal cord is at risk of physical damage which would lead to more severe disabilities; therefore children born with spina bifida have to undergo surgery as soon as possible.

The incidence of spina bifida is very low, but can be reduced further if prospective mothers take folate supplements (0.5 mg/day) for at least 1 month before conception and for at least the 1st trimester of pregnancy. Some countries fortify staple foods (e.g. bread) with folate to make sure all prospective mothers get sufficient folate. This is controversial because it is a form of mass medication that does not give people (e.g. men) who do not want or need folate supplements any choice in the matter.

In a recent study in China, women giving birth to children with neural tube defects (e.g. spina bifida) had lower (9.6 nmol/L) folate levels in their blood than women who gave birth to 'normal' children (14.0 nmol/L) (data from Zhang et al. (2008) *Biomedical and Environmental Sciences*, **21**, 37–44). This illustrates the importance of folate in the mother's diet to assure sufficiently high levels in the mother's blood to transfer sufficient to the embryo to facilitate neural tube development.

Fats

As already discussed at the beginning of this chapter, recent work has shown that the amount of fat in a mother's diet can influence gene expression in her offspring. This is a very important finding because it shows that the mother's nutritional status can lead to changes in the embryo/fetus that can have effects which last beyond the gestational period and can influence the child, adolescent and adult up until the end of his/her life.

Experiments carried out by scientists at the National Research Centre for Growth & Development, New Zealand, have already been introduced at the beginning of this chapter. Their findings are very important in the context of the effects of diet on offspring; for this reason I will discuss them in greater detail here. Female rats (dams) were given a normal diet (i.e. controls), a high-fat diet throughout their life (including pregnancy and lactation), or a high-fat diet just during pregnancy and lactation. The offspring (pups) were weighed at birth and when weaned were fed either a normal or a high-fat diet and were weighed at time intervals until they reached maturity. Pups from both high-fat diet group dams had lower body weights and lower blood lipid levels than the controls. As the pups grew up the ones from high-fat diet dams increased in weight and became obese and their blood lipid levels were high, even if they had been fed a normal diet. This is very interesting because it shows that the mother's diet determined the body status (i.e. fat or thin) of the offspring rather than the offspring's diet. Taking this idea one stage further tells us that the mother's diet must have switched on a process (perhaps via genes) that determined her offspring's weight. There is no reason to believe that this would not be the case in humans too. Don't forget, all of the rats in this study were normal (i.e. not fat) before the experiment began; the only change that any of them was subjected to was a different dietary fat profile (see Howie et al. (2009) in Further reading).

Effects of food chemical contaminants

It is well known that many chemicals in the mother's diet cross the placental barrier and enter the fetus's circulatory system, and we assume that these chemicals have the same effects on the fetus as they would on adults. Mothers in the past knew that if they drank alcohol their fetus would calm down and did not kick as much as it might normally. This is because alcohol has an anaesthetic/calming effect and is readily transferred from the mother to her child via the placenta. Incidentally, this is the origin of gripe mixture that used to be given to infants to calm them - gripe mixture (no longer available) was a sugar alcohol concoction; also in days gone by mums would drink stout (a dark beer) before nursing because it calmed their child - alcohol from the stout was transferred from the mother's circulatory system via her milk to the baby. We now know that alcohol can cause serious neurological birth defects (fetal alcohol syndrome) and therefore no sensible mum drinks alcohol during pregnancy.

Despite our understanding of chemical transfer from mother to fetus across the placenta we know surprisingly little about the effects of dietary chemical contaminants on embryo/fetus development. As with other dietary chemical exposures it is very difficult to prove cause/effect relationships.

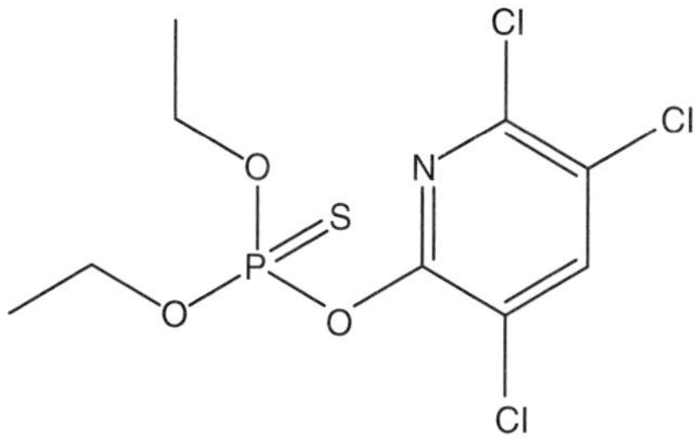

Figure 13.5 The organophosphorus insecticide chlopyriphos.

There are, however, several good examples of chemical effects in the children of chemically exposed mothers (e.g. female farmers who use pesticides during pregnancy).

Case report – adverse effects on the children of mothers exposed to the insecticide chlopyriphos

Four children in Michigan, USA, born to three mothers, presented in 1995 with multiple structural deformities (e.g. severe brain deformities) and other growth abnormalities (e.g. deformed external genitalia). Two of the children had the same mother but had normal brothers and sisters. The other two had different mothers and they too had normal siblings. There was no history of genetic abnormalities in their families. The mothers of all of the children had been exposed to the insecticide chlorpyriphos (Figure 13.5), a known teratogen, during the 1st trimester of their pregnancies, either in their workplace or at home. Chlopyriphos was used to treat insect infestations (e.g. fleas in carpets originating from household pets) (see Sherman (1996) in *Further reading* for the full case report).

The effects seen in the children of pesticide-exposed mothers strongly suggest a cause/effect relationship. There is no reason why an embryo/fetus exposed to pesticides originating from the mother's diet (rather than via her occupation) would not incur the same effects. The key determinant of effect is dose, and it might be that the diet-derived pesticide dose would be too low to result in an effect on the developing child. There are no proven examples of dietary chemical contaminants causing effects on embryo/fetal development, but this does not mean that there are no effects.

Effects of microbiological contaminants

The only food-borne pathogens associated directly with fetal damage are *Listeria monocytogenes* (see Chapter 3, *Listeria*) and *Toxoplasma gondii* (the parasite that causes toxoplasmosis; see Chapter 5, *Toxoplasma*).

Severe *L. monocytogenes* infection in adults leads to systemic bacteria that are able to cross the placental barrier and infect the fetus. Fetal infection with *Listeria* can lead to fetal death and/or abortion. Similarly toxoplasmosis in pregnancy can result in *T. gondii* crossing the placental barrier and infecting the fetus which can also lead to fetal death and/or abortion.

There is a high rate (19–63% in infected mothers) of abortion following *Listeria* infection in pregnant mothers; for this reason women are advised not to eat food associated with *Listeria* contamination (e.g. soft cheeses like Brie) during pregnancy.

Effects on ova and sperm

So far we have discussed only direct effects on the developing embryo/fetus (i.e. maternal dietary components or contaminants received during pregnancy). However, effects on sperm and ova can also cause changes that result in abnormalities of the developing child. In order for chemical exposures to affect sperm and/or ova in such a way that this would lead to deformed (including biochemical changes) children, the chemical must change the DNA in some way. This change could either be direct (i.e. a chemical change to DNA – mutation) or indirect (e.g. a change that alters gene expression). It is well known that mutagens result in birth defects, which is why mothers and fathers who have received cancer chemotherapy (most anticancer drugs are mutagenic) are advised not to have children. Since sperm are produced at a high rate it is likely that a short time after the mutagenic insult the new sperm produced will have no genetic defects. The situation with ova is quite different because a woman's ova are all pre-formed and are simply waiting in a queue to be released (i.e. ovulation which occurs every 28 days post-puberty and pre-menopause). If ova are exposed to mutagens it is likely that they will all be damaged and thus the risk of deformed offspring being produced from them is great; in fact too great for it to be advisable for many women who have received cancer chemotherapy ever to have children.

The example of cancer chemotherapy can be extrapolated to exposure to any mutagens. The sperm are likely to recover, but the ova might harbour mutational changes for their lifetime. The difference between cancer chemotherapy and food-borne mutagens is, of course, the dose. The latter will be very much lower than the former which means that the risk of dietary chemicals resulting in severe malformations of offspring is very low.

Chemicals that indirectly affect sperm and ova by modifying gene expression are far more worrying in the context of embryo/fetal abnormalities. However, we know very little about this; indeed what we do know is the subject of a great deal of argument and heated debate amongst scientists. It will be many years before we understand the significance (if any) of such potential effects.

Chemicals that influence gene expression without altering the chemical nature of nucleic acids work by epigenetic effects (Greek επί (*epi*) meaning on or upon). For example, exposure to the female hormone 17β-estradiol (see Chapter 9, *Estrogen receptors – ERs*) influences gene expression without changing DNA irreversibly – 17β-estradiol binds to ERs which then bind to specific sites on DNA and up-regulates genes up-stream of the bound 17β-estradiol/ER complex. So, any chemical (e.g. a xenoestrogen) that binds to an ER can have a similar (albeit lesser) effect than the natural ligand, 17β-estradiol. Therefore such chemicals could alter gene expression in sperm or ova which might result in the development of a zygote, then embryo, then fetus, then infant, then adult that harbours the effects of this gene up-regulation.

There are no unequivocal examples of such effects at the moment, but scientists are finding evidence for multigenerational effects resulting from parental exposure to chemicals that have epigenetic mechanisms of toxicity.

My laboratory is studying New Zealand veterans of the Malaya (now Malaysia) Emergency (1948-1960) who were exposed to dibutylphthalate (DBP) used as an acaricide (applied to clothing) to prevent infestation by ticks that carry Bush Typhus. DBP inhibits the synthesis of the male hormone testosterone and therefore can result in cellular feminisation by reducing testosterone levels. Chemicals that either mimic estrogens (i.e. xenoestrogens) or reduce testosterone levels can cause developmental retardation of the genitals (e.g. hypospadias and cryptorchidism) in exposed males, or perhaps boys born to exposed mothers – this is discussed in greater detail in Chapter 9. Interestingly, we found that the sons of New Zealand Malaya Campaign veterans exposed to DBP had a greater incidence of hypospadias than the general population. These data are very new and somewhat controversial, but they suggest that sperm can be affected following exposure to chemicals with an epigenetic mechanism of toxicity, in such a way that the resultant children have a greater risk of abnormality ... now that is interesting! There is, of course, no reason why dietary exposure – providing the dose is high enough – could not have the same effect. DBP was commonly used as a plasticiser in food contact plastics (e.g. in cling films) and remains one of the most prevalent environmental contaminants worldwide and so we are all likely to have been exposed to it (see Carran & Shaw (2012) in *Further reading*).

Take home messages

- Infection with *L. monocytogenes* and *T. gondii* can lead to fetal infection and fetal death/abortion.
- The implications of food chemicals (both natural and contaminants) is a very new field of interest and there are few hard and fast rules.
- It is becoming clear that the food we eat can affect the growth and development of our children.
- Food additives (e.g. colours) and contaminants (e.g. environmental chemicals like xenoestrogens) might have a role to play in the multigenerational effects of food.

Further reading

Carran M & Shaw IC (2012) New Zealand Malayan war veterans' exposure to dibutyl-phthalate is associated with an increased incidence of cryptorchidism, hypospadias and breast cancer in their children. *New Zealand Medical Journal*, **125** (No. 1358). ISSN 11758716.

Curley JP, Mashoodh R & Champagne FA (2011) Epigenetics and the origins of paternal effects. *Hormones and Behaviour*, **59**, 306-314.

Howie GJ, Sloboda DM & Kamal T (2009) Maternal nutritional history predicts obesity in adult offspring independent of postnatal diet. *Journal of Physiology*, **587**, 905–915.

Sastry B (1999) Techniques to study human placental transport. *Advances in Drug Delivery Reviews*, **38**, 17–39.

Shaw IC *et al.* (2009) The effect of dietary endocrine disruptors on the developing fetus. In: Shaw IC (ed.) *Endocrine Disrupting Chemicals in Food*. Woodhead Publishing, Cambridge, pp. 3–35.

Sherman JD (1996) Chlopyriphos (Dursban)-associated birth defects: report of four cases. *Archives of Occupational and Environmental Health*, **51**, 5–8.

Chapter 14
Organic Food

Introduction

In the context of farming, 'organic' means produced without the use of artificial fertilisers or pesticides – it conjures up a vision of the halcyon farms of the past. However, organic farming is now big business, with a significant price premium paid by consumers of certified organic produce.

In this chapter I will discuss the pros and cons of organic farming. There is no doubt that organic farming has a lower environmental impact than 'conventional' farming because many agrochemicals (e.g. man-made pesticides) are not used and instead of using artificial fertilisers animal waste and composted herbage are returned to the land, which is an excellent way of recycling. In addition, there is a price premium on organic food because it is sought-after by some consumers. I will not discuss these two positive aspects (one to the environment, the other to the farmer) of organic production in detail because they are outside the scope of this book; instead I will concentrate on the direct risks and benefits of organic food to the consumer.

There is a great deal of controversy about the nutritional value of organic food. Organic enthusiasts claim that it has higher nutritional value than 'conventionally' produced food; others say there is no difference between organic and conventional food. In this chapter I will discuss laboratory studies that have compared the nutrient content of both types of food – you might be surprised at their findings!

Regulations covering organic farming are now enshrined in legislation in most countries. This is necessary to make sure that what the consumer expects of the organic food they purchase is what the producer is delivering. Such legislation defines the term 'organic' and lays out the dos and don'ts of organic farming (e.g. which pesticides are allowed). Some countries have policing schemes which involve testing organic produce for residues of chemicals that are not permitted (e.g. man-made pesticides) as a means of checking that farmers abide by the organic production rules.

Finally, I will discuss the question, 'Is organic food safer than conventionally produced food?' You might be surprised that there is not a simple answer to this question.

Food Safety: The Science of Keeping Food Safe, First Edition. Ian C. Shaw.
© 2013 John Wiley & Sons, Ltd. Published 2013 by John Wiley & Sons, Ltd.

What does 'organic' mean?

The word 'organic' means derived from an organism – a plant or animal hav-ing an *organ*ised physical structure. An organic molecule is a molecule from a living organism. The definition is further clarified by chemists as a hydro-carbon (i.e. a molecule containing both hydrogen and carbon atoms). Lipids (e.g. palmitic acid), sugars (e.g. glucose) and DNA bases (e.g. thymine) (Figure 14.1) are all organic molecules even though they contain other atoms (e.g. N and/or O) in addition to C and H.

'Organic' has now taken on another, rather misleading, meaning. It used to denote a sustainable farming methodology that employs only natural processes – for example, organic fertilisers (i.e. manure from living organisms, e.g. cattle). In fact 'organic' is now so often used in the context of vegetables and meat produced by traditional and 'natural' farming methods that many

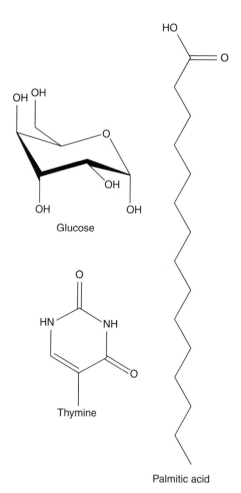

Figure 14.1 Three organic molecules: the sugar glucose, the DNA base thymine and the lipid palmitic acid – they all contain carbon and hydrogen atoms which is the chemist's definition of 'organic'.

people would not know the real meaning of the word. Most chemists find the phrase 'organic food' rather strange because by definition all food is organic!

The idea of organic farming conjures up an idealistic vision of a slow pace of life with farmers growing their crops without the use of harsh chemicals that might harm our fragile environment. On this halcyon organic farm the farmer strolls through verdant fields to tend his animals that he knows individually by name. This might be somewhat embellished, but can be distilled down to the ideal of happy animals, unadulterated crops and low environmental impact. Most consumers would aspire to this, I think – I certainly do. But is organic farming like this and are its products pure and unadulterated? I will address these questions later, but first we will explore the history and philosophy of organic farming.

The history and philosophy of organic farming

It might surprise you to learn that organic farming has its origins in the 1930s when under the leadership of Sir Albert Howard, a UK agricultural scientist, a group of farmers and consumers rejected modern agrochemical techniques and introduced a system of holistic and natural animal and plant husbandry in which the waste products from cities and farms were used as fertilisers for crops. Howard published his book *An Agricultural Testament* in 1940 in which he outlined the principles of organic farming, most of which are still adhered to today.

Organic farming eschews man-made fertilisers and chemical pest control; instead biological pest control methods are used and natural fertilisers (animal manures) replace man-made chemicals (e.g. nitrates and phosphates). Of course, the natural fertilisers also contain high concentrations of phosphates and nitrates which is why they work, and so their immediate environmental impact is likely to be similar to the man-made alternatives, but they are recycled animal and plant waste products, unlike man-made fertilisers which are continually added to the land and result in nutrient build-up.

The biological pest control methods that organic farmers replace chemical pesticides with definitely have a much lower environmental impact. For example, instead of insecticides organic farmers might use diversified farming and crop rotation methods to minimise insect attack. However, organic farming methods permit the use of specific 'organic' insecticides (I will discuss these later), for example, nicotine derived from the tobacco plant (*Nicotiana tobaccum*; i.e. natural). Nicotine might be natural, but it is still very toxic indeed and its residues in food would be harmful to consumers ... but more of this later. In addition, organic farmers advocate the use of natural pest predators to control pest infestation of their crops – this, of course, means that there is no chemical risk to the environment.

Organic farming became very fashionable in the 1980s and '90s and continues to attract much attention from consumers. It is now commonplace to find shelves devoted to organic produce in supermarkets and well-attended farmers markets that specialise in organic crops. Many people will pay extra for organic crops because they believe that they are helping to reduce the environmental impact of farming and that they are eating food that is better for them.

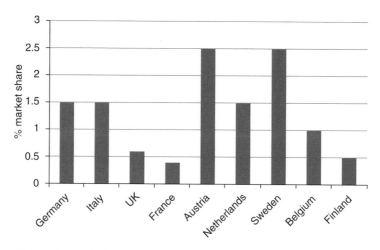

Figure 14.2 The market share of organic food in Europe. (Data from Kristensen & Thamsborg in Kyriazakis & Zervas (eds) (2001) *Organic Meat and Milk from Ruminants*. EAAP Publication No. 106, Athens, p. 6.)

The methods used by organic farmers tend to be more labour intensive than those used by 'conventional' farmers. For example, many organic farmers weed their crops by hand because they cannot use herbicides. This means that the cost of organic production is greater than 'conventional' crop production which, in turn, translates into more expensive food. It is arguable that the higher price of organic food reflects not only the increased production costs, but also the value added by the organic brand.

Organic farming is becoming big business in many countries in the developed world and has even gained the approval of President Barak Obama's White House, with the First Lady, Michelle Obama, creating an organic vegetable garden in 2009 - an endorsement indeed!

Demand for organic food

Organic food is a niche product which has a relatively small market share (Figure 14.2) in the developed world. It is a product that people are often passionate about and organic devotees will go to great lengths to source organic products.

Organic farming methods

As Sir Albert Howard advocated in the 1930s, organic farming should be a system of holistic and natural plant and animal husbandry in which waste products are returned to the soil for utilisation as nutrient materials. Organic philosophy requires that waste materials (i.e. organic fertilisers, e.g. animal manure) are applied to fields to improve both the soil's structure and its moisture-holding capacity and nourishes soil life, which, in turn, provides a nourishing environment for crops to grow in (this is the holistic approach).

Chemical fertilisers, on the other hand (according to organic philosophy), feed the plants directly rather than feeding the environment in which they grow.

Pest control is also very clearly prescribed and must be achieved by preventative methods rather than applying pesticides when there is a pest problem. These preventative methods include:

- growing diverse crops rather than large acreages of monoculture;
- crop rotation to stop pests becoming established in the field ecosystem;
- planting pest-deterrent species alongside crops (e.g. some insect pests are deterred by marigold so if marigolds are grown with crops they might deter pest insects);
- growing plants that attract predators of insect pests interspersed amongst crop fields;
- releasing sterile male insects to inhibit pest reproduction;
- releasing pest predators.

All of these methods have little impact on the environment while having an inhibitory impact on pest species, so helping the organic farmer to grow his crops efficiently and at a profit.

These are the ideals of organic farming, but what do organic farmers do that is different from conventional farming? I will address this by first looking at organic farming legislation and then exploring the means by which organic farmers control pests and how they fertilise their crops.

Organic farming legislation

Most countries that grow food organically have authorities or organisations that lay down the rules for growing organic crops and often accredit farmers and their land for organic production. They often authorise accredited organic farmers to use a recognised symbol on their produce which signifies that the 'organic rules' have been followed.

Accreditation for organic production requires that the farmer uses only organic farming methods and that his land or previous crops have not been treated with conventional pesticides or fertilisers for a specified period of time (often 5 years). The latter requirement is to ensure that residues of pesticides do not remain in the soil and so will not end up as residues in the 'organic' crop. Therefore, the history of the land in which an organic crop is grown is as important as the methods used to grow the crop.

The EU has comprehensive legislation governing organic production and I will use it here as an example. Other jurisdictions have similar legislation. The EU has a Council Regulation (EC 834/2007) that governs 'production and labelling of organic products' (this replaces an earlier Regulation (EC 2092/91) which was enacted in 1991). EC 834/2007 is a long, detailed document which spells out very clearly what is expected of organic farmers. On the whole it follows Howard's original organic maxim of recycling nutrients and doing as little harm to the environment as possible. For example, Clauses 12 and 15 state:

(12) Organic plant production should contribute to maintaining and enhancing soil fertility as well as to preventing soil erosion. Plants should preferably be fed through the soil eco-system and not through soluble fertilisers added to the soil.

(15) In order to avoid environmental pollution, in particular of natural resources such as the soil and water, organic production of livestock should in principle provide for a close relationship between such production and the land, suitable multiannual rotation systems and the feeding of live-stock with organic farming crop products produced on the holding itself or on neighbouring organic holdings.

The USA has its 2008 Farm Bill (officially called the Food, Conservation and Energy Act 2008) which includes provisions in its 673 pages of small print for organic production.

In order to ensure that farmers are complying with their country's legislation or are following the rules governing the use of organic symbols on their products, policing schemes are often imposed. These take the form of food and crop surveillance where samples of organic produce are taken and analysed for residues of chemicals (e.g. pesticides) that are not permitted for use in organic production. If residues are found the farmer is traced and serious questions are asked. If the farmer is found to have contravened the regulations he might be prevented from using organic produce logos, barred from selling organic crops or even prosecuted. Such policing schemes are important to maintain the integrity of the organic food brand and to ensure that the consumer is getting a fair deal.

The UK's Pesticide Residues Committee (formerly called the Working Party on Pesticide Residues – WPPR) carries out extensive pesticide monitoring schemes in a wide range of UK produced and imported foods (see Chapter 7, *Pesticide residues in food – assessing risk to the consumer and making sure farmers use pesticides properly*) and includes organic foods in its monitoring. Surprisingly, pesticide residues are sometimes found in organic food, albeit usually at very low concentrations (i.e. well below MRLs). The question is, where did the residues come from? And did the farmer contravene the organic production legislation? I will use a UK example to illustrate how a regulator might deal with the issues relating to pesticide residues in organic food later in this chapter.

Organic fertilisers

The underlying organic principle relating to fertilisers is that they must be worked on by soil bacteria to release nutrients that might then be used by plants. They must not be the primary nutrient source – superphosphate is a direct nutrient source; plant compost is indirect. For example, the New Zealand Biological Producers & Consumers Council (NZBPCC) lists the following permissible fertilisers:

- gypsum (calcium sulphate; $CaSO_4$);
- elemental sulphur alone or combined with bentonite (a clay);
- feldspar (e.g. potassium aluminium silicate; $KAlS_3O_8$);

- limestones (calcium hydroxide; $Ca(OH)_2$);
- rock minerals (e.g. chalk – calcium carbonate; $CaCO_3$);
- seaweed and fish products;
- unrefined rock or sea salt (sodium chloride; $NaCl$);
- dolomite (calcium magnesium carbonate; $CaMg(CO_3)_2$);
- glauconite (a complex iron silicate);
- rock phosphate.

Since some rocks also contain toxic heavy metals (e.g. cadmium) the NZBPCC often requires soil analysis to determine levels of heavy metals before accreditation can be given.

The New Zealand regulatory example above is similar to those used in other parts of the world.

Organic pest control

The most important organic principle of pest control is to protect natural predators (e.g. ladybirds (*Coccinella* spp.)) by allowing natural field ecosystems to flourish – the happy natural predators then kill the crop pests (e.g. ladybirds eat aphids). This can be helped by planting shelter belts and hedges because many natural predators live in these field-edge environments.

If natural pest control methods fail the NZBPCC lists the following means of pest control:

- biological controls (e.g. natural predators);
- diatomaceous earth;
- herbal sprays (e.g. rhubarb leaf spray which contains toxic oxalic acid – see Chapter 8, *Oxalic acid and rhubarb*);
- homoeopathic preparations;
- mechanical controls (e.g. traps);
- natural purgatives (e.g. sea water);
- pheromones (see Chapter 7, *Pesticides*);
- potassium-based soaps;
- Stockholm tar (also called pine tar; dry distilled pine wood containing phenols);
- thermal sterilisation;
- water (salt and/or fresh);
- gas saturation (e.g. CO_2 atmosphere to kill insects in stored grain);
- hydrogen peroxide (H_2O_2 – a powerful oxidising disinfectant that breaks down to water; $H_2O_2 \rightarrow H_2O + \frac{1}{2}O_2$);
- sulphur burning (this forms toxic sulphur dioxide which dissolves in water to sulphuric acid; $SO_2 + H_2O + O_2 \rightarrow H_2SO_4$);
- vegetable oils;
- waterglass (sodium silicate; Na_2SiO_3).

Organic weed control

There are just two organic methods of weed control – mechanical (i.e. weeding) and thermal (e.g. flame thrower). No herbicides are permitted.

Animal health remedies

When farm animals are ill farmers have a duty of care to treat them or eutha-nase them if they are in extreme pain. For 'conventional' farmers this means calling out the vet who will prescribe an appropriate veterinary medicine (see Chapter 7, *Veterinary medicines*) and all being well the animal will recover. Since medicines are usually man-made chemicals they are not permitted under the organic rules. For this reason, farmers have only three options available to them if their animals get sick: homoeopathic remedies, plant-based remedies (e.g. garlic drenches) or euthanasia. There is much contro-versy about the efficacy of homoeopathic and herbal remedies and the latter might contain toxic components that are equally, or even more, harmful to consumers than veterinary medicines. This is a huge debate that I do not have space to go into here; an important factor to consider though is that non-intensively reared organic animals are less likely to become ill and so are less likely to need medication.

Food processing

The organic principles apply to the processing of organically produced crops and meat if the final food product is to be labelled 'organic'. There are lists of approved processing chemicals (e.g. preservatives) in just the same way as the organic regulators' list of approved organic pest control methods and organic remedies for farm animals.

The International Federation of Organic Agriculture Movements (IFOAM) only allows the use of the following food additives:

- Calcium carbonate
- Sodium carbonate
- Potassium carbonate
- Sulphur dioxide
- Potassium metabisulphite
- Lactic acid
- Carbon dioxide
- Ascorbic acid (vitamin C)
- Tocopherols (vitamin D)
- Lecithin
- Citric acid
- Potassium citrate
- Calcium citrate
- Tartaric acid
- Sodium tartrate
- Potassium tartrate
- Calcium phosphate
- Ammonium phosphate
- Agar
- Carragenan
- Locust bean gum
- Guar gum
- Arabic gum
- Xanthan gum
- Pectin
- Potassium chloride
- Calcium chloride
- Calcium sulphate
- Ammonium sulphate
- Nitrogen
- Oxygen

There are a lot of approved organic food additives (Table 14.1) and some are not 'naturally' produced (e.g. the preservative potassium metabisul-phite), but it is important that appropriate food additives are added to

Table 14.1 Numbers of additives permitted in organic and conventional foods. The numbers vary from country to country and change over time, but these data make the point that far fewer additives are used in organic than in conventional foods. (Data from Bavec & Bavec (2007) *Organic Production and Use of Alternative Crops*. Taylor & Francis, London, p. 17.)

	Number of additives permitted	
Additive	Organic food	Conventional food
Antioxidants	11	55
Colours	1	48
Gelling, thickening and stabilising agents	12	74
Flavours	Not allowed	19
Preservatives	3	50
Acids	6	28
Sweeteners	Not allowed	11
Anticoagulants	Not allowed	10
TOTAL	33	295

manufactured food to ensure its safety; for example, to prevent the growth of pathogens by acidification with lactic acid, because pathogens grow just as well on organic food as they do on conventionally produced food. Even though the list of approved organic food additives seems long it is very much shorter than the list of conventional food additives – there are almost 300 approved conventional food additives compared to 31 on the organic list. Compare the organic additives with those discussed in Chapter 11; you will see a great difference in the type of molecules used and that some of the organic additives are based on chemicals naturally produced in foods (e.g. lactic acid is naturally produced as a preservative in fermented meats like salami).

Is organic food better for you?

'The organic label is not a health claim, it is a process claim' (quote from Koumba in Kyriazakis & Zervas (eds) (2001) *Organic Meat and Milk from Ruminants*, EAAP Publication No. 106, Athens, p. 62), but some consumers assume that organic food is more healthy for them. Organic production is certainly better for the environment and usually kinder to farm animals, but whether it is better for the consumer is a subject of much debate.

As a means of comparing organic and conventionally produced food, I will select a few measures of chemicals in specific examples of both food types in order to make a comparison. This does not fully answer the question I posed, but it does give an indication of whether the differences show that organic food might be beneficial or not.

Table 14.2 Milk composition from organic and conventional cows showing that there is little difference between the two milk types. (Data from Pirisi *et al.* in Kyriazakis & Zervas (eds) (2001) *Organic Meat and Milk from Ruminants*. EAAP Publication No. 106, Athens, p. 144.)

Parameter	Concentration (g/100 g)	
	Conventional	Organic
Total solids	16.94	17.11[2]
Fat[1]	6.25	6.46[2]
Protein	5.79	5.61[2]
Casein	4.20	4.44[3]
Whey protein	1.41	1.35[2]

[1] g/100 mL.
[2] Not significantly different.
[3] Significant difference; $P < 0.05$.

Nutrients

There are surprisingly few published studies that compare the nutrient content of organic and conventional foods, and some of the studies published are the subject of considerable controversy. One of the problems of making such comparisons is that the 'normal' variability of nutrient levels in foods (both organic and conventional foods) is great and so it is difficult to decide what is the control value with which to make our comparison. Bearing this in mind we can only accept large differences as being meaningful; but then the question is, what is large? When you look at the milk example I have used below remember that the nutrient ranges are not given.

Milk is an important staple food in many parts of the world and therefore is a good example to illustrate the differences (if there are any) between organically and conventionally produced milk. The only statistically significant difference noted in a study of organic milk compared to conventional milk was in the casein (milk protein) concentration; there was 5.7% more casein in organic milk – clearly there is very little difference between the two types of milk (Table 14.2).

Scientists at Rutgers University in the USA carried out a major study on organic versus conventional food and published a now famous report in 1948 (the Firman Bear Report – Dr Firman Bear is its lead author, hence its name) which shows some large differences in specific nutrients (e.g. thiamine – vitamin B_1) between organic and conventional foods (Table 14.3); in general, organic food was shown to have higher nutrient levels. There has been significant debate about the reliability of the data in the Firman Bear Report and therefore, once again, controversy reigns about whether organic food contains higher levels of nutrients than conventional food or not. I suspect it will be a long time before scientists agree on these important issues. In the meantime a small minority of consumers believe that organic food is better for them and maintain this important niche market which is avidly defended by its supporters.

Table 14.3 Differences between levels of calcium, iron and thiamine in organic and conventional produce. These controversial data show very much higher levels of all of these nutrients in organic than conventional vegetables. (Data from Bear *et al.* (1948) *Proceedings of the Soil Science Society of America*, **13**, 380–384.)

	Calcium (mEq/100 g)		Iron (mEq/100 g)		Thiamine (mEq/100 g)	
Crop	Conventional	Organic	Conventional	Organic	Conventional	Organic
Cabbage	17.5	60	20	94	2	13
Lettuce	16	71	9	516	1	169
Tomatoes	4.4	23	1	1,938	1	68
Spinach	47.5	96	19	1,584	1	117

Chemical residues

National surveillance schemes are carried out in many countries to measure and assess the health implications to consumers of residues of pesticides, fertiliser components (e.g. nitrate - NO_2^-) and veterinary medicines in a wide range of foods (see Chapter 7). Organic foods are sometimes included in these surveillance schemes as a means of policing the integrity of organic farmers and ensuring that consumers are buying the product they want (e.g. pesticide residue free). Surprisingly, residues of pesticides are quite often found in organic food, albeit usually at very low levels. The question is where did the residues come from? There are several possible answers:

- illicit use of pesticides by an 'organic' farmer;
- spray drift (e.g. a neighbouring farmer sprays his field and the wind carries the pesticide onto the organic farmer's crop - this should not happen because most countries' pesticide legislation bars spraying in high winds);
- contaminated land (e.g. pesticides remain in the soil from past use - this should not be the case because organic certification usually requires soil analysis to check for contamination before a farm is accredited);
- contamination during post-harvest storage (e.g. organic grain stored in silos previously used for pesticide-treated grain);
- mixing of organic and pesticide-treated conventional grain during processing.

The UK regulatory authority monitored pesticide residues in organic and conventionally produced breads between 1988 and 1996 (see Chapter 7, *Pesticide residues in food - assessing risk to the consumer and making sure farmers use pesticides properly*, Figure 7.1) and found that 15% (n = 53) of the organic breads analysed contained measurable pesticide residues compared with 12% (n = 512) for white bread and 27% (n = 375) for wholemeal bread - most organic breads are wholemeal and therefore conventional wholemeal bread is the best comparator. The pesticide residues found in all of the breads were below their MRLs. Residues found in organic breads were etrimfos and/or pyrimiphos-methyl (post-harvest OPs), both of which, of course, are not allowed in organic agriculture. We will never know why the residues were

Table 14.4 Pesticide residues in organic and conventional fruit and vegetables in Switzerland 1980–1983. (Data from Lampkin N (2002) *Organic Farming*. Old Pond Publishing, Ipswich, p. 564.)

	Percentage of samples with pesticide residues	
	Conventional	Organic
Number of samples	856	173
No residues detected	61	97
Residues <MRL	33	3
Residues >MRL	6	0

present, but their presence illustrates the problems associated with maintaining the integrity of the organic standard.

Studies in Switzerland between 1980 and 1983 showed that organic fresh fruit and vegetables have a considerably lower frequency of residues detected compared to conventional produce – in the Swiss study 6.2% of samples contained pesticide residues above MRLs whereas no MRL exceedances were found in organic fruit and vegetables (Table 14.4); in fact only approximately 3% of fruit and vegetables tested had measurable residues. This is very much lower than the percentage residues found in the UK bread study outlined above and in Chapter 7, but again illustrates that organic produce can contain pesticides.

A similar study in Germany looked at vegetables in 1985 and found that pesticide residues were detected in 20% ($n = 25$) of organic and 55% ($n = 132$) of conventional produce. Clearly, just because a product is organic does not mean it is necessarily pesticide residue free.

Natural toxins

As discussed in Chapter 8 (*Why produce natural toxins?*), natural toxins are often pesticidal and are the plant's inbuilt mechanism to protect itself from attack by insects or fungi. Therefore if a plant is left to its own devices it will respond to insect and fungal attack by producing phytoalexins (i.e. natural pesticides). On the other hand, if the plant was sprayed with insecticides and/or fungicides it would not need to activate its own defence mechanisms. For this reason, organic crops (i.e. not sprayed with conventional pesticides) are more likely to have higher levels of natural toxins than conventionally grown crops. It is important to remember that organic farmers can use selected methods to minimise pest infestation (see above) and if these are successful the plant might not need to respond by synthesising its own pesticides.

I have already discussed two examples of natural toxins with elevated levels in organic vegetables (see Chapter 8, *Cucurbitacins from the cucumber family* and *Furocoumarins in parsnips, parsley and celery*). The health implications of these two toxins are quite different; the furocoumarins in parsnips are carcinogens and therefore have a chronic effect (i.e. after long-term exposure) whereas the cucurbitacins in courgettes (zucchini) are acute toxins. Without knowing the toxin doses that consumers of organic parsnips and courgettes are exposed to and their frequency of exposure it is impossible to assess the health implications, but clearly the likelihood of exposure is greater to the consumers of organic vegetables compared to the majority of people who eat conventionally produced vegetables because the levels of both natural pesticides are likely to be higher in organic produce.

There are many more phytoalexins (Table 14.5), many of which have not been measured in organic and conventional crops, but it is possible that their levels will be higher in organic produce for the reasons outlined above.

We still do not understand many of the complex responses of plants to their growing conditions; this combined with the dearth of data on levels of natural toxins in organic produce means that we can draw no firm conclusions about the health implications at this stage.

Nitrate residues

Fertilisers are used by organic and conventional farmers and the elements delivered to crops by both organic and conventional fertilisers are the same (e.g. N as NO_3^-; P as PO_4^{3-}), but the delivery methods are different. For example, conventional fertilisers usually contain a direct source of N (e.g. ammonium nitrate, NH_4NO_3) whereas organic fertilisers deliver nitrate indirectly via N-rich composts or manure; the nitrate is formed by soil bacteria and only then is it taken up by the crop – remember, the philosophy of organic fertilising is that the soil must 'work' on the fertiliser to release its nutrients. This difference between conventional and organic fertilisers means that crops grown organically receive a gentle trickle of nutrients (e.g. NO_3^-), whereas conventional crops get a burst of nutrients when the fertiliser is applied.

The different nutrient delivery systems might explain the differences in fertiliser component residues (e.g. NO_3^-) found in organic and conventional crops. For example, nitrate concentrations are usually lower in organic than conventional crops (Figure 14.3).

When a plant is exposed to high nitrate levels in soil it simply absorbs the nitrate for use in amino acid and protein synthesis later (Figure 14.4). If, however, it is exposed to a gentle trickle of nitrate it is more likely to utilise it straight away rather than storing it.

Nitrate is carcinogenic to mammals (including humans) because it forms nitrite in the gut and the nitrite reacts with molecules in the intestine to form carcinogenic nitrosamines (see Chapter 7, *Nitrate – NO_3^-*). For this reason, food with high nitrate concentrations might increase the risk of colon cancer; on this basis organic vegetables might be safer than conventional vegetables.

Myths and facts about organic food

There is much heated debate about organic food between its staunch supporters and others who either don't care or can't understand why organic food usually costs more than its conventional counterpart and don't want to pay a premium for something that they don't believe is better. Unfortunately,

Table 14.5 Examples of phytoalexins in food plants – their concentrations have not been studied in both organic and conventional crops, but it is likely that they are modulated by fungicide use and so might be higher in organic crops. This illustrates the uncertainty about the effects of organic production on natural toxin levels. (If you would like to see more examples of phytoalexins look at Harborne JB & Baxter H (eds) (1996) *Dictionary of Plant Toxins*. John Wiley & Sons Ltd., Chichester.)

Crop	Phytoalexins
Sweet potato (*Ipomoea batatas*)	Ipomeamarone
Avocado (*Persea americana*)	1-Acetoxy-2-hydroxyheneicosa-12,15-diene-4-one
Oats (*Avena sativa*)	Avenalumin I

Table 14.5 *(Continued)*

Crop	Phytoalexins
Sunflowers (including the Jerusalem artichoke, *Helianthus truberosa*)	Ayapin
Chicory (*Cichorium intibus*)	Chicoralexin
Sugar beet (*Beta vulgaris*)	Betavulgarin
Radish (*Raphanus sativus*)	Brassinin
Celery (*Apium graveoleus*)	Columbianetin

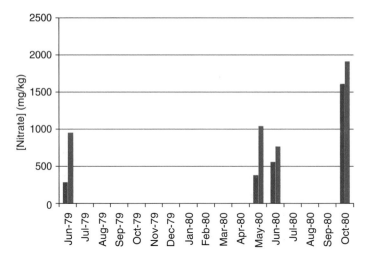

Figure 14.3 Nitrate concentrations in organic (grey) and conventional (green) lettuce crops in a year's farming cycle. You can see that the nitrate concentrations vary considerably, but that the conventional lettuces always have higher levels. (Data from Tempereli *et al.* (1982) *Schweizerische Landwirtchaftliche Forschung*, **21**, 167–196.)

much of the debate relies on belief or faith rather than scientific facts; there simply aren't good robust scientific studies that give us definitive answers about the differences (if any) in the risks either type of food poses to consumers' health. At the moment we simply don't know if organic foods are better for us than their conventionally produced counterparts, or whether conventional foods are best. One thing is certain, the debate will continue. In the meantime there is one certainty, that is that organic production is better for the environment than conventional production because it uses less pesticides and no harsh fertilisers.

Take home messages

- Organic food is a niche market (approx. 0.4–2.5% market share in Europe).
- Organic farming is better for the environment because it does not use harsh chemicals (e.g. man-made pesticides).
- There is controversy about whether organic produce contains more nutrients than conventional produce.
- Some organic foods contain pesticide residues.
- Some organic produce contains higher levels of natural toxins (e.g. cucurbitacins in courgettes).
- There is some evidence that organic food is better for its consumers (e.g. low nitrate residues), but this topic remains controversial.
- There are several examples of food safety issues attributed to organic produce (e.g. cucurbitacins in New Zealand organic courgettes).

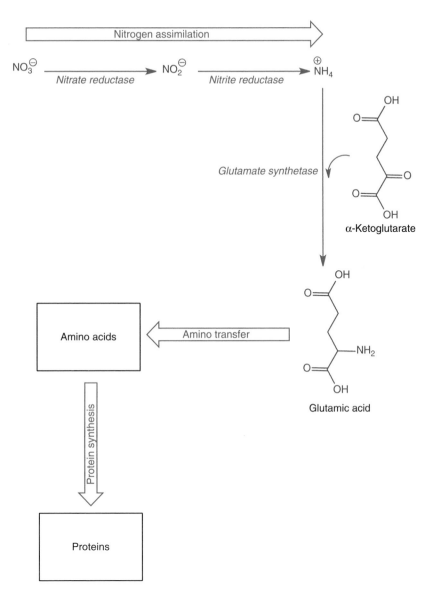

Figure 14.4 The synthesis of proteins from nitrate in plants. (If you want to read more about this see Garrett RH & Grisham CM (2010) *Biochemistry*, 4th edn. Brooks/Cole, USA, Chapter 25, Nitrogen Acquisition and Amino Acid Metabolism, pp. 769–776.)

Further reading

Bavec F & Bavec M (2007) *Organic Production and Use of Alternative Crops*. Taylor & Francis, London. Chapter 1 is an excellent introduction to organic agriculture.

Bear FE, Toth SJ & Prince AL (1948). Variation in Mineral Composition of Vegetables. *Proceedings of the Soil Science Society of America*, **13**, 380–384. This is an old report but is historically very important and well worth a read.

IFOAM (1996) *Basic Standards for Organic Agriculture and Food Processing*. Contact: IFOAM-secretary@oln.comlink.apc.org.

Lampkin N (2002) *Organic Farming*. Old Pond Publishing, Ipswich. Chapter 15, The Wider Issues, has an excellent section (pp. 562–573) on chemical residues in organic compared with conventional produce.

Lockeretz W (ed) (2007) *Organic Farming – An International History*. CAB International, Oxford.

New Zealand Biological Producers & Consumers Council. *Certified Bio-Gro Organic Production Standards*. Contact: NZBPCC, PO Box 36-170, Auckland 9, New Zealand.

Chapter 15
Food Allergy

Introduction

Allergy to food is gaining increasing importance in our society. This might be because it is getting more common, or it might be that consumers are getting more interested in such issues, which have led them to look for symptoms of allergy and respond by eliminating 'causative' foods from their diets. It is even becoming fashionable to have a food allergy and to eat special diets. Indeed, gluten-free diets have become fashionable as can be seen by the proliferation of advertising by food outlets and manufacturers. This might not equate to an increased frequency of the debilitating gluten allergy, coeliac disease, but might reflect the consumer's own diagnosis of the effect of gluten-containing foods on their wellbeing. I will address these issues in this chapter.

Despite my rather sceptical introductory paragraph, there are many well documented food allergies (including coeliac disease); most are due to the body's immune system rejecting specific proteins in a particular food (e.g. wheat in coeliac disease) which leads to the sufferer becoming ill if they eat the food. In this chapter I will explain how the immune system responds to proteins in food (allergens) and why this leads to the symptoms of food allergy in some but not all consumers.

The immune system is very complex, but an understanding of its basic function is important if you are to understand food allergies. Specialised blood cells (B- and T-cells) cooperate to synthesise antibodies that bind to specific food allergens (e.g. the wheat protein, gluten in coeliac disease) and initiate the production of inflammatory chemicals that lead to an inflammatory response (e.g. wheezing) – this is the mechanism of food allergy. I will discuss the bare bones of this very complex process in this chapter.

Finally, I will explore some common food allergies (e.g. gluten (coeliac disease), milk, peanuts and seafood) and discuss why food allergies appear to be getting more common.

Food Safety: The Science of Keeping Food Safe, First Edition. Ian C. Shaw.
© 2013 John Wiley & Sons, Ltd. Published 2013 by John Wiley & Sons, Ltd.

What is an allergy?

Put simply, an allergic response is when the body rejects a chemical that it is challenged with. The chemical might be a protein or glycoprotein (a protein molecule with sugar chains attached) associated with the surface of a cell that makes up a tissue. This is what happens when a transplanted tissue (e.g. heart) is rejected by a recipient. Similarly, rejection of the glycoprotein markers on the surface of red blood cells (blood groups A, B, AB and O are determined by red blood cell surface glycoproteins) occurs when the wrong blood group is transfused. The previous examples are unwanted immune responses, but there are many beneficial immune responses that are important in combating diseases. For example, the body rejects the surface proteins of a virus following a viral infection – this is immunity and is the mechanism underlying immunisation. When someone is immunised, dead bacteria, inactivated viruses or fragments of bacterial cell walls or viral coats are injected which elicits an immune response. If the person is exposed to the bacterium or virus after being immunised, they will already have the ability to make antibodies to immobilise the organism, i.e. they are immune to the bacterium or virus.

An allergic response or rejection can occur when the body is challenged with almost anything that has a specific molecular shape that the immune system 'sees' as foreign. A food allergy occurs when a person eating a particular food elicits an immune response to a molecular component of the food.

Before I discuss food allergies, you need to understand the basics of the science of immunology which underpins our understanding of the allergic response.

The basics of immunology

I will only cover the rudiments of immunology here – just enough for you to understand food allergy. If you want to read more see Roitt *et al.* (2001), Abbas & Lichtman (2009) or Calder & Field (2002) in *Further reading*.

The science of immunology began in 1796 in Gloucestershire, England, when Dr Edward Jenner found that farm workers who had been in contact with cattle did not contract smallpox. He took this observation further and inoculated people with a crude cattle serum and found that these people also did not get smallpox – this was the first vaccination. It was many decades before the science underpinning vaccination was unravelled, but we now know that some of Jenner's cattle were suffering from cow pox (vaccinia – hence vaccination) and that the surface proteins on the vaccinia virus are very similar to those on the smallpox virus; so similar, in fact, that the body's immune system could not tell the difference, and so challenge with vaccinia led to protection against (i.e. immunity to) smallpox. But what *is* immunity?

Immunity and the immune response

Specialised cells in the blood (lymphocytes – a form of white blood cell), bone marrow and some other tissues (e.g. soft connective tissue) are able to synthesise specific proteins that recognise particular large molecules – or small molecules (i.e. haptens) bound to large molecules (e.g. proteins). The proteins the cells synthesise are immunoglobulins (antibodies) and the molecules they recognise are antigens. The immunoglobulins each has a unique protein structure which interacts specifically with the surface shape of the antigen (like a lock and key). The immunoglobulin binds to the antigen and initiates a series of cellular reactions that results in destruction of the antigen (there are multiple mechanisms by which this is achieved, but I will not go into them here – see later in this chapter for more detail). The structure of the immunoglobulin is remembered by specific immune system cells so that if the body is re-challenged with the same antigen, the cell remembers and synthesises the correct antibody quickly in order to quickly dispatch the foreign molecule. This is how vaccines work – you are given a dose of an antigen (e.g. a killed virus or its surface coat) which leads to the synthesis of antibodies and the development of memory cells. If you are exposed to the virus later in life the memory cells recognise the viral surface proteins and make antibodies quickly enough to destroy the virus before you are able to develop the disease that it causes.

There are two types of lymphocytes called B-cells and T-cells. In simple terms, T-cells bind to antigens and send a chemical signal (e.g. interleukin-2) to stimulate B-cells to make antibodies against the specific antigen.

Immune responses can be at four levels:

- **Type I** (Figure 15.1) – this is the immune response that is responsible for allergies, including food allergies. It involves a specific immunoglobulin called immunoglobulin E (IgE). IgE is synthesised by B-cells in response to challenge by an antigen (e.g. pollen or a chemical in food), the IgE binds to the antigen and the complex migrates to mast cells in soft connective tissue (e.g. mucous membranes of the intestine or respiratory tract) where it binds to the cell surface, the mast cell surface receptors which recognise the non-allergen binding end of IgE (this part of its structure is the same in all IgEs). When the IgE/antigen complex binds to the mast cell surface receptors it initiates the synthesis or release of specific signal chemicals (e.g. histamine) which dilate blood vessels, constrict the bronchial air passages in the lungs or cause tissue swelling. The result is difficulty in breathing, sneezing, wheezing and swelling, i.e. a typical allergic response (if you suffer from hay fever or asthma you will be very familiar with these symptoms). A typical allergen that might initiate an IgE response is pollen. The result is hay fever which is treated with antihistamines which counteract the histamine response and therefore prevent the symptoms (e.g. sneezing); if the allergen is found in food (e.g. peanuts) this leads to a food allergy with symptoms very similar to hay fever, but sometimes much more extreme (see *Type III (anaphylaxis)*).

  The mast cells with IgE bound are able to quickly recognise future challenges with the specific antibody and therefore respond more quickly the second, third, fourth, etc. time around. The antibody binds across two IgE molecules on the mast cell surface to initiate a process called degranulation; this is when pre-synthesised inflammatory chemicals (e.g. histamine – stored as granules in the mast cell cytoplasm) are released from the mast cell to initiate the inflammatory response.

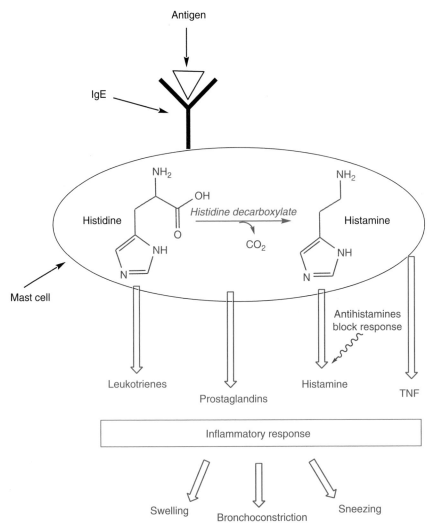

Figure 15.1 Type I immune response. IgE is synthesised by B-cells in response to an antigen (e.g. a chemical in food), the antigen binds to the IgE and the complex migrates to a mast cell located in soft connective tissue (e.g. mucous membranes of the respiratory or gastrointestinal tract). Here it stimulates the release of inflammatory chemicals including histamine which is synthesised from histidine, prostaglandins, leukotrienes and tumour necrosis factor (TNF). The histamine is exported from the mast cell and initiates an extracellular histamine response, leading to symptoms including swelling.

Whether or not someone initiates an IgE response to a particular allergen is, at least in part, inherited, but if they do their mast cells 'remember' the allergen (see above) that initiated the response, and when challenged again IgE is synthesised much more quickly and in greater amounts which means that later antigen challenges can lead to more severe symptoms (this is called sensitisation). In some cases an individual can get so sensitive to a particular allergen that their response is extreme (anaphylaxis), which leads to severe bronchoconstriction and swelling which makes breathing difficult and in very severe cases breathing is impossible and the person can die – this, however, is rare providing treatment with adrenalin (which causes bronchiodilation) is quick. Some people who are hyperallergenic (e.g. to peanuts) carry a syringe of adrenaline (sometimes called a 'pen') with them so that they can respond quickly to the early symptoms following peanut ingestion and prevent anaphylaxis.

The Type I immune response is responsible for food allergies. I will discuss specific food allergies later in this chapter.

- **Type II** – in a type II response B-cells produce immunoglobulin G (IgG) or immunoglobulin M (IgM) that is directed against cell surface markers (e.g. on bacteria); it is the body's response to infection. When the IgG binds to the cellular antigen the antibody/antigen complex initiates a complex cascade response (the complement system) which kills the target cell or macrophages (literally 'big eaters' – they are specialised lymphocytes) that recognise the invading cell surface bound immunoglobulin and engulf and kill the cell. The type II response can also be directed against other (e.g. mammalian) cells or large molecules (e.g. proteins) produced by cells. The type II response is not involved in food allergy; I have described it here so that you can see how immune responses to food fit into the totality of the highly complex immune system.

- **Type III** (anaphylaxis) – this occurs when someone is sensitised (i.e. their response is rapid and extreme – see below) to a particular antigen and they are re-exposed to the same antigen. B-cells produce IgG which forms a complex with the antigen which is deposited on the walls of small blood vessels and triggers the complement system. This results in an inflammatory response leading to damage to the blood vessels. This response is also called anaphylaxis.

- **Type IV** – this is also called a *delayed response* and is mediated by T-cells which have cell surface binding sites that interact directly with an antigen. The delay is simply because the T-cells take longer to migrate to the site of the antigen than immunoglobulins. A T-cell response against alien mammalian cells is the mechanism of transplant rejection – this is treated with immunosuppressant drugs (e.g. Azathioprine). The type IV immune response is not involved in food allergy, but is included here for completeness.

Sensitisation

Sensitisation (Figure 15.2) is very important in the development of food allergies and therefore you need to understand the basics of this phenomenon before you can begin to understand why some people are allergic to specific foods.

If someone is challenged repeatedly with a particular allergen they might become more sensitive to the allergen; thus, with each challenge they develop a greater and greater immune response. This occurs because IgE bound to the surface of mast cells recognises (remembers) the structure of the antigen and is able to elicit an immediate and bigger response when re-exposed. This does not occur in everyone exposed to a particular antigen. There is still uncertainty about why some people become sensitised whereas others do not; the reason is, at least in part, genetically determined (i.e. inherited).

Sensitisation is important in food allergy because it is possible to become sensitised to a particular food allergen following repeated exposure; this means that the sensitive person's response becomes more severe with each exposure.

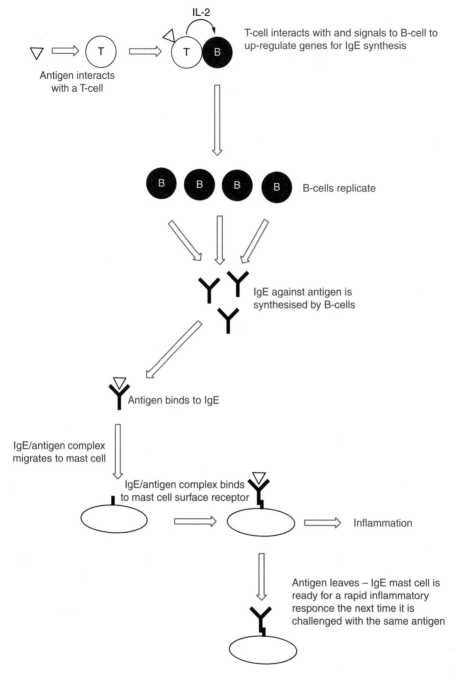

Figure 15.2 Type I immune response leading to sensitisation. ∇, Antigen; IL-2, Interleukin-2; Y, IgE.

Interestingly, sensitisation might begin *in utero* (i.e. while a baby is in the womb); when mother eats she transfers some components of her food via the umbilical cord to her developing child (see Chapter 13). This is the way that nutrients are supplied to the embryo and fetus, but many other chemicals are transferred from mother to baby at the same time. If allergens (e.g. peanut allergens) are transferred, the developing child might become sensitised to the allergens and be born with the makings of a food allergy (e.g. peanut allergy). As the growing child eats food (e.g. peanuts) containing the allergens to which she/he was sensitised *in utero*, she/he might become more severely allergic to the food until such time that the child undergoes a full blown anaphylactic reaction.

Food allergies

Approximately 2% of the adult population suffers from food allergies; the incidence in infants and children is greater because the incidence of some food allergies diminishes with age. The most common food allergy is to peanuts.

Several foods are particularly associated with allergies (e.g. peanuts and milk); although in theory it is possible to develop allergies to any food this does not happen. It is possible that food allergies are related to the frequency of consumption of a particular food (but this is not the only determinant; otherwise potatoes and rice would be major food allergens), or *in utero* sensitisation to particular allergens or, perhaps, the ease with which an antibody response is initiated by a molecule found in food.

We don't fully understand the reasons why food allergies develop in response to particular foods and not others; indeed this is the subject of a great deal of research currently. I will discuss only the most common food allergies, and explore their immunological mechanisms and what is known about the antigens involved.

Some food allergy facts and figures

- In the USA there are >12,000,000 food allergy sufferers; of these, 3,000,000 (25% of the total) are children.
- Approx. 6% of under 3 year olds have a food allergy.
- In the USA 300,000 people need to be rushed to hospital each year because of a severe food allergic response.
- Milk, eggs, peanuts, tree nuts (e.g. walnuts), wheat, soy, fish and shellfish account for 90% of all food allergies.

(Data from www.foodallergy.org.)

The genetics of allergy

A detailed discussion of this important subject is far beyond the scope of this book, but it is important to know that to some extent allergies are genetically determined. People who suffer from a particular allergic disease (e.g. asthma) are more susceptible to other allergies (e.g. to food) because their cells have inherited the genetic makeup to produce the antibodies associated with allergy (e.g. IgE). Since allergic reactions to different allergens go via the same basic immunological mechanisms it is understandable why someone with a propensity to elicit an allergic response is more likely to develop allergies to a wide range of allergens. I have mentioned this here because allergy to foods is likely to involve broader symptoms than those associated with acute food allergy (e.g. breathing difficulties and tissue swelling). For example, a food allergy might be manifest as asthma-like symptoms or eczema (i.e. an allergic skin disease). No one knows what proportion of asthma and eczema is caused by food allergy, but it is a likely cause and therefore food allergy is probably much more common than is often thought.

Food allergens

I will discuss allergies to specific foods below; even though the foods responsible for particular allergies might be very different (e.g. shellfish and milk) there are some important common features (e.g. they are proteins) of the allergens from these very different foods.

Food allergens (Table 15.1) are usually proteins that either interact directly with intestinal mast cells (i.e. they are not absorbed) to initiate a gastrointestinal response, or are absorbed and result in systemic IgE synthesis with a consequent mast cell-mediated systemic response (e.g. bronchoconstriction). The absorption of proteins is unusual because they are usually metabolised by gastric proteases to protein fragments (peptides) or individual amino acids and then the amino acids are absorbed. However, very small amounts of intact proteins, or perhaps allergenic protein fragments, can be absorbed and it is these allergens that initiate the systemic allergic response seen in food allergy sufferers.

Table 15.1 Allergens responsible for some common food allergies. The allergen nomenclature is derived from the Latin name for the food species and denotes the specific form of a particular protein type that is responsible for initiating an immune response, e.g. α-lactalbumin is a class of proteins found in milk, *Bos d 4* is the specific amino acid sequence of α-lactalbumin that results in an allergic response in some people.

Food	Animal or plant species	Protein	Allergen
Cow's milk	*Bos domesticus*	α-Lactalbumin	*Bos d* 4
		β-Lactoglobulin	*Bos d* 5
Egg	*Gallus domesticus*	Ovomucoid	*Gal d* 1
		Ovalbumin	*Gal d* 2
Cod	*Gadus callarias*	'Allergen M'	*Gad c* 1
Shrimp	*Metapenaeus ensis*	Tropomyosin	*Met e* 1
Peanut	*Arachis hypogaea*	Vicilin	*Ara h* 1
		Conglutinin	*Ara h* 2

Milk allergy

There are two important dietary problems humans might have following cow's milk consumption:

(1) Inability to metabolise milk sugar (lactose) to its component monosaccharides (galactose and glucose) because of an inherited lack of the enzyme lactase. This leads to gastrointestinal disturbances due to the excess lactose present. This is a serious problem in suckling infants. This dietary problem is not immunologically based and therefore is not a food allergy – it is a food intolerance.
(2) Allergy to milk proteins (α-lactalbumin and/or β-lactoglobulin). This is immunologically based and is a food allergy.

The above two milk-related dietary problems must not be confused. I will discuss only the milk protein allergies here.

Milk allergies occur mainly in infants and children and usually subside by adolescence – 19% of children outgrow their cow's milk allergy by age 5 years and 79% by age 16 years (data from www.foodallergy.org). It is a significant problem early in life because many children are fed cow's milk or cow's milk-based formulae. Most children with cow's milk allergy develop gastrointestinal symptoms, but some have skin-associated reactions (e.g. eczema) and fewer have more serious systemic effects (e.g. respiratory problems) (Table 15.2).

Milk contains three main proteins: casein, α-lactalbumin and β-lactoglobulin. There are significant differences in the molecular structures of bovine and human caseins (especially in the type and number of sugar units attached (caseins are glycoproteins)) which explains why some human allergies to cow's milk involve antibodies to bovine caseins. On the other hand, the amino acid sequences of the lactalbumins and lactoglobulins from cows and humans are similar, but the small differences in amino acid sequences lead to differences in the tertiary structure (i.e. folding) of the protein, which appears to cause the human immune system to elicit a response to both bovine α-lactalbumin and β-lactoglobulin. Differences in the tertiary structures of a protein means that the conformation (shape) presented to the immune system is different even though the amino acid sequence of the two proteins might be very similar; therefore, an immune response might result in the synthesis of an antibody (e.g. IgE) to the foreign protein's conformation rather than its primary amino acid sequence. This is perhaps easier to understand if you consider a very simple hypothetical example.

Table 15.2 Cow's milk allergy symptoms in infants. (Data from Kagan (2003) *Environmental Health Perspectives*, 111, 223–225.)

Symptoms	Proportion of cases (%)
Gastrointestinal	≈ 100
Skin-associated (cutaneous)	50–70
Respiratory	20–30

Below are two amino acid sequences (primary structures) for the same sections of two very similar large proteins:

PROTEIN A
-Gly-Gly-Lys-Gly-Thr-Cys-Val-Phe-Val-Ala-Leu-Val-Val-Cys-Ser-Gly-Glu-Gly-Gly-

PROTEIN B
-Gly-Gly-Lys-Gly-Thr-Glu-Val-Phe-Val-Glu-Leu-Val-Val-Glu-Ser-Gly-Lys-Gly-Gly-

I have marked the differences between the two sequences in green. Remember, this is just a small part of our hypothetical protein molecules and therefore the small number of different amino acids account for only a tiny proportion of the total primary structure of the proteins. Despite this, they would lead to a significant difference in the way the two proteins might fold (i.e. their tertiary structure).

This is not the place to go into protein structure in detail, but put simply it is the primary structure of a protein that determines the protein's overall shape (i.e. tertiary structure) because interactions between amino acid residues determine how the protein folds. Amino acids can be attracted to each other by electrostatic interactions; when positively charged (e.g. lysine – $R = -CH_2CH_2CH_2CH_2NH_3^+$) and negatively charged amino acid (e.g. glutamic acid – $R = -CH_2CH_2COO^-$) residues align; by hydrophobic interactions (e.g. van der Waal's forces) when two hydrophobic (i.e. water repelling) amino acids (e.g. valine – $R = -CH(CH_3)_2$ and phenylalanine – $R = CH_2-C_6H_5$) interact; by hydrogen bonds between appropriate amino acids (e.g. two serines – $R = -CH_2OH$; a hydrogen bond forms between the –OH oxygen of one serine and the –OH hydrogen of the other), or by the formation of a disulphide bridge between two cysteines ($R = -CH_2SH$ which forms $-CH_2S-SCH_2^-$) (Figure 15.3). (If you want to read more about amino acids and protein folding see Chapters 5 and 6 in Garrett & Grisham (2010) listed in *Further reading*.)

Our two hypothetical closely related proteins (i.e. the same proteins from two different species) have key amino acid differences that will lead to different tertiary structure (Figures 15.4 and 15.5; Plate 15.1), which means that they might well elicit different immune responses. For example, if protein A is human α-lactalbumin and protein B is bovine α-lactalbumin the difference might be sufficient for a human consuming cow's milk to 'see' the bovine α-lactalbumin as foreign and so synthesise IgE in response. This is a likely explanation of human allergy to cow's milk.

The specific cow's milk allergens are termed *Bos d* 4 (α-lactalbumin) and *Bos d* 5 (β-lactoglobulin) – the Latin name for cattle is *Bos domesticus*, hence the allergens' terminology. Antibodies in blood can be measured using a binding assay in which blood serum is mixed with an allergen (e.g. *Bos d* 4) and the degree of binding determined. Sera from people who suffer from cow's milk allergy contain IgEs that bind *Bos d* 4 (Figure 15.6) which proves that bovine α-lactalbumin (i.e. *Bos d* 4) is an important allergen in the aetiology of cow's milk allergy. Interestingly, some of the sera also bind human α-lactalbumin, albeit at a very low response, which shows that the IgE against *Bos d* 4 cross-reacts with human α-lactalbumin to a very minor extent.

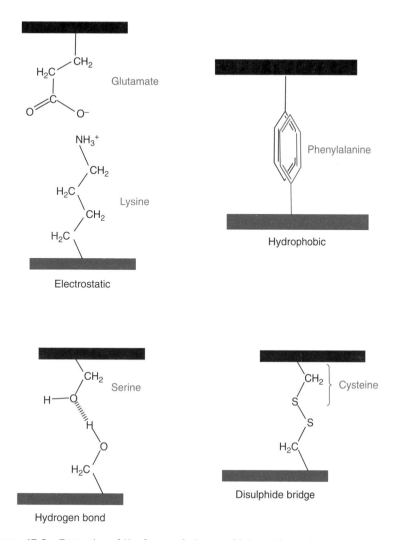

Figure 15.3 Examples of the four main types of interaction between amino acid residues in a protein's primary structure that determine how the protein folds (i.e. its tertiary structure).

Peanut allergy

The peanut (*Arachis hypogaea*) is a member of the pea family (Leguminosae) – it is not a nut. It produces aerial flowers that as they mature turn towards the ground and push the developing seed pod into the ground (peanuts are sometimes called groundnuts for this reason); the mature 'nuts' are harvested from the ground. Peanuts have very high levels of protein (more per gram than beefsteak) and fats and are often used in food processing to increase protein levels in the finished product. This presents a huge problem for peanut allergy sufferers.

Approximately 8% of children and 1-2% of adults suffer from peanut allergy; it is the commonest food allergy. Unlike many allergies, it does not

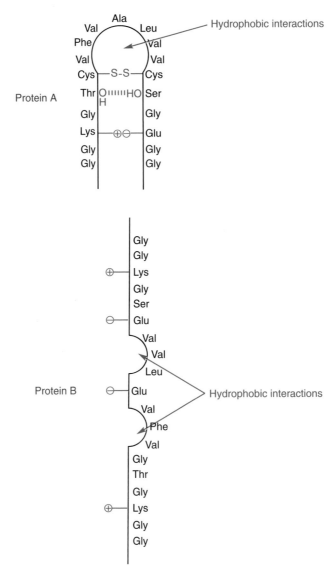

Figure 15.4 Hypothetical proteins A and B have small differences in their overall primary structure (amino acid sequence) which leads to a significant difference in their tertiary structures (folding) because of the different amino acid R-group interactions. This means that the immune system 'sees' the proteins as different and therefore might elicit an immune response against the foreign protein. This is the basis of allergy to cow's milk α-lactalbumin in humans. There are several ways these hypothetical proteins could fold; I have shown just one example to illustrate the differences that folding can make.

diminish as its sufferer gets older; indeed it often gets worse and it is usually much worse than other food allergies. The only way to 'treat' peanut allergy is to avoid eating peanuts, but this is becoming more difficult as peanut products (e.g. peanut meal) are increasingly used in processed foods. Some peanut

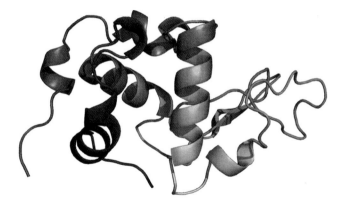

Figure 15.5 The tertiary structure of α-lactalbumin showing the importance of folding in determining its shape: molecular weight = 14 kDa; concentration in cow's milk ≈ 1 g/L. (Molecular structure from http://upload.wikimedia.org/wikipedia/commons/6/69/Protein_LALBA_PDB_1a4v.png.) (To see a colour version of this figure, see Plate 15.1.)

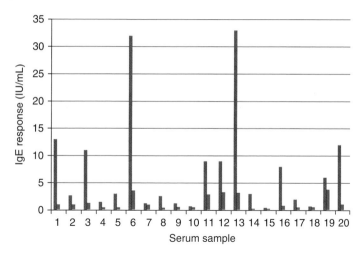

Figure 15.6 IgE binding assays for bovine (*Bos d* 4) (grey) and human (green) α-lactalbumin in serum samples from people suffering from cow's milk allergy showing that some allergic people (i.e. numbers 1, 3 ,6, 11, 12, 13, 16, 19 and 20; 45% of total, *n* = 20) elicit a significant IgE response to bovine α-lactalbumin. (Data from Maynard *et al.* (1999) *Food and Agricultural Immunology*, 11, 179–189.)

allergy sufferers respond severely to traces of peanuts in their food – even traces resulting from a food being processed in the same factory as peanuts. This has led to accurate labelling to indicate if peanuts are used in a particular processed food, or even if peanuts are processed in the same factory that the food was manufactured in.

Surprisingly, despite a great deal of research, the identity of the allergen responsible for peanut allergy has not been unequivocally proved. However, some very interesting recent work (see Rabjohn *et al.* (1999) in *Further reading*) has provided strong evidence that there are several protein allergens present in peanuts that cross-react with (i.e. bind to) antibodies present in the blood of peanut allergy sufferers. The peanut protein allergens are glycoproteins (i.e. proteins with sugar chains attached) of molecular weights approximately 63.5 kDa (Ara h 1 – vicilin) and 17 kDa (Ara h 2 – conglutinin) – *Ara* is derived from the name of the peanut genus (*Arachis*).

Vicilin and conglutinin are storage proteins and so are present at high concentrations in peanuts; they have no human counterparts and so on ingestion the human body 'sees' them as foreign. Why it is possible to elicit such a severe allergic response to these proteins is uncertain because we are exposed to myriad foreign proteins every day, but don't respond to them in the same way as some people do to vicilin and conglutinin.

There is a great deal still to learn about the whys and wherefores of peanut allergies, but for now the only way to treat the disorder is to COMPLETELY remove peanuts from the sufferer's diet – and this is very difficult indeed.

Soy allergy

Like peanuts, soy (also called soya, soya bean or soybean) is a member of the pea family (Leguminosae) – there is some dispute in the literature about the botanical origins of soy; it is referred to as either *Glycine ussuruensis* or *G. max* in most texts. Soy is a very important food that has been used in Eastern cuisine for thousands of years and is increasingly being used in Western food particularly as a protein source – soy contains about 35% protein. You might be surprised if you look in the ingredients of everyday foods to find soy amongst the key ingredients when a cheap source of protein is needed. For example, bread flour is often 'improved' with soy flour – look at the ingredients list and warning on a major brand of bread I found in my local supermarket:

> *Wheat flour, water, baker's yeast, salt, soy flour, vegetable oil, emulsifiers (471, 472e, 481), acidity regulators. CONTAINS WHEAT AND SOY*

The warning is necessary because some people are allergic to soy or wheat (see *Gluten allergy (coeliac disease)*). It is, perhaps, not surprising that one can develop an allergy to soy since it has such high protein content and proteins are the culprit allergens in most food allergies.

Soy is also used to make soy 'milk'; this is basically soy homogenised in water to form a white suspension/emulsion which some people use instead of milk. Soy milk and soy-based formulae are sometimes used to feed babies and infants – they are important for children with lactose intolerance or cow's milk allergy. The use of soy product to feed babies might lead to their sensitisation and the development of soy allergy, which might persist into their adolescent and adult lives.

Approximately 1–6% of infants suffer from soy allergy; however, most soy allergy children are able to tolerate soy by the time they go to school. There is evidence that the incidence of soy allergy is increasing in the Western world – this is not surprising since the production and therefore consumption of soy

Table 15.3 Soy allergens. The glycinins are large storage proteins made up of six subunits which are allergenic in their own right; three of the subunits (Gly m 1A, 1B and 2) have been identified. (Data from Helm *et al.* (2000) *International Archives of Allergy and Immunology*, **123**, 205–212.)

Allergen	Molecular weight (kDa)	Protein function
Glycinin	65	Storage protein
β-Conglycinin (vicilin)	?	Storage protein
Gly m 1A	7.5	Glycinin subunit
Gly m 1B	7	Glycinin subunit
Gly m 2	8	Glycinin subunit

is also increasing (48% increase in acreage of soy grown in the USA from 1971–1996). Clearly we need to understand soy allergy better so that we can attempt to minimise its potentially increasing impact on health.

In order to understand soy allergy we need to ask the important question, 'Which protein(s) in soy lead to an allergic response?' The answer to this question is far from simple and still not fully understood.

There have been five allergens reported in soy (Table 15.3) – they are all proteins called glycinins. (NB the *Gly* stem of their names relates to the soy genus *Glycine*; it has nothing to do with the amino acid glycine which is also abbreviated to Gly.)

- β-Conglycinin (also called vicilin)
- Glycinin
- *Gly* m IA
- *Gly* m IB
- Kunitz trypsin inhibitor

Glycinin is a large storage protein composed of six different subunits (i.e. the subunits have different protein structures) and it is likely that some of the antibodies in soy-allergic people are directed against these glycinin subunits (e.g. *Gly* m 1A and *Gly* m 1B), but this is an area of current research and is not yet fully understood.

Soy is also used as a source of oil for cooking and food processing, but allergy to soy oil is unusual because it does not usually contain the protein allergens.

Nut allergies

Allergies to tree nuts are common; the self-reported (i.e. reported by sufferers, but not confirmed by a physician) incidence of tree nut allergy in Canada is 1.14% (*n*=9,677; see Table 15.4). Tree nuts are nuts derived from trees (obviously!) as distinct from pseudo-nuts from other plants (e.g. peanuts). The following tree nuts have been associated with allergy in humans:

- Brazil nuts
- Cashews (not a true nut, but classified here for convenience)

Table 15.4 Self-reported (i.e. reported by sufferers, but not confirmed by a physician) incidence of some food allergies in Canada showing that tree nut allergy is relatively common. (Data from Ben-Shoshan *et al.* (2010) *Journal of Allergy and Clinical Immunology*, http://www.sciencedirect.com/science?_ob=MiamiImageURL&_cid=272425&_user=103118&_pii=S0091674910005373&_check=y&_origin=&_coverDate=30-Jun-2010&view=c&wchp=dGLbVlt-zSkWz&md5=be0ef98a63c0e36fab8b75ee7fab10cc/1-s2.0-S0091674910005373-main.pdf.)

Food	Incidence of allergy (%, $n = 9,677$)
Shellfish	1.42
Tree nuts	1.14
Peanuts	0.93
Fish	0.48
Sesame	0.09

- Chestnuts
- Hazelnuts (also called filberts)
- Macadamia nuts
- Pecans
- Pine nuts (not a true nut, but classified here for convenience)
- Pistachios
- Walnuts
- Almonds
- Coconut

People with tree nut allergies are often allergic to more than one species of nut (e.g. hazelnuts and walnuts) because many tree nuts have allergens in common even though the plants are not closely related botanically. For example, the storage proteins legumins, vicilins (also found in peanuts) and 2S albumins (S refers to the sedimentation rate when centrifuged – the larger the number, the bigger the protein) are common to several species and antibodies are found to them in tree nut allergy sufferers. In addition, some non-storage proteins are also common to several nut species and their antibodies are also found in tree nut-allergic people, including prolifins, lipid transfer proteins and members of a family of proteins called *Bet v* 1. This is another area of very active research and we have much still to learn about the mechanisms and cross-reactivities of the tree nut allergies.

The common allergens between several tree nuts associated with allergy are vicilin storage proteins (Table 15.5); they are very similar to the vicilins (e.g. *Ara h* 1; see Table 15.1) associated with peanut allergy. It appears that the molecular structure of the vicilins is particularly immunoactive in some people; this also explains why many nut allergens cross-react, including peanuts and tree nuts in some people.

Table 15.5 Tree nut vicilin storage protein allergens – note the allergen naming system is based on the Latin name of the nut tree. (Data from Barre *et al.* (2008) *Molecular Immunology*, **45**, 1231–1240.)

Nut	Tree	Allergen
Walnut	*Juglans regia*	*Jug r* 2
Hazelnut	*Corylus avellana*	*Cor a* 11
Cashew	*Anacardium orientale*	*Ana o* 1

Table 15.6 The most commonly consumed seafoods and freshwater fish (*) in the USA in rank order. (Data from Wild & Lehrer (2005) *Current Allergy and Asthma Reports*, **5**, 74–79.)

Seafood	Consumption (kg/person/year)
Shrimp	1.5
Tuna	1.3
Salmon	0.92
Alaska pollock	0.54
Catfish*	0.52
Cod	0.25
Clams	0.21
Crab	0.20
Flatfish	0.18
Tilapia*	0.16

Seafood allergies

Approximately 2.3% of the USA population are allergic to seafood. This equates to some 6.6 million people (data from Sicherer *et al.* (2004) – see *Further reading*) and therefore seafood allergy is an issue that deserves attention. Seafood is consumed worldwide, but the amounts consumed vary very much from country to country and between geographical regions within a country – the consumption rate often depends on proximity to the sea. In the USA, the average *per capita* consumption of seafood was 7 kg/year in 2002 whereas in New Zealand the *per capita* consumption was 27 kg/year in 2005. The large difference in consumption of seafood in the USA compared to New Zealand is unlikely to be explained by the different years the surveys were carried out, but rather because most large cities in New Zealand are coastal, whereas many large cities in the USA are far from the sea.

Seafood includes anything that is caught in the sea (Table 15.6) and can be divided into the following categories:

FISH
● Vertebrate finned fish (e.g. cod, salmon, tuna)

SHELLFISH
- Crustaceans (e.g. shrimp, crab, lobster)
- Molluscs (e.g. squid, scallop, mussels, oysters)

Seafood allergies in some people are confined to one particular species – I am allergic to skate (*Raja batis*) and nothing else – which suggests that some people respond only to one allergen that is present in only one species.

It is also important to consider freshwater (i.e. river and lake) fish because in some countries they are consumed (e.g. catfish in the USA; Figure 15.6). I will look at the two categories of seafood and the allergies they might cause separately and will include some data for freshwater fish under 'fish'.

Shellfish

A key protein involved in muscle contraction is tropomyosin which is found in all animals with musculature. Tropomyosins vary in protein structure (i.e. amino acid sequence) between taxonomic groups even though their functions are essentially the same. These protein structure variabilities are due to evolution of the protein and selection of the best molecular structure for purpose in a particular species. The tropomyosins of invertebrates (including shellfish) have a molecular weight range 38–41 kDa and show significant amino acid sequence homology between species.

The tropomyosins are the main allergens in seafood. Interestingly some people who are allergic to seafood also show allergic responses to insects and arachnids (e.g. spiders) because the tropomyosins of insects and arachnids are remarkably similar to those from *Crustacea* (e.g. shrimps and crabs). This is not the place to discuss other allergies, but it is interesting that some people with shellfish allergies also have asthma and that asthma can be caused by dust mites (arachnids which have allergens of molecular weight 33–36 kDa).

The shellfish allergens are named in the same way as the other allergens we have discussed in this chapter (Table 15.7). For example, a commonly consumed shrimp species is the black tiger shrimp (*Penaeus monodon*); its major allergen is a 40 kDa protein termed *Pen m* 1.

Fish

There is very little, if any, cross-reactivity between fish and shellfish allergies, or between fish allergy and allergy to insects. This is because the fish allergens are very different to the tropomyosin allergens of the invertebrates. However, people allergic to fish are sometimes allergic to frogs too; this is because frog and fish allergens are the same group of proteins – the parvalbumins (approx. molecular weight 12 kDa) (Table 15.7).

Parvalbumins are globular proteins important in controlling calcium ion (Ca^{2+}) concentrations in muscle cells. Ca^{2+} is very important in muscular contraction which is why its influx and efflux and its free concentration in muscle cells is carefully controlled – the parvalbumins bind and release Ca^{2+}, thus controlling the free Ca^{2+} concentration in the cytoplasm.

Table 15.7 Allergens from commonly consumed seafoods compared with allergens from insects, mites and frogs showing the similarity between the molecular weights of the fish and frog allergens and the shellfish and insect/mite allergens. This suggests that the fish and frog allergens are closely related proteins (i.e. parvalbumins) and that the shellfish, insect and mite allergens are also closely related (i.e tropomyosins), but are very different to the fish and frog allergens. This explains why people allergic to shellfish can also be allergic to insects and mites – inhalation of mite allergens is thought to be a cause of asthma – and people allergic to fish can also be allergic to frogs. (Data from Wild & Lehrer (2005) *Current Allergy and Asthma Reports*, **5**, 74–79.)

Invertebrate	Allergen	Molecular weight (kDa)	
Fish			
Baltic cod (*Gadus callarias*)	*Gad c* 1	12.3	
Atlantic cod (*G. morhua*)	*Gad m* 1	11.5	
Atlantic salmon (*Salmo salar*)	*Sal s* 1	14.1	Parvalbumins
Frog			
Edible frog (*Rana esculenta*)	*Ran e* 1	11.9	
	Ran e 2	11.7	
Shellfish			
Brown shrimp (*Penaeus aztecus*)	*Pen a* 1	36	
Black tiger shrimp (*P. monodon*)	*Pen m* 1	38	
Lobster (*Homarus americanus*)	*Hom a* 1	33	
Crab (*Charybdis feriatus*)	*Cha f* 1	34	
Squid (*Todarodes pacificus*)	*Tod p* 1	38	
Abalone (*Haliotis midae*)	*Hal m* 1	49	Tropomyosins
Insects and mites			
American dust mite (*Dermatophagoides farina*)	*Der f* 10	33	
European dust mite (*D. pteronyssinus*)	*Der p* 10	36	
American cockroach (*Periplaneta Americana*)	*Per a* 7	37	
Silverfish (*Lepisma saccharina*)*	*Lep s* 1	36	

*NB the silverfish is an insect not a fish.

The parvalbumins are not destroyed by heat or broken down by mammalian digestive proteases (e.g. trypsin in the stomach), which means that they survive digestion and enter the intestine where they can be absorbed across intestinal mucosal cells or cause a localised inflammatory response in the intestine.

There is a great deal of protein sequence homology between the fish parvalbumins; this means that there is cross-reactivity between allergens from different fish, i.e. if you are allergic to cod you will probably be allergic to salmon too. The response to different parvalbumin allergens might differ because of their slightly different molecular structures which means that the IgEs directed against them might not interact quite as well with a slightly different parvalbumin. This is why, for example, people very allergic to cod might be less sensitive to salmon and *vice versa*.

Gluten allergy (coeliac disease)

Gluten is an important protein found in wheat and is responsible for many of wheat's culinary properties. If you take a handful of wheat flour and mix it with water until it becomes a pliable solid mass, then wash it gently under running water, you will eventually be left with a stringy, elastic substance that you can pull into long strands before it breaks; this is gluten. Gluten is important in bread and cake making because when it is cooked it changes its form and traps bubbles of carbon dioxide generated either by yeast fermentation (e.g. in bread making) or baking powder (e.g. in cake making; baking powder is a mixture of sodium hydrogen carbonate and tartaric acid that liberates CO_2; Figure 15.7), so making the product light and spongy.

There is a broad spectrum of gluten allergies, from a full blown and serious allergic response (called coeliac disease) to an individual's perception that eliminating gluten from their diet makes them feel better. The latter has evolved into a dietary fad that has led to eating gluten-free food becoming trendy in some Western countries – whether gluten-free food is beneficial or not is the subject of much debate. I will discuss only coeliac disease here.

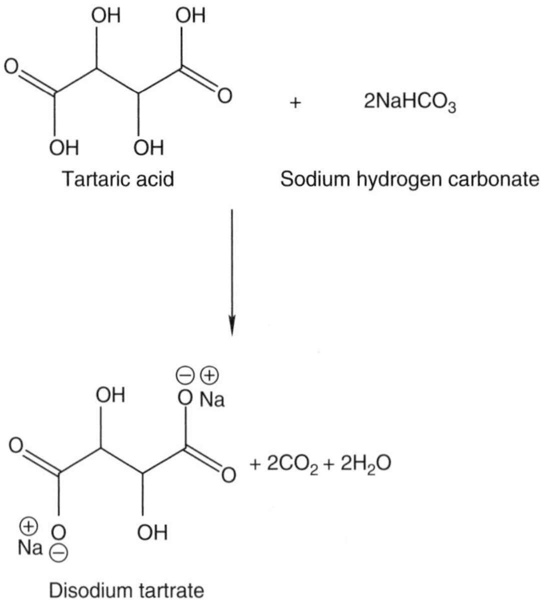

Figure 15.7 The generation of CO_2 from baking powder (a mixture of tartaric acid and sodium hydrogen carbonate) during cooking – CO_2 bubbles are trapped in heat-denatured gluten to give cakes their light spongy texture.

Coeliac disease

Coeliac (also spelled celiac) disease is reasonably common; 1 in 1,750 (0.06%) people have the disease in the USA. The disease is caused by a reaction to wheat leading to an enteropathy (disease of the gut) – 'coeliac' is from the Greek κοιλιακός (*koiliakós*) meaning abdominal and was first described by the ancient Greek physician Aretaeus Cappadocia in the 2nd century AD, but his description was lost until the late 1800s.

To be precise, the immune response in coeliac disease is to a modified gluten, not gluten itself, so the disease is, strictly speaking, not an allergy to gluten *per se* ... you'll see why later.

Wheat gluten proteins cause coeliac disease in genetically susceptible people. Within the gluten protein group are the prolamins (they are rich in the amino acid proline, hence their name); they can be separated into subclasses by electrophoresis (a laboratory method used to separate molecules based on their differential charges). The protein responsible for coeliac disease is a pro-lamin called gliadin. Other cereals have prolamins (e.g. horden in barley; seca-lin in rye), but gliadin is only found in wheat and therefore only wheat causes coeliac disease.

Gliadin is a glycoprotein rich in the amino acids proline and glutamine; it is resistant to proteases in the gut and so is not broken down on its passage through the digestive tract. A region of the gliadin molecule interacts with specific cells in the intestine (enterocysts) and causes them to relax their connections (tight junctions) with neighbouring cells; this allows large molecules to enter the circulatory system through the leaky cell junctions. One of the large molecules that enters the circulatory system by this gliadin-assisted mechanism is gliadin itself – in short, gliadin modifies gut cells to facilitate its own absorption. When in the circulatory system gliadin elicits immune responses at two levels (Figure 15.8):

(1) Innate response
  (a) A region of the gliadin molecule interacts with lymphocytes and stim-ulates them to release interleukin-15.
  (b) Interleukin-15 attracts inflammatory cells which release chemicals (e.g. histidine) that initiate an inflammatory response.
(2) Adaptive response
  (a) A 33 amino acid fragment (33mer) of gliadin is modified by the intestinal enzyme transglutaminase (tTG) which either removes the amino group from glutamine to form glutamate or crosslinks the gli-adin glutamine residue to the lysine of the tTG molecule (Figure 15.9). Formation of glutamate substituted gliadin or, more likely, the cross-linked 33mer-tTG complex leads to an immune response.
  (b) The modified 33mer interacts with T-cells which results in IgE produc-tion which in turn leads to an inflammatory response.

Both the innate and adaptive responses lead to an inflammatory response in the intestine which results in malabsorption of food nutrients, which, in turn, leads to some of the symptoms of coeliac disease (e.g. diarrhoea). The adaptive response also leads to the formation of autoantibodies (i.e. anti-bodies against the body's own molecules) to tTG because antibodies against the crosslinked gliadin-tTG complex cross-react with native tTG because a

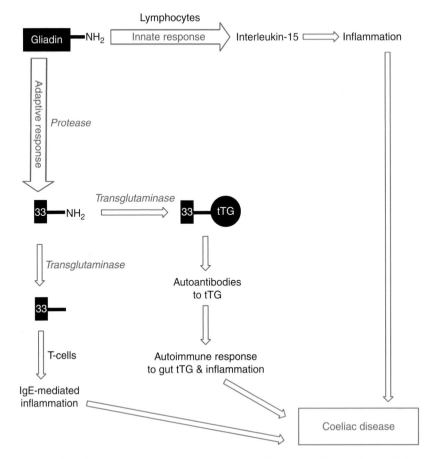

Figure 15.8 The immune responses triggered by the wheat gluten gliadin that lead to coeliac disease. The *innate response* involves a direct interaction between gliadin and lymphocytes causing an inflammatory response mediated by interleukin-15. The *adaptive response* is mediated by a 33-amino acid residue (33mer) fragment of gliadin which either forms a complex with the gut enzyme transglutaminase (tTG) leading to autoimmunity or tTG catalysed deamination of the glutamine residue in the 33mer to form a glutamate residue; the modified 33mer initiates an IgE-mediated inflammatory response. All of these routes lead to intestinal inflammation and coeliac disease.

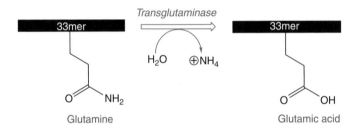

Figure 15.9 Deamination of the glutamine residue in the gliadin-deriver 33mer to form glutamic acid – this initiates an IgE-mediated inflammatory response in coeliac disease.

large proportion of their protein structure is common. Since tTG is a gut enzyme, the immune response occurs in the intestine, which further exacerbates the enteric symptoms of coeliac disease.

In summary, coeliac disease is complex and not fully understood, autoimmunity to a natural protein (tTG) brought about by exposure to a wheat protein (gliadin) and an IgE-mediated inflammatory response to gliadin together lead to a debilitating intestinal disorder triggered by consumption of wheat or wheat products. The only way to treat coeliac disease is to remove all wheat products from the diet.

Allergy to eggs

Egg allergy is the second most common food allergy worldwide – its prevalence is 1.6–3.2% of the world population. In some developed countries, egg allergy in infants is more common than milk allergy.

Most people who develop allergy to eggs have antibodies to one or more of four egg white proteins – ovomucoid, ovalbumin, ovotransferrin or lysozyme. Some people develop antibodies to egg yolk proteins, but this is much rarer and is not well understood. The egg yolk proteins to which antibodies have been detected in egg-allergic people are α-livetin, apovitellenins I and VI and/or phosvetin. Some people might even develop allergy to both egg white and egg yolk proteins and particularly sensitive people might also be allergic to chicken meat because low levels of the egg proteins are present there too (Table 15.8).

By far the most common immune system target in egg allergy is the egg white protein ovomucoid, or *Gal d* 1 as it is often called – '*Gal d*' is derived from the Latin name for the domestic fowl (*Gallus domesticus*).

Table 15.8 Egg antigens – the functions of these proteins in the egg are either storage (i.e. to be used by the developing chick) or to protect the stored proteins from bacterial or enzyme degradation. (Data from Mine & Yang (2008) *Journal of Agricultural Food Chemistry*, **56**, 4874–4900.)

Allergen	Molecular weight (kDa)	Protein function
Egg white proteins		
Ovomucoid (*Gal d* 1)	28	Serine protease inhibitor – prevents enzymes breaking down egg storage proteins
Ovalbumin (*Gal d* 2)	45	Storage protein
Ovotransferrin (*Gal d* 3)	76–77	Iron binding and antimicrobial
Lysosyme (*Gal d* 4)	14.3	Antibacterial
Egg yolk proteins		
α-Livetin (*Gal d* 5)	65–70	Binds ions, fatty acids and hormones
Phosvitin	35	Metal chelating
Apovitellenins I	9.5	Liporotein lipase inhibitor
Apovitellenins VI	170	Binds lipids

As with other food allergies, the only certain way to treat egg allergy is to avoid eggs; however, this is very difficult because eggs and egg products are used in many foods and beverages. For example, egg white is even used to clear ('fine') wine and therefore people with egg allergy have to look carefully at the label on wine bottles to see whether eggs have been used during processing. Below is a warning from a bottle of wine from my wine cellar.

This wine was made using fining agents which contain egg and/or milk products. Traces may remain.

Allergen cross-reactivity

If two allergens have similar molecular structures, or have a section of their molecule (e.g. part of a protein allergen) that is similar to another allergen, they might both bind to the same IgE – this is called cross-reacting. If two allergens cross-react they will both elicit an allergic response. So, if you are exposed to one allergen you might develop sensitivity and if exposed to the second allergen you might develop an allergic response.

There are several examples of allergen cross-reactivity. Some people with hay fever (i.e. allergy to pollen) caused by a specific pollen (e.g. oak) are also allergic to certain foods (e.g. apple) because the oak pollen allergen to which the hay fever sufferer has elicited an immune response has aspects of its molecular structure in common with certain apple proteins. Therefore, when an oak pollen hay fever sufferer eats an apple they might develop an allergic response, not because they have IgE against apple proteins, but because their anti-oak IgE 'thinks' the apple protein is oak allergen.

Another interesting example of allergen cross-reactivity relates to latex (a natural or synthetic polymer used to make rubber) allergen and molecules found in bananas. This means that people with allergy to latex might also be allergic to bananas.

Banana/latex allergy

On the face of it this is very strange. Why would allergy to latex (rubber) be so closely associated with banana allergy?

Latex has very many uses, from surgical gloves to gumboots. Latex can be made from the natural juices of the rubber tree (*Hevea brasiliensis*) by cutting the tree's bark and catching the white fluid (latex) that drips out, or it can be made synthetically. Natural rubber comprises a huge (molecular weight (MW) 100–1,000 kDa) isoprene (2-methyl-1,3-butadiene) polymer (Figure 15.10) – rubber tree latex contains isoprene monomers that polymerise slowly on leaving the tree to form a solid rubbery mass. In addition, natural rubber contains hundreds of other trace components (including proteins) derived from the rubber tree juices. Synthetic latex is simply polymerised isoprene of approximately the same molecular weight as natural latex.

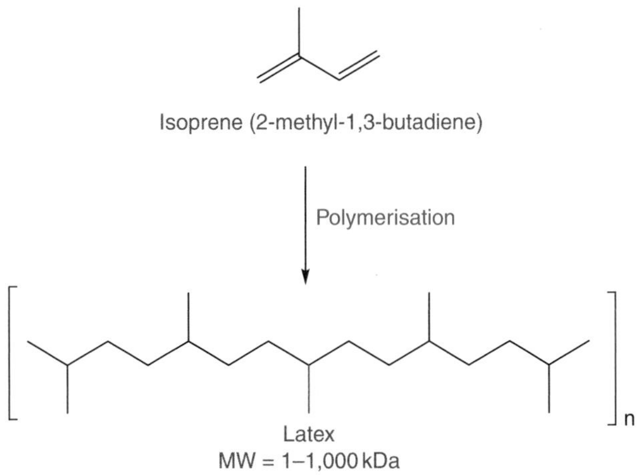

Isoprene (2-methyl-1,3-butadiene)

Polymerisation

Latex
MW = 1–1,000 kDa

Figure 15.10 Latex can be synthesised by polymerising isoprene monomer. Natural latex is very similar to synthetic latex, but also contains proteins, fatty acids and resins from the rubber tree.

Table 15.9 Cross-reactivity of natural latex allergy with foods. (Data from Perkin (2000) *Journal of the American Dietetic Association*, **100**, 1381–1384.)

Food	Percentage of latex allergy sufferers also allergic to this food
Banana	18.3
Avocado	16.3
Shellfish	12.2
Kiwi fruit	12.2
Fish	8.1
Tomato	6.1
Watermelon	<5.0
Peach	<5.0
Carrot	<5.0

Interestingly, while allergy to natural latex is common, allergy to synthetic latex hardly ever occurs. This means that the isoprene polymer is not the cause of the allergy, but instead other components of the rubber tree juice are the allergenic culprits. Rubber tree latex comprises isoprene polymer (95% dry mass), proteins, fatty acids and resins (5% dry mass combined) – it is very tempting to blame the proteins for natural latex's allergenicity. Read on ....

Fifty to 70% of latex allergy sufferers also have allergies to fruits, nuts and grains, with a disproportionate number showing a link with allergy to bananas (Table 15.9). Much research has been carried out to try to work out the reason for this link. We still are not absolutely certain, but the finger is pointing very firmly at a plant enzyme called chitinase as the common allergen.

Chitin is a polymer of N-acetylglucosamine that is an important component of the exoskeleton of insects (and other creatures such as crabs and shrimps) and the cell walls of fungi. Chitinase breaks down chitin to its component amino sugars and is an important plant defence against insects and fungi. Plants produce chitinase to kill insects that eat them or suck their juices or fungi that infect them.

Banana plants (*Musa sapientum*) and rubber trees both produce chitinase and if someone becomes sensitised to one they will respond to the other when exposed. Therefore, a person who regularly wears natural latex surgical gloves might become sensitised to latex chitinase which will be manifest as a food allergy when they eat a banana. The fact that many plants produce chitinase explains the cross-reactivity between latex allergy and a wide range of nuts, fruits and vegetables – banana is, however, the commonest latex allergy cross-reactivity.

Cross-reactivity between latex and shellfish also occurs. This is because shellfish have chitin exoskeletons and utilise chitinase metabolically. Similarly, latex/fish cross-reactivity has also been reported; presumably some fish have chitinase to digest their shellfish prey.

Food additives allergy

All of the food allergies I have discussed so far are thought to be caused by specific proteins in the particular food. However, some of the food additives used in food processing (see Chapter 11) cause allergic responses. When absorbed, the food additives might bind to proteins, thus forming a hapten (the additive) carrier (the protein) complex which is large enough to initiate an immune response (e.g. cause the synthesis of IgE by B-cells) – remember, small molecules are not immunogenic and so must be bound to a large molecule to elicit an immune response. The yellow food colour tartrazine (E102; see Chapter 11) can cause an allergic response and therefore foods coloured with it are associated with some food allergies.

Food additive allergies are quite rare – they comprise only about 10% of all food allergy cases (i.e. occur in about 0.2% of the population as a whole).

Why is the incidence of food allergies increasing?

Increased food allergen intake

The incidence of some food allergies (e.g. peanut allergy) is steadily increasing while others (e.g. shellfish allergy) are remaining constant; this might at first be difficult to explain – peanut allergy doubled in children over the 5-year period 1997-2002 (USA data from www.foodallergy.org). These allergy statistics might be influenced by the consumption of specific foods containing allergens increasing with a concomitant increase in the incidence of associated allergies. If this were the case, at a population level we might be being exposed to more peanut allergens, resulting in increased sensitisation. On the other hand, the consumption of other foods that might be associated with food allergies might remain unchanged over time and thus consumers' exposure to their allergens also remains unchanged, and, therefore, the incidence of allergies to these foods also remains unchanged.

In utero transfer of food allergens

Transfer of allergens from mother to developing child *in utero* is also probably important in the increasing incidence of some food allergies (e.g. peanut allergy). The increasing consumption of peanuts means that the developing embryo/fetus is also being exposed to the allergens by placental transfer; thus, they too become sensitised and in later life develop food (e.g. peanut) allergy. The importance of the mother's diet in conferring peanut allergy is supported by the fact that a high proportion of children with peanut allergy display symptoms on their first exposure to peanuts, i.e. they must have been sensitised before they first ate peanuts.

Excretion of food allergens in mother's milk

Transfer of allergens in mother's milk also occurs and might lead to sensitisation of the suckling child. For example, if a mother eats peanuts and excretes the *Ara h* 1 and *Ara h* 2 in her milk her child will also receive a dose of the antigen. Subsequent sucklings following the mother having eaten peanuts will result in the child getting successive doses of peanut allergens that might result in sensitisation. When the child eats his or her first peanut he/she could rapidly develop allergic symptoms because his/her mast cells will recognise the peanut allergens and will degranulate, thus resulting in a full inflammatory response.

A cautionary note

In addition to the above 'scientific' explanation, there is also the possibility that some people think they are allergic to certain foods when they are not. As discussed above, it has become trendy to eat gluten-free food on the grounds that gluten is responsible for its consumer's apparent ill health. When the person stops eating foods containing gluten they might feel better; whether this is explained by a food allergy or has a psychosomatic explanation is often not possible to ascertain; these issues must be borne in mind when considering data on the increasing incidence of food allergies.

Take home messages

- The immune system is complex which makes food allergies appear complex, but they are in fact quite simple because all they involve is the body eliciting an immune response to a specific protein in a food (e.g. peanuts).
- Some allergies to different foods (e.g. tree nuts) involve the same or very similar proteins that are common to the foods. This is why some people are allergic to several foods.

- Exposure to food allergens *in utero* or during infancy (i.e. sensitisation) can lead to food allergy later in life.
- Food allergies are more common in youngsters – they often disappear by adolescence.
- Food allergies can only be treated by avoiding the allergenic food, but this can be difficult because some allergenic foods are very widely used in processed foods (e.g. eggs).

Further reading

Since food allergy is the subject of a great deal of current research, I have included a longer *Further reading* list than in other chapters. I want you to see for yourself the conjecture, controversy and uncertainty that prevail.

Abbas AK & Lichtman AH (2009) *Basic Immunology*. Saunders Elsevier, Philadelphia. Chapter 1 is an excellent introduction to the immune system.

Calder PC & Field CJ (eds) (2002) *Nutrition and Immune Function*. CABI Publishing, Oxford & New York. Chapter 15 by Elizabeth Opara is an excellent account of food allergy and its underlying immunology.

Garrett RH & Grisham CM (2010) *Biochemistry*, 4th edn. Brooks/Cole, USA. Chapters 5 and 6 give a detailed account of protein structure and the importance of amino acids in determining conformation.

Maynard F, Chatel JM & Wal JM (1999) Immunological IgE cross-reaction of bovine and human α-lactalbumin in cow's milk allergic patients. *Food and Agricultural Immunology*, **11**, 179-189.

Perkin JE (2000) The latex and food allergy connection. *Journal of the American Dietetic Association*, **100**, 1381-1384.

Rabjohn P *et al.* (1999) Molecular cloning and epitome analysis of the peanut allergen Ara h 3. *Journal of Clinical Investigations*, **103**, 535-542. This paper describes the identification of peanut allergens.

Roitt I, Brostoff J & Mahe D (2001) *Immunology*. Mosby, London.

Sicherer SH, Munoz-Furlong A & Sampson HA (2004) Prevalence of seafood allergy in the United States determined by a random telephone survey. *Journal of Clinical Immunology*, **114**, 159-165.

Chapter 16
Food Legislation

Introduction

In order to ensure that food is safe to eat (i.e. fit for purpose, in legal jargon) it is necessary to grow and manufacture it in such a way that contamination with harmful bacteria, viruses, parasites, prions and toxic chemicals is as low as possible; or at least low enough to cause no harm to the consumer. In an ideal world this should be the case because any diligent food producer would not want to harm their customers. However, in these days of big business and profit margins, it is, perhaps, understandable that shortcuts might be taken to make that little bit more profit on a particular food line. Even with such commercial pressure, no food producer or manufacturer would want to harm their customers and so even if they did decide to take a manufacturing shortcut they would do this thinking that it would not result in harm to their customers. Despite all of this we should not leave the producers and manufacturers to decide what is an acceptable risk for their customers because they have a conflict of interest. This is where food laws are important; they set a level playing field that ALL producers and manufacturers MUST comply with, and, if they don't, legal action can be taken against them. The need to comply with food legislation – and to have legislation to comply with – has assumed greater importance following the BSE saga (see Chapter 6) as a means of increasing public confidence in the food we eat.

All developed countries have food legislation, but some developing countries have little or no effective food regulation. Food-borne illness is rife in many developing countries and so, arguably, this is where our legislative effort should be focused. It is, however, very important to keep the value of legislation in perspective; many people in the developing world are more concerned about where their next meal will come from rather than whether it complies with food safety regulations. This is a sobering thought that should remain in our minds when we explore food laws in this chapter.

International food laws are also important because they set standards for international food trade that ensure that if we import food from another country that that food complies with basic standards

Food Safety: The Science of Keeping Food Safe, First Edition. Ian C. Shaw.
© 2013 John Wiley & Sons, Ltd. Published 2013 by John Wiley & Sons, Ltd.

(e.g. pesticide residues below MRLs – see Chapter 2, *Maximum Residue Level (MRL) and Maximum Limit (ML)*).

Food legislation has been with us for a surprisingly long time. It began in the UK with the baking laws of 1155 which covered adulteration of bread with, for example, sand to make it heavier. In 1641 England introduced laws relating to the inspection of bakeries – bread is a staple food and so early laws concentrated on it. In 1785, America (it did not become the USA until 1789) introduced the Prevention of Food Adulteration Act, which was followed in 1886 with the first regulations preventing the adulteration of margarine. Comprehensive food acts are very much more recent (i.e. late 20th century).

It is all very well having a food law in place, but if it is not being obeyed it is of little use. For this reason food laws are usually accompanied by policing schemes to check that producers and manufacturers are complying with the law. Food surveillance is the most common form of policing; it involves sampling food or produce and measuring parameters covered (e.g. pesticide residues) by the food law of the particular country. If the parameter measured does not comply with the standard set in the food law, the producer is liable. I will discuss the different types of surveillance schemes used around the world in this chapter.

Our problem now is that food production is global and food is moved around the world to provide consumers with out-of-season produce, and to allow food production in countries where it is cheaper as a means of maintaining competitive food prices. This presents food legislation with a huge challenge. When we import food from another country should we assume that that country has good food laws and efficient surveillance schemes? You might think the answer to this important question is yes, but that is certainly not the case for many countries that can produce food more cheaply. For example, some milk produced in China in 2009 contained toxic levels of melamine that resulted in deaths of children in China and appeared in milk-containing products (e.g. confectionery) exported to other countries (e.g. New Zealand). Melamine was added because it gives a positive reaction in a test used to measure protein levels; melamine artificially inflated the apparent protein levels, so resulting in the milk complying with statutory levels for milk proteins and allowing it to be exported. Clearly international compliance with legislation must be the focus for a safe food future.

Before we can understand the legislation relating to food we must first appreciate how laws are created and be familiar with the terms used by the legal profession to describe the different levels of legislation. Legal processes vary from country to country and so do the legal terms used, but there are several important processes and terms that are common pretty well worldwide. I will outline them here.

Legal processes – how laws are made

I will outline the British system here, but its general principles apply to most democratic countries around the world. Legislation begins in parliament with informal discussion about an issue (e.g. food safety) about which it soon becomes clear some rules need to be set (e.g. to minimise the risk to consumers). A member of parliament (MP) gets the ideas together in the form of a Draft Bill (also called a Green Paper in the UK) which is presented as a speech to Parliament. Following this, the ideas are further debated and reformulated into a second draft called a Bill (this is also called a White Paper in the UK) which incorporates useful ideas that came out of the parliamentary debate. The Bill is then sent for further discussion in committee (i.e. outside the confines of parliamentary debate) during which time the implications of the Bill becoming law are discussed in detail and any potential problems ironed out. This can take months or longer. The amended Bill is then taken back to parliament for debate and subject to modifications and acceptance following a parliamentary vote (this is termed Passage of a Bill) it becomes an Act of Parliament (usually termed an Act). The Act is then sent to the monarch for approval – this is termed the Royal Assent and means that the Act is added to the British Statutes and at that point becomes law (Figure 16.1). When an Act is on the Statutes the authorities (e.g. Food Standards Agency in the UK) can prosecute people who infringe the Act and the courts use the Act to determine whether the accused is guilty or not. If the accused is found guilty an appropriate penalty will be given. Such penalties are laid down in the Act as guidance to the courts.

If we consider a food-related example, the process becomes clearer. A restaurant sells a customer a hamburger, and the consumer contracts *E. coli* O157 (see Chapter 3) and becomes ill. He visits his doctor and the doctor realises, when he takes the case history of his patient, that he has seen several similar cases in the past few weeks. The doctor reports his findings to the food safety authority because *E. coli* O157 is a notifiable disease (i.e. the doctor who makes the diagnosis must report the case to the authorities) under the country's Food Act. The food safety authority notes that several doctors have reported similar clusters of cases, and, following further investigation, all of the affected patients had eaten at the same restaurant, and all had eaten a hamburger. At this point the food safety authority would visit the restaurant and if they found *E. coli* O157 contamination would probably prosecute under the Food Act. The authority is likely to have emergency powers under the Food Act to temporarily shut down the restaurant while they determine the source of the *E. coli* O157 (i.e. to prevent further cases). If they find that the problem was that the chef had not cooked the hamburgers thoroughly (see Chapter 3) this would almost certainly lead to the chef being found guilty of misconduct under the Act because the Act clearly states that food should be fit for purpose and contain nothing that would cause harm to its consumer. This would lead to a fine (laid down in the Food Act) and perhaps result in him losing his job (i.e. his employer's response to his conviction).

The above discussion of laws and how they are made applies only to democratic countries; dictatorships simply impose laws without public debate, but such issues are far outside the remit if this book (and my knowledge).

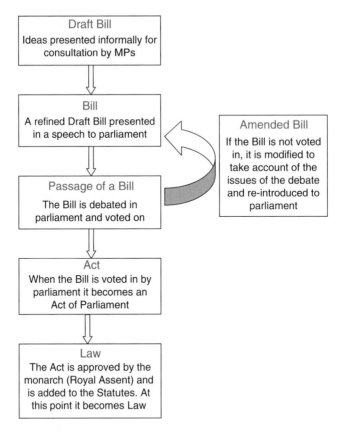

Figure 16.1 The process of creating law in Britain – many countries have very similar approaches because their legal processes are based on the British system. The major difference is that in countries without a constitutional head (e.g. a monarch) the Act becomes law as soon as parliament votes it in.

A very brief history of food law

It might surprise you to learn that the first food 'law' was introduced in England in 1155 when the Worshipful Company of Bakers (Baker's Guild) was instituted. The Baker's Guild was responsible for the Assize of Bread and Ale (1266) (an assize is a decree of the court, i.e. law) which governed the manufacture of bread and, in particular, prohibited selling short measure or adding sand to the dough to increase its weight. Violation of the Assize was punishable by an appropriate time in the pillory – a wooden device, usually in the town centre, that secured the hand and neck through a series of holes (Figure 16.2) – and sometimes whipping or stoning while the violator was restrained. So, the concept of punishing food adulterators has been with us for a thousand years. Interestingly the idea of a baker's dozen (i.e. 13) probably arose from the Assize of Bread and Ale to ensure that the baker did not sell short measure – clearly 13th century bakers took the law very seriously (so would you if the punishment was being stoned in the pillory in public!).

Figure 16.2 A man being stoned in a pillory – in this case stoned refers to a multifarious array of unsavoury missiles including a rat. (From the *Newgate Calendar* (1732).)

Since 1155 there has been a plethora of food-related legislation worldwide covering food issues as diverse as sugar (The Sugar Act, American Colonies, 1764), corn (British Corn Laws, 1815–1846) and even Maraschino cherries (defined by the US Food & Drug Administration, 1912), but there was no comprehensive food legislation anywhere in the world until the mid to late 20th century, although there were government departments (e.g. te US Food Administration, set up in 1917) that had responsibility for food, but they worked under multiple acts and regulations.

Food legislation around the world

It would be impossible in a single chapter to review comprehensively food legislation from around the world and therefore I will outline the approaches to legislation in the countries Britain, the USA and New Zealand to give you an idea of the similarities – and differences – between the food laws of three quite different jurisdictions.

Food legislation in the USA

The USA has a federal food agency (Food and Drug Administration (FDA)) that implements high-level food legislation across the 50 states and each state has its own state-specific legislation. The FDA, which has responsibility for public safety in respect of drugs, food, cosmetics, medical devices and clinical trials of new medical products, has been part of the Department of Health and Human Services since 1979 and is one of the oldest food legislations in the world.

State legislation in relation to food is variable, but the general ethos that food should not cause its consumer harm runs across the legislation of all states. The definition of what might cause harm varies from state to state. For example, permitted pesticide residues in food are lower in California than other states which reflects California's 'pure green' philosophy.

I will not discuss the 50 states' individual food legislations, but will give a brief overview of the USA's federal food legislation. You will see when you read the overviews of legislation for other countries below that the general principles are international.

Instead of the Acts of Parliament of the UK, the USA has US Codes – to all intents and purposes Codes are the same as Acts. They originate as discussions in the US Congress and go through a process of debate and committees to refine them before they are agreed and become Codes (i.e. US laws). The Codes are brought together in a huge document called the Code of Federal Regulations (CFR). The Codes must be followed by the individual states, but how they do this with respect to their own State Legislation is up to them – hence the state to state variability. The Code provides the lowest common denominator. All states must at least comply with the Code, but can enhance the Code with their own particular state laws that go over and above the federal requirements laid down in the Code (e.g. California has more stringent pesticide regulations than the corresponding Code requires). The US Code that covers food is US Code Title 21 – Food & Drugs. The Code is divided into 26 Chapters covering all aspects of food and medicines legislation. Chapter 26 – *Food Safety* is also very varied and covers, very comprehensively, aspects of food safety.

US Code Title 21 – Food & Drugs; Chapter 26 – Food Safety

To give you an idea of the breadth of issues covered in Chapter 26 – *Food Safety*, here are the titles of the sub-sections (denoted in the Code by §):

§ 2101 Findings (see later)
§ 2102 Ensuring the safety of pet food
§ 2103 Ensuring efficient and effective communications during a recall
§ 2104 State and Federal cooperation
§ 2105 Enhanced aquaculture and seafood inspection
§ 2106 Consultation regarding genetically engineered seafood products
§ 2107 Sense of Congress [this sub-section is about what the US Congress must provide to ensure that food safety regulations can be met, e.g. funding for inspectors]
§ 2108 Annual report to Congress
§ 2109 Publication of annual reports
§ 2110 Rule of construction

§ 2101 covers the 'findings' of the US Congress with respect to the importance of food safety both from public health and economic standpoints. The 'findings' are effectively the reasons for assuring safe food; they are as follows:

(1) the safety and integrity of the United States food supply are vital to public health, to public confidence in the food supply, and to the success of the food sector of the Nation's economy;

(2) illnesses and deaths of individuals and companion animals caused by contaminated food–

   (A) have contributed to a loss of public confidence in food safety; and

   (B) have caused significant economic losses to manufacturers and producers not responsible for contaminated food items;

(3) the task of preserving the safety of the food supply of the United States faces tremendous pressures with regard to–

   (A) emerging pathogens and other contaminants and the ability to detect all forms of contamination;

   (B) an increasing volume of imported food from a wide variety of countries; and

   (C) a shortage of adequate resources for monitoring and inspection;

(4) according to the Economic Research Service of the Department of Agriculture, the United States is increasing the amount of food that it imports such that–

   (A) from 2003 to 2007, the value of food imports has increased from $45,600,000,000 to $64,000,000,000; and

   (B) imported food accounts for 13 percent of the average American diet including 31 percent of fruits, juices, and nuts, 9.5 percent of red meat, and 78.6 percent of fish and shellfish; and

   (C) the number of full-time equivalent Food and Drug Administration employees conducting inspections has decreased from 2003 to 2007.

As you can see the Congress' 'findings' cover a wide range of implications relating to food safety, including what the effects of not assuring safe food might be, trends in food imports and the number of food inspectors. They set the scene for the need for US food to be safe.

§ 2107 'Sense of Congress' covers what the US Congress must provide (i.e. resources) to facilitate food safety programmes. For example, clause 1 states that:

> It is vital for Congress to provide the Food and Drug Administration with additional resources, authorities, and direction with respect to ensuring the safety of the food supply of the United States.

Sub-chapter B – *Food for Human Consumption* lays down the requirements relating to specific food safety issues. For example, Part 170 – *Food Additives* details the regulations that specifically apply to the safety testing and approval of chemicals used as colours, flavours and preservatives in food (see Chapter 11). The section headings under Sub-part B of Part 170 will give you a good idea of the content of the legislation:

Sub-part B – Food Additive Safety

170.20 General principles for evaluating the safety of food additives

170.22 Safety factors to be considered

170.30 Eligibility for classification as generally recognised as safe (GRAS*)

170.35 Affirmation as generally recognised as safe (GRAS)

---

*GRAS is an approach to accepting that chemicals that have been used for a long time without any adverse effect do not need to be subjected to toxicity testing before they are permitted for use in food (and other products intended for human use), i.e. these compounds are pre-approved for food use. An example of a GRAS preservative is benzoic acid.

170.38 Determination of food additive status
170.39 Threshold of regulation for substances used in food contact articles

Food legislation in the UK

There are two important Acts relating to food in the UK: the Food Act 1984 and the Food Standards Act 1999. The Food Act covers the definition of food and the basic, and very comprehensive, principles of food safety (including surveillance, inspection and penalties), whereas the Food Standards Act legislates for the agency – the Food Standards Agency (FSA) – which is responsible for food safety in the UK.

The Food Act 1984

The UK Food Act 1984 opens with the grand words:

An Act to consolidate the provisions of the Food and Drugs Acts 1955 and 1982, the Sugar Act 1956, the Food and Drugs (Milk) Act 1970, sections 7(3) and (4) of the European Communities Act 1972, section 198 of the Local Government (Miscellaneous Provisions) Act 1982, and connected provisions.

[26th June 1984]

Be it enacted by the Queen's most Excellent Majesty, by and with the advice and consent of the Lords Spiritual and Temporal, and Commons, in this present Parliament assembled, and by authority of the same, as follows:-

PART I
FOOD GENERALLY
Composition and labelling of food

1. - (1) A person is guilty of an offence who –
    (a) Adds any substance to food,
    (b) Uses any substance as an ingredient in the preparation of food,
    (c) Abstracts any constituent from food, or
    (d) Subjects food to any other process or treatment, so as (in any such case) to render the food injurious to health, with intent that the food shall be sold for human consumption in that state.

It is a very comprehensive Act that set (in 1984) the scene for a significant increase in attention to food safety in the UK. This book is not the place to go into great detail about food legislation, but rather to give food safety scientists and aspiring food safety scientists an overview that encompasses the breadth and philosophies of legislations from different countries. The best way to do this, I think, is to begin by listing the sub-sections of legislative documents to give an idea of their subject matter and breadth of content; the UK Food Act 1984 is divided into Parts and Sections. The Sections are grouped under headings that define their scope. I will list the Sections and headings – their titles are self explanatory and illustrate very well the comprehensive nature of this important Act:

Food Act 1984

Part I - FOOD GENERALLY

Composition and labelling of food
Food unfit for human consumption
Hygiene
Control of food premises
Ice-creams, horseflesh and shellfish
Food poisoning

Part II - MILK, DAIRIES AND CREAM SUBSTITUTES

Milk and Dairies
Special designations of milk and their use
Compulsory use of designations for specific areas, and licenses for
  specific areas
Cream substitutes

PART III - MARKETS

PART IV - SALE OF FOOD BY HAWKERS

PART V - SUGAR BEET AND COLD STORAGE

Ministerial functions as to sugar beet
Cold storage

PART VI - ADMINISTRATION, ENFORCEMENT AND LEGAL PROCEEDINGS

Administration
Sampling and analysis
Enforcement
Legal proceedings
Appeals
Compensation and arbitration

PART VII - GENERAL AND SUPPLEMENTAL

Acquisition of land, and order to permit works
Inquiries and default
Protection
Subordinate legislation
Notices, forms and continuances
Expenses and receipts
Interpretation and operation

I think you will agree that the Act covers almost everything conceivably related to food safety. To illustrate this still further I have listed below the Sections under PART VI - ADMINISTRATION, ENFORCEMENT AND LEGAL PROCEEDINGS - *Sampling and analysis* which shows how food safety is assured by sampling and analysis for contaminants and the powers of the inspectors:

PART VI - ADMINISTRATION, ENFORCEMENT AND LEGAL PROCEEDINGS
Sampling and analysis

Section 76 Public analysts
77 Facilities for examination
78 Powers of sampling
79 Right to have samples analysed
80 Samples taken for analysis
81 Sampling of milk
82 Sampling powers of Minister's inspectors
83 Minister's power of direction
84 Where division not practicable
85 Examination of food not for sale
86 Quarterly reports of analysis

At the outset of this section I showed an extract from the beginning of the Food Act 1984 which shows the overriding philosophy of the Act, i.e. that nothing shall be added to or taken away from food that makes it unfit for human consumption - as an aside consider for a while the legal problems associated with natural toxins (see Chapter 8) that are harmful to consumers, but have not been added or taken away from the food - a conundrum that often has to be dealt with outside the Act.

Finally, the working of Acts generally, and the Food Act in particular, requires that specific procedures (e.g. taking samples of food for analysis) are clearly laid down. This is done in a series of Schedules included as addenda to the Act. There are 11 Schedules in the UK Food Act 1984; for example, Schedule 7 covers sampling and goes into great detail about how and when samples can be taken, who (i.e. the Public Analyst) is allowed to analyse them, how the samples should be transported and stored, how the results should be reported and the right of the 'owner' of the sample to have an independent analysis. To illustrate the detail of a Schedule, here is an extract from the beginning of Schedule 7 - SAMPLING, Part I:

MANNER IN WHICH SAMPLES TAKEN OR PURCHASED FOR ANALYSIS ARE TO BE DEALT WITH
1. The sampling officer shall forthwith divide the sample into three parts, each part to be marked and sealed or fastened up in such a manner as its nature will permit, and shall -
   (a) With respect to one part of the sample comply with paragraph 2 to 8[1], and
   (b) Deal with the remaining parts in accordance with paragraph 9[2].

Food Standards Act 1999

The Food Standards Act simply (although very importantly) provides the UK administration and expertise for operating under the Food Act 1984 and

---

[1] These paragraphs outline the ways in which samples can be obtained (e.g. by purchasing from a shop) and that the person from whose premises the sample is taken (e.g. a shop owner) must be informed that a sample has been taken.
[2] This paragraph states that the sample will be divided, one sub-sample for official analysis and the other for future analysis (e.g. in the event of a dispute).

development of new food safety policy. The opening paragraph of the Food Standards Act makes this very clear:

> An Act to establish the Food Standards Agency and make provision as to its functions; to amend the law relating to food safety and other interests of consumers in relation to food; to enable provision to be made in relation to the notification of tests for food-borne diseases; to enable provision to be made in relation to animal feedingstuffs; and for connected purposes.
>
> [11ᵗʰ November 1999]

The detail and language of the Food Standards Act 1999 is similar to that of the Food Act 1984 and lays down in great detail the way in which the FSA will be set up, staffed and operate. Section 1 of the Act defines the FSA thus:

1. (1) There shall be a body to be called the Food Standards Agency or, in Welsh, yr Asiantaeth Safonau Bwyd (referred to in the Act as "the Agency") for the purpose of carrying out the functions conferred on it by or under this Act.
   2. The main objective of the Agency in carrying out its functions is to protect public health from risks which may arise in connection with the consumption of food (including risks caused by the way in which it is produced or supplied) and otherwise to protect the interests of consumers in relation to food.
   3. The functions of the Agency are performed on behalf of the crown.

The FSA's main job is to ensure sampling and analysis of food by the appropriate agencies (e.g. local authorities), collate and report on the results, communicate findings to the public and advise Ministers of any food-related issues. Section 12 (1) of the Act makes this clear:

> 12. (1) The Agency has the function of monitoring the performance of enforcement authorities in enforcing relevant legislation.

In summary, the FSA is an agency that co-ordinates food monitoring and makes sure everyone is aware of the results and their implications in the hope that food safety disasters (like BSE; see Chapter 6) don't reoccur.

Food legislation in New Zealand

The New Zealand Food Act 1981 preceded the UK's Food Act 1984 (remember the UK Food Act 1984 is derived from a series of preceding Acts including the Food and Drugs (Milk) Act 1970) and is based on, and refers to, New Zealand's Food Hygiene regulations 1974. The Act has been amended four times (1985, twice in 1996, 2002) since being enacted in 1981; most of the amendments enable particular administrations to be set up and/or function (e.g. the New Zealand Food Safety Authority – Amendment No. 26 (2004)).

The Act is divided into four parts (including examples of Sections under each Part) as follows:

From this brief look at the New Zealand Food Act it is clear that it includes similar provisions to both the USA's Code and the UK's Food Act and in general its purpose is to ensure food safety by regulating its production and sale, carrying out surveillance with authority given to officers to sample food, regulation and assurance of analysis and reporting of results, and powers to prosecute offenders.

As mentioned above, New Zealand amended its Food Act 1981 in 2002; this amendment (Food Amendment Act 2002) paved the way for the creation of the New Zealand Food Safety Authority (NZFSA) by removing responsibility of some aspects of food safety from the Ministry of Health and transferring it to the Chief Executive of the Ministry which the Prime Minister determines will be responsible for the Food Act 1981. This might seem rather covert, but it provided the means of creating a new agency (rather like the UK's FSA) that has 'independent' authority over food safety. The NZFSA became that 'independent' agency that did not 'belong' to a ministry and this was seen to be independent of fiscal influences related to the sale of food. In 2010 the New Zealand government removed the 'independence' of the NZFSA and included it in the Ministry of Agriculture and Forestry (MAF). This change did not require a further amendment of the Act because the wording of the Food Amendment Act 2002 is sufficiently vague to allow the Prime Minister to alter responsibilities; this is clear from the following quote from Part 1 4 (1) of the Food Amendment Act 2002. I have only included the text for "Ministry" because this is the important wording that gave the Prime Minister the authority to change responsibility for the Food Act 1981 and thus the authority to set up the 'independent' NZFSA, then, at a later date, transfer it to MAF:

4. **Interpretation**
   (1) Section 2 of the principal Act[1] is amended by repealing the definition of "Director-General", "Minister", and "officer", and inserting, in their appropriate alphabetical order, the following definitions:

   "designated officer"
   "Director-General"
   "Minister"
   "Ministry" means the department of State that, with the authority of the Prime Minister, is responsible for the administration of this Act
   "officer"

A quick foray into the Food Acts from three countries shows that food safety thinking (and legislation) in three quite different jurisdictions (i.e. the USA, UK and New Zealand) are quite similar; this is because food safety issues worldwide are also quite similar (the difference is usually in the frequencies of food-borne illnesses rather than the illnesses themselves).

Policing food legislation

For food laws (or any laws for that matter) to be effective they must be policed to identify any violations and allow action to be taken against the violators. Such action reinforces the law and shows others the consequences of not obeying the law. Consider driving your car; if you speed you might get caught by the police, and if you get caught you will almost certainly be fined. If this unfortunate situation arises, you are likely to tell your friends about it, and this, in turn, is likely to make them think twice about exceeding the speed limit. Therefore, policing works.

There are many ways that food laws can be policed, but they fall into three main categories:

1. Food surveillance schemes – random samples of food and produce are taken at the point of sale or from the farm and are analysed for residues of chemicals and/or microbiological content. If levels/microbiological count exceed the limits laid down in the law, court action might be taken.
2. Total diet surveys (TDSs) – samples of meals that consumers actually eat are taken and analysed for selected residues or microbiological contamination. Typically a group of consumers (i.e. a statistically significant number; e.g. 1,500 for the New Zealand TDS) are recruited and asked to prepare duplicate meals for a prescribed period of time. They eat one meal as normal and the other is sent to the laboratory for analysis. This does not allow immediate action under food legislation, but it does give an indication of parameters that might exceed lawful intakes.
3. Enforcement sampling – samples are taken when there is concern that a food safety problem exists. This might lead to food recalls if results show unacceptable residues or microbiological counts.

---

[1]Principal Act means the Food Act 1981.

Food surveillance schemes and TDSs are not used to prevent exposure to hazards in the food being sampled at, or around, the time of sampling because the time from sampling to results analysis is often very long. These methods are used to assess population exposure and determine overall risk to populations rather than individuals. They sometimes lead to future action though; for example, high levels of the carbamate pesticide phorate found in carrots in the UK in 1995 (see Chapter 7, *OP and carbamate residues in food*) as part of routine surveillance led to changes in the regulations for use of phorate to minimise the risk of consumer exposure to unacceptable levels of the pesticide in the future.

Enforcement sampling is immediate. Samples are taken, analysed urgently and the results acted upon quickly. If unacceptable levels, for example, of a pathogen (e.g. *Listeria monocytogenes*) are found, the regulatory authority might close the manufacturer or restaurant until the problem is solved and withdraw food from shops to prevent consumer exposure. Such action is often a response to cases of a particular food-borne illness (e.g. listeriosis) being linked to a particular food outlet or manufacturer (see Chapter 3, *Listeriosis case example*). Enforcement sampling might result from regular surveillance if high levels of a particular chemical or microbe are found. This was the case in the phorate example outlined above.

Food surveillance schemes

I have outlined the principles of food surveillance schemes above. In this section I will illustrate a good food safety scheme with the UK's pesticide monitoring programme.

The programme is run by the Pesticide Residues Committee which is an independent committee given authority under the Food Act 1984 and the Food Standards Act 1999 to sample food, analyse the samples for pesticides, report the results to the FSA and communicate its findings to the public.

The pesticide monitoring programme runs continuously and develops a sampling plan each year which includes the following foods types:

- Staples – bread, milk and potatoes – are analysed every year because they are consumed in large quantities and thus any contamination would lead to many people being exposed regularly.
- Frequently consumed fruit and vegetables (e.g. carrots) – analysed on a 5-year cycle.
- Infrequently consumed fruit and vegetables (e.g. figs) – analysed on a 10-year (or longer) cycle.
- Meat and other animal products (e.g. butter).

There are many excellent food surveillance schemes around the world; the UK's pesticide monitoring scheme is just one good example.

National surveillance

Most countries in the developed world (and many in the developing world) run surveillance schemes to ensure that their food safety regulations are being applied effectively to food produced within their countries; these are national surveillance schemes (NSSs). NSSs are policing schemes; they involve sampling home (i.e. within the country)-produced food and analysing the samples for contaminants (e.g. pesticides used in farming) or additives (e.g. preservatives) to make sure that farmers and food manufacturers are obeying the law. If contaminants are found at levels above statutory limits (e.g. MRL; see Chapter 2, *Maximum Residue Level (MRL) and Maximum Limit (ML)*) or approved substances (e.g. preservatives) are found in food that they should not be in, or at levels above those allowed, the food source is traced and appropriate action taken. This is usually a caution for a first offence, but can lead to significant fines or a prison sentence for repeat offenders.

Import surveillance

A great deal of our food is imported. The days of eating only locally produced food in season are long gone – most people expect to eat out of season fruit and vegetables (e.g. strawberries) which creates a vibrant import/export market worldwide. Interestingly, as we become more aware of the need for a sustainable lifestyle we are beginning to question the environmental cost of transporting non-essential food items around the world; this, combined with carbon taxes, is likely to reduce non-essential food trading that involves long journeys. Despite this, at the moment there is a vibrant world trade in food and therefore it is necessary for importing countries to assure their consumers that foreign food is safe. This is done via import food surveillance schemes. These schemes, in principle, are the same as any food monitoring programmes (e.g. NSSs), but samples are taken only from food entering a country from another country.

A good example of import surveillance is the UK's Veterinary Residues Surveillance Scheme which includes imported food. Any violations are reported to the authorities of the exporting country with the expectation that the problem will be dealt with and no further violations will occur. For example, in 2007 the UK Veterinary Residues Committee found unacceptable levels of nicarbazim (a coccidiostat; see Chapter 7, *Antiprotozoal drugs (e.g. Imidocarb)*) in 1/100 imported pâté samples taken for analysis. The nicarbazim-containing sample was from France, and so the UK authorites 'talked' to their French counterparts who informed them that the pork from which the pâté was manufactured came from the Netherlands; so the UK authorities 'talked' to their Dutch counterparts with a view to understanding the cause of the problem and leaving the Dutch regulators to ensure that it did not recur. This illustrates the importance of import surveillance schemes and further illustrates the international origins of seemingly simple foods like pâté.

Total diet surveys

Total diet surveys (TDSs) are another way of checking that the food we eat does not contain unacceptable residues. A TDS differs from other food sampling schemes by sampling the food as it is eaten rather than sampling raw food from farmers or food stores (e.g. supermarkets). In a TDS a large group of consumers are asked to prepare their meals as normal, but in duplicate. They eat one meal and the other is sent to the lab for analysis. TDSs are good because the food preparation and cooking process might have an effect on residues and this is accounted for when TDS samples are analysed. It is usually difficult to use TDS data for legislative surveillance though, because no one can be certain that an unacceptable residue level (e.g. of a pesticide) was a result of the food producer's or food manufacturer's illegal action – it might be something to do with the cooking process or perhaps cross contamination in the kitchen. Despite this, data from TDSs can be used to initiate surveillance monitoring of a food that is found to have unacceptable residues – you could consider the TDS a screening method in the food legislation world.

Microbiological surveillance

The surveillance schemes discussed above are either national or import monitoring schemes and relate only to chemical residues or food additives. Microbiological surveillance is very important indeed, at both a national and a local level – remember, nearly all acute food-related illness is caused by bacteria or viruses. Making sure that we are not exposed to pathogens in food is a very important aspect of food surveillance.

Microbiological surveillance involves sampling food and culturing it to see if any pathogenic bacteria are present. This is often via local authorities sampling food from food shops, food manufacturers or restaurants. If the results show local contamination (e.g. of a restaurant) the local authority will deal with their findings under the provisions of the food legislation. For example, they might close a restaurant until the source of contamination has been found and dealt with. On the other hand, if the contamination has national or international significance (e.g. contamination in a factory that manufactures food for national consumption or for export) the results are reported to the national authority for action. This might lead to a recall of the contaminated product and will result in an investigation of the manufacturer that might lead to a prosecution.

Does food legislation reduce risks to consumers?

There is no doubt whatsoever that food safety legislation makes food safer. Food producers, manufacturers and sales outlets (including restaurants) are well aware of the implications of non-compliance. The policing schemes (e.g. NSSs) which are usually random and secret (i.e. where and when samples will be taken is not disclosed in advance) and the legal ramifications (e.g. prosecution), with their associated publicity, ensure that producers, manufacturers and retailers are not tempted to break the law. Severe non-compliance can put the perpetrator out of business and perhaps in prison – these are very good reasons to obey the food laws. Despite all of this, food producers, manufacturers and retailers do disobey legislation and do end up in court, but the number of non-compliance cases is very small indeed when you consider the amount of food grown, manufactured, sold and consumed every day.

Case example – non-compliance follow-up

In 2007 the UK authorities sampled royal jelly as part of its food surveillance scheme and found unacceptable residues of the antibiotic chloramphenicol.

Royal jelly is a substance produced by the salivary glands of worker bees; it is fed to bee larvae and causes them to metamorphose into queen bees. Royal

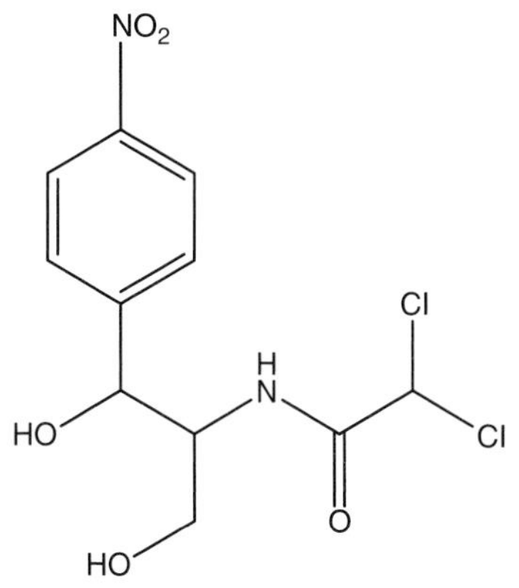

Figure 16.3 Chloramphenicol.

jelly's chemical composition, while extensively investigated, is not fully understood, and the mechanism by which it causes a generic bee larva to turn into a queen bee is still not known. Because of these intriguing properties, royal jelly is marketed as a dietary supplement (legally classified as a food) because some people believe that it has beneficial health properties. Royal jelly is expensive and is a lucrative market.

Chloramphenicol (Figure 16.3) is an antibiotic that has been banned for use in food-producing animals in many jurisdictions (e.g. the EU) because it is linked to a specific, very serious blood disease called aplastic anaemia.

Eighteen of 71 samples (25%) of royal jelly had chloramphenicol residues in the range 0.33–21 µg/kg – the EU statutory limit is 0.3 µg/kg. The samples analysed in the study were purchased from shops and via the internet. Eleven of the samples were sourced from the UK (10 were bought off the internet and one from a shop). The remaining seven samples were sourced from outside the UK. The UK-sourced royal jelly samples containing chloramphenicol residues had levels in the range 0.93–21 µg/kg. Eight of the UK-sourced samples were found to have originated from outside the UK (i.e. the company selling them had bought the raw product from overseas); three were from the USA, five from China, one from Turkey and one could not be traced. The UKs Chief Veterinary Officer (CVO; the British government official responsible for legislation relating to animal products) wrote to senior officials of the royal jellies' countries of origin expressing his concern and asking to be kept informed of the outcome of investigations. In the meantime, the UK Food Standards Agency worked through local authorities to withdraw remaining stocks of chloramphenicol-contaminated batches and notified purchasers of the contaminated products that they were unsafe to consume.

The relevance of national food legislation in a global food market

Many people now want food out of season (e.g. strawberries in winter); international trade in foods makes this possible. During the northern hemisphere winter, southern hemisphere countries (e.g. Australia) grow and export strawberries to the north. Such produce is expensive, but it fulfils the needs of the consumers who are often prepared to pay a premium for a treat out of season. At the other end of the spectrum there is a severe price pressure on food. Consumers shop around for cheaper food; food produced in countries where agricultural and manufacturing labour is cheap (e.g. China) can undercut prices from countries where labour is more expensive. Both of the above scenarios have contributed to the steadily growing international food trade. The international trade in food makes robust food safety legislation essential if we are to minimise food-borne illness due to imports from countries that do not have strict food safety regulations in place. This is a very real concern and is the reason that most countries with robust food safety laws apply import surveillance to make sure that imported food complies with their national standards. In addition, some countries might only allow imports (e.g. manufactured food products; e.g. canned goods) if the exporter can demonstrate compliance with the importer's food laws and will undergo periodic inspections of their production and manufacturing facilities. For example, some supermarkets use overseas producers and often impose very strict production (e.g. control of pesticide use) and manufacturing rules (e.g. in relation to food hygiene) to ensure that the products comply with legislation of the country in which the supermarket intends to sell the product. The exporting manufacturer is often willing to submit themselves to the strict manufacturing regime imposed by the importer because it means they will receive financially lucrative business.

A quick look in my pantry (in New Zealand) revealed a supermarket 'own brand' can of Italian plum tomatoes that clearly stated 'Made in China'. I must admit when I bought the product I assumed it was Italian! This illustrates the multinational nature of food production and marketing and shows why agreed international food safety standards are important.

Take home messages

- Most developed and many developing countries have comprehensive food safety legislation.
- Compliance with the legislation is policed by surveillance schemes in which food samples are taken and analysed for residues, microbiological contaminants and/or additives.
- Food producers, manufacturers and retailers who do not comply with legislation are subject to prosecution which can lead to fines or even prison sentences.
- The compliance policing schemes are important because they warn would-be offenders of the consequences – this reduces non-compliance.

Further reading

New Zealand Food Act 1981. http://www.legislation.govt.nz/act/public/1981/0045/13.0/DLM48687.html.

UK Food Act 1984. http://www.legislation.gov.uk/ukpga/1984/30/contents.

UK Food Standards Act 1999, Chapter 28 – The Food Standards Agency. http://www.legislation.gov.uk/ukpga/1999/28/contents.

US Code; Title 21 – Food and Drugs. http://www.law.cornell.edu/uscode/usc_sup_01_21.html.

Index

Page numbers in *italics* denote figures, those in **bold** denote tables.

Food Safety: The Science of Keeping Food Safe, First Edition. Ian C. Shaw.
© 2013 John Wiley & Sons, Ltd. Published 2013 by John Wiley & Sons, Ltd.